STATE AGRICULTURAL COLLEGE.

THE NEW HALL.

THE new hall which was erected through the appropriation of $30,000 by the Legislature of 1869, was nearly completed by the opening of the term of 1870. Its total cost was $34,550. The excess of cost over the appropriation was met by the sale of swamp lands.

Anticipating further growth of the College, and necessity, perhaps, for a new building for ladies or gentlemen, the dining-hall, kitchen, washing and ironing rooms, etc., have been made large enough for perhaps double the number of students that can now be received. The cost, therefore, of another structure will be considerably lessened.

The building has a supply of water from the Red Cedar river, steam engine, boiler, force pump, a cistern holding two thousand barrels, a tank in the attic, affording head of water, holding twenty-seven barrels, and hose on each floor, to be used in case of fire. The hall is heated by steam, has a large ventilating flue also heated, gas-pipes, cooking range, steam fixtures for cooking, washing, etc. These items and the furnishing of the house cost $13,075. About $7,000 of this expense was met by the sale of swamp lands, leaving the institution in debt about $6,000.

U. S. LAND GRANT.

During the year 1870 the sum of $2,779 89 was received on account of the Congressional grant of lands. This is the first of any income from that source. The interest accruing from the sale and investment of the proceeds is in charge of the State Board of Agriculture. The sales and investments are by legislative enactment committed to the Agricultural Land Grant Board, consisting of the Governor, the Auditor General, Secretary of State, State Treasurer, Attorney General, and Commissioner of the State Land Office.

OFFICERS.

Prof. W. J. Beal of Chicago has been appointed Professor of Botany in the College. He came highly recommended by Dr. Asa Gray, of Harvard University, author of text-books on botany, by Prof. Louis Agassiz, and others, and has proved himself worthy the testimonials of these distinguished men.

The teaching force of the institution now consists of the President, and Professors of Agriculture, Chemistry, English Literature, Entomology, Botany, and a Superintendent of the Horticultural Department. A large part of the duties of the Professor of Agriculture and of the Superintendent of the Horticultural Department is with the students at their work.

The Professors work hard and enthusiastically for the College, and some of them are becoming widely known. The College is in constant danger of losing Professors, from the fact that salaries in this State are much lower than is frequently offered them in other States. Men competent to fill chairs in Agricultural Colleges are hard, and, for some places, almost impossible to find; and other institutions tempt officers who have gained their knowledge and skill through much study, observation, failure, and success in this College, to give them the benefit of this experience.

STUDENTS.

The number of students during the term of 1870 has been 132, distributed in classes as follows: Seniors, 12; Juniors, 18; Sophomores, 16; Freshmen, 38; Preparatory, 36; Special Course, 2; Ladies, 10.

The average age of the students is as follows: Seniors, 22 5-12; Juniors, 21 1-18; Sophomores, 20; Freshmen, 18 7-18; Preparatory Class, 18 9-35; Ladies, 18 2-5; Special Students, 20.

The applications of ladies for admission were in many instances rejected for lack of room for them. The applications of gentlemen to take courses with special analytical chemistry as one study, have been rejected, and still are so, for want of room in the chemical laboratory.

Students of only fifteen years of age, the minimum legal age for entering, have been discouraged from entering as a general thing. Few at that age, however well fitted to enter, are competent to go through the severe discipline of the first two years of study.

GRADUATES.

I do not see why the Agricultural College may not well be proud of its graduates; for in the first place not one of them is addicted to bad habits, so far as we have means of knowing, but all of them are esteemed as upright, honorable, industrious, intelligent citizens.

The graduates of 1870 consisted of twelve gentlemen from nine different counties of the State, with an average age of 22 5-12 years. The whole number of graduates is 56. Two of these died in the army, and one at his home on his farm.

Of the forty-one living graduates of the classes previous to the one just graduated, sixteen, or more than two-fifths, are farmers. Five others are superintendents of gardens and orchards in State institutions, thus showing more than one-half of all the graduates as farmers or horticulturists. Yet the statement is not unfrequent in the papers that our graduates do not go upon farms. An error once stated keeps its place in the public prejudice. The very last report of the Department of Agriculture contains the statement, in an abstract, that not one in four go to farming. A later Shiawassee paper says the same, and assigns as a reason that educated persons will not do manual labor, thus making it necessary that the farmers should always be an ignorant class, from which professional men, manufacturers, and mechanics will be distinguished yearly more and more, as education which *cannot* belong to farmers becomes more and more their portion. One Washtenaw paper, in speaking of the fact that two out of the forty-three who were to graduate from the University in 1869 intended becoming farmers, adds, "A better showing than the Agricultural College can make." This is only one of many misrepresentations afloat. This College figures in a Pennsylvania report on the organization of Agricultural Colleges, in Governor Andrew's message to the Massachusetts Legislature, and in Appleton's new American Cyclopedia, as for some cause suspended, and not in operation.

Seven of the graduates are connected as officers with Agricultural Colleges. One is Professor of Botany in Cornell University; one is Professor of Agriculture and Analytical Chemistry in Wisconsin State University; one is Instructor of Botany and Superintendent of the Gardens in the Iowa State Agricultural College; one holds a like position in the Maine State Agricultural College; another a similar position in our College, where another is employed as Professor of Zoölogy and Entomology, and another as Assistant Secretary. It is not often that any college or school of law or medicine can show, as this does, more than one out of every ten of its graduates occupying places in the faculties of colleges, and those scattered so widely over the land.

Five of the graduates are engineers, surveyors, and machinists, making twenty-eight out of forty-one of the living graduates previous to those just graduated, who are engaged in pursuits supposed to be in keeping with the objects of the institution.

The students of the College are mostly from the country,—are mostly the sons of farmers of limited means, and are mostly largely dependent on their own efforts for the securing of an education; and although, as is natural, the attention of most observers and writers on education is given to schools of literature, law, medicine, divinity, and the like, since they have long had the eye of the world, there are still many who look with interest on an experiment honestly prosecuted, with the view of educating young men, not out of the class of farmers, but to remain in it. At this College young men keep up their habits of labor by three hours' daily work; they keep up familiarity with all farm operations and hardships; they study in connection with their work, and so let the one exercise illustrate and add interest to the other. If an experiment of the sort can succeed, this would seem to be the method of making it.

The graduates who are farmers stand high as farmers in the communities where they live. The necessarily expensive ways of preparing fertilizers, at times, of spreading them, of seeding, harvesting, and other operations which the prevalence of experimenting requires to be practiced at the College, has not led any graduate into expensive ways or unapproved theories of farming.

NEW CHEMICAL LABORATORY NEEDED.

A new chemical laboratory is indispensable to the further proper growth of the institution. The present laboratory is in a building most of which is used for other purposes, as lecture rooms, library, museum, and offices. It is a room poorly lighted, while such a room should have abundance of light. It is very poorly ventilated, so that the fumes become not only extremely disagreeable to all the occupants of every room in the building, but very deleterious to the health. The art of ventilating chemical laboratories, and that of the construction of the conveniences and necessary furnaces, and all the appointments of such a building, have been greatly advanced since the time of the construction of this one.

It is *too small,* and cannot be enlarged. It has been so crowded heretofore that applicants for special courses, embracing analytical chemistry, are refused admission to the College. Now, there is no room with the utmost crowding for the regular class in chemistry. Full one-third of the class are unprovided for.

A plan for a chemical laboratory has been made, after careful examination of nine of the best laboratories of the United States, and the study of the plans of the new structures in Berlin. Yet, I believe, the plan presented does not call for near the cost of the cheapest of any permanent structure in any one of these institutions. Nothing not essential to the building as a good working laboratory, with room for lectures to the class, has been added to the plan. It has been submitted to competent builders, and it is believed that $10,000 or $10,500 will cover all the cost of building and furnishing, ready for use.

Unusual attention is given to chemistry and its application, in the College. These, with chemical physics, occupy the student with a daily exercise, for two full years. For one half year in analytical chemistry they work three hours daily in the laboratory, using Dr. Kedzie's "Hand-Book." The one hundred and fifty substances analyzed by the class of 1870 included mineral poisons, mineral paints, ores, ashes of plants, plaster of Paris, marls, soils, commercial articles, mineral waters, etc. The class made a volumetric analysis of ten acids and ten alkaline substances. All made at least two quantitative analyses, some of them five, of marls, ores, etc.

LADIES AS STUDENTS.

Applications for admission of ladies have been and still are frequent and urgent. The Faculty admitted a few, who occupy rooms on the floor of the Steward's family, or in private houses. They studied chemistry, botany, horticulture, floriculture, trigonometry, surveying, entomology, book-keeping, and other branches. Their progress in study was rapid, and their improvement marked.

Work was furnished them when it could be; they prepared seed for the ground, cut potatoes, transplanted tomatoes and flowering plants, pruned shrubbery, gathered small fruit, did some work in the green-house, and many other kinds of work.

The experiment of having women as students has worked so successfully that there would be no hesitation in admitting them if there were a hall for them.

Should provision be made for them, they should occupy the present new hall, as it contains kitchen, dining-room, washing and ironing rooms, etc., and should do the work of the hall. Another hall, without dining-room, etc., being merely a building of students' rooms, could be built for the young men.

Many ladies would find our course of study agreeable and useful. They would find a knowledge of scientific principles comprising as much additional interest and delight to them in the practice of floriculture, the care of gardens, ornamental shrubs, and orchards, in the operations of the kitchen, and in their general reading, as it does to men. Women are frequently left in circumstances where they would highly prize some knowledge of agriculture.

The applications of chemistry to women's work are so many that a half year's course of daily lectures would not be too long a one. Among these applications are, cooking, preserving of fruits, utilization of materials usually wasted, cleansing by acids and soaps, bleaching, manufacture of soaps of different kinds, disinfection, fermentation, and neutralization of poisons A course of lectures on dairying is now given every year.

Women are turning their attention more and more to studies such as are taught here. Some would like the out-of-door labor, some the aid which the compensation for their labor would afford them in acquiring an education; and it is to be regreted that they cannot avail themselves of the same privilege here that is offered the young men.

OTHER THINGS.

Our experiments have been continued on from those published in the report of 1869, pages 53 to 101. They were planned and carried out after consultation with practical farmers and scientific men, and it is gratifying to the College that they have received the highest praises from some of the best known practical farmers and most esteemed writers in the country, for their accuracy and practical value. Their scope has been somewhat enlarged the present year.

There are many things the College lacks. It is without a horse-barn, proper work-shops, has no wood-shed, and its buildings for the pigs are wretched. Several of its other buildings are mere shanties. It has but a few of the best varieties of swine, sheep, and cattle; the herds are small and by no means what they ought to be. Many of its implements are of the best kinds in use, but it lacks variety in them, and has no museum of implements of any extent. The College should be a place where a more extensive study of these things can be made. Its library and museum need increasing, and permanent improvements need to be furthered more rapidly, the green-house needs perfecting and enlarging, and houses need to be built for the officers of the College.

But, as in the past, so for the present, it is expected that these improvements must wait, except so far as they can be met without further appropriation than pays for student and other necessary labor. The farm teams and hands were very much occupied in excavating and grading the past season, having excavated for the new hall, and having taken down quite a hill from between the halls and the new farm-house. Yet, besides the care of the crops and this excavating and grading, the students have laid two car-loads of tile in underdrains, enlarged the granaries, relaid the barn floor in plank, laid in ceton and coal-tar, and logged and underbrushed to some extent in the south part of the farm.

SECRETARY'S ACCOUNT.

DR.

To Receipts for the year beginning December 1st, 1869, and ending November 30th, 1870.

1870.			
Dec. 1.	Cash at sundry times from the State Treasurer, on account of current expenses		$20,000 00
	Cash from the State Treasurer, on account of construction of the new hall		5,000 00
	Cash from the State Treasurer, on account of the Agricultural College interest fund (the U. S. grant)		2,779 89
	Cash from sales of swamp lands, to-wit:		
	Big Marsh	$717 75	
	Detached lots	602 27	
			$1,320 02
	Cash from sales of produce, to-wit:		
	Farm, stock account	$2,316 04	
	Farm, other produce	529 39	
	Garden sales	260 02	
			$3,105 45
	Cash on account of College bills against students and others		7,134 60
	Cash from all other sources (farm account)		120 83
			$39,460 79

CR.

1870.		
Dec. 1.	By cash paid at sundry times to Joseph Mills, Treasurer	$39,460 79

NOTE—The receipts from students and others on account of College bills add almost nothing to the real income of the College. The money is paid out again for them, and the law of the State, 1861, Act No. 188, provides that the rate of board " shall be so estimated that no profit shall be saved to the institution, and as near as possible at the actual cost."

WARRANT STATEMENT

For the year ending December 1, 1870.

Number.	Date.		To Whom Drawn.	Object.		Amount.
	1869.					
1284	Dec.	1	Sanford Howard...	Secretary's Office, Lansing—See 1417...		$ 50 00
1285	"	3	M. Miles...........	Salary................................		300 00
1286	"	8	T. C. Abbot........	Salary................................		187 50
1287	"	8	Edwards & Cooper	New Hall—contract....................		1,000 00
1288	"	8	Geo. T. Fairchild..	Library—periodicals..................		35 00
1289	"	21	Geo. T. Fairchild..	Salary................................		60 00
1290	"	21	R. C. Kedzie.......	Salary................................		50 00
1291	"	22	A. J. Cook.........	Salary................................		50 00
1292	"	22	R. Haigh, Jr.......	Office of President and Secretary:		
				Services as Assistant Secretary.......		30 00
1293	"	22	S. S. Rockwell.....	New Hall—See 1451....................		907 00
1294	"	27	M. Miles...........	New Farm House—See 1297............		375 00
1295	"	30	John Davis & Co...	New Hall—steam fitting...............		500 00
	1870.					
1296	Jan.	4	R. C. Kedzie.......	Salary................................		75 00
1297	"	4	M. Miles...........	New Farm House—account covers:		
				Warrant 1294........$375 00		
				Warrant 1297........ 836 34		
				Warrant 1298........ 93 43	$1,304 77	
				Expenditures:		
				N. C. Chapman—contract for construction $875 00		
				Plastering, etc....... 224 81		
				Digging well........ 16 75		
				Wood 49 50		
				Oil and turpentine.. 3 30	$1,169 36	
				Furniture	71 98	

WARRANT STATEMENT—Continued.

Number.	Date.	To Whom Drawn.	Object.	Amount.
	1870.			
			Farm Department:	
			Current expenses............ $63 43	$836 34
1298	Jan. 4	M. Miles..........	Farm Department—See 1297..........	93 43
1299	" 4	Sanford Howard...	Salary..........	41 66
1300	" 4	Edwards & Cooper	New Hall contract..........	1,000 00
1301	" 4	T. C. Abbot..........	Salary..........	187 50
1302	" 8	J. W. Robinson....	Horticultural Department:	
			Current expenses..........	29 15
1303	" 11	R. Haigh, Jr.......	Museum:	
			Insect cases..........$5 00	
			Current expenses.......... 2 00 $7 00	
			Office of President and Secretary:	
			Office Books..........$17 00	
			Advertising 7 00	
			Stamps and Stationery.. 21 00 $45 00	52 00
1304	" 11	C. E. Bessey.......	Horticultural Department:	
			Greenhouse repairs..........$ 3 02	
			G. H. current expenses.......... 10 74	13 76
1305	" 11	R. Haigh, Jr.......	Office of President and Sec'y—See 1391	200 00
1306	" 11	S. S. Rockwell.....	Boarding Hall—account covers:	
			Warrant 1269..........$100 00	
			Warrant 1306.......... 18 81	
			Cash receipts.......... 8 38 $127 19	
			Expenditures:	
			Current expenses.......... $127 19	18 81
1307	" 11	S. S. Rockwell.....	New Hall—See 1451..........	1,200 00
1308	" 14	Will W. Tracy.....	Salary..........	75 00
1309	" 18	R. Haigh, Jr........	State Board Expenses:	
			H. G. Wells..........$10 75	
			J. W. Childs 23 30	
			A. C. Prutzman.......... 10 90	44 95
1310	" 20	Edwards & Cooper	New Hall—contract..........	500 00
1311	" 20	A. J. Cook..........	Salary..........	200 00
1312	" 28	Geo. T. Fairchild..	Salary..........	65 00

WARRANT STATEMENT—CONTINUED.

NUMBER.	DATE.		TO WHOM DRAWN.	OBJECT.	AMOUNT.
	1870.				
1813	Jan.	28	S. S. Rockwell	New Hall—See 1451	$600 00
1814	"	28	R. Haigh, Jr.	Office of President and Sec'y—See 1391	200 00
1815	"	28	C. E. Bessey	Horticultural Department:	
				Services in Greenhouse	50 00
1816	"	28	M. Miles	Farm Department—See 1822	465 00
1817	"	28	M. Miles	Farm Department—See 1822	200 00
1818	Feb.	1	John Davis & Co.	New Hall—steam fitting	1,000 00
1819	"	3	C. E. Bessey	Horticultural Department:	
				Current expenses	6 06
1820	"	3	C. E. Bessey	Horticultural Department:	
				Services in Greenhouse	70 00
1821	"	3	Edwards & Cooper	New Hall—contract	400 00
1822	"	7	M. Miles	Farm Department—account covers:	
				Warrant 1816 $465 00	
				Warrant 1817 200 00	
				Warrant 1822 495 79 $1,160 79	
				Expenditures—Farm Department:	
				Implements $512 17	
				" freight 47 71	
				Lumber 55 33	
				Hardware 9 37	
				Current expenses 330 20 $954 78	
				Farm House:	
				Furniture $149 23	
				Current expenses 56 78 $206 01	495 79
1823	"	7	R. Haigh, Jr.	Office of President and Sec'y—See 1391	100 00
1824	"	7	C. C. Stowe	Farm House:	
				Current expenses	263 67
1825	"	7	T. C. Abbot	Salary	187 50
1826	"	9	Will W. Tracy	College General Account—wood	119 37
1827	"	9	Will W. Tracy	Horticultural Department:	
				Greenhouse $21 00	
				Current expenses 16 68	37 68

WARRANT STATEMENT—Continued.

Number.	Date.		To Whom Drawn.	Object.			Amount.
	1870.						
1328	Feb.	9	Will W. Tracy	Salary			$60 00
1329	"	9	Will W. Tracy	Horticultural Department—See 1460			25 00
1330	"	9	Sanford Howard	Salary			41 66
1331	"	12	S. S. Rockwell	New Hall—See 1451			600 00
1332	"	12	R. Haigh, Jr.	Office of President and Sec'y—See 1391			200 00
1333	"	17	Edwards & Cooper	New Hall—contract			1,000 00
1334	"	17	Edwards & Cooper	New Hall—contract in full			2,662 79
1335	"	18	O. Hosford	State Board—expenses			85 00
1336	"	18	D. Carpenter	State Board—expenses			63 70
1337	"	18	John Davis & Co.	New Hall—steam-fitting			500 00
1338	"	18	G. W. Harrison	College General Account—insurance			81 50
1339	"	21	M. Miles	Farm Department—See 1461			200 00
1340	"	21	M. Miles	Salary			150 00
1341	"	23	Geo. T. Fairchild	Salary			250 00
1342	"	28	R. C. Kedzie	Salary			250 00
1343	"	28	A. J. Cook	Salary			125 00
1344	"	28	Will W. Tracy	Salary			245 00
1345	"	28	Sanford Howard	Salary			41 68
1346	"	28	S. S. Rockwell	Salary			150 00
1347	"	28	C. C. Stowe	Salary			125 00
1348	"	28	E. H. Hume	Salary			90 00
1349	"	28	Thomas Dawson	Horticultural Department			47 33
1350	Mar.	3	S. S. Rockwell	New Hall: Steam-fitting—See 1451			150 00
1351	"	7	S. S. Rockwell	Boarding Hall—See 1393			100 00
1352	"	7	Will W. Tracy	Horticultural Department:			
				Greenhouse—current expenses			49 70
1353	"	7	Will W. Tracy	Horticultural Department:			
				Greenhouse—			
				Repairs	$14 95		
				Plants	32 65	$47 60	
				Implements	6 50		
				Current expenses	79 57	86 07	133 67

WARRANT STATEMENT—Continued.

Number.	Date.	To Whom Drawn.	Object.	Amount.
	1870.			
1854	Mar. 9	M. Miles	Farm Department:	
			Books $ 5 90	
			Hardware 47 00	
			Current expenses 113 29 $166 19	
			Farm House—repairs 38	
			College General Account 3 50	$170 07
1855	" 9	S. S. Rockwell	Office of President and Secretary:	
			Furniture—	
			Office of Assistant Secretary $78 08	
			Office of President 13 23	91 31
1856	" 9	S. S. Rockwell	Boarding Hall:	
			Current expenses $521 77	
			New Hall—construction 12 00	533 77
1857	" 9	R. Haigh, Jr.	Office of President and Secretary:	
			Services as Assistant Secretary	15 00
1858	" 9	J. W. Robinson	Salary	95 43
1859	" 9	S. S. Rockwell	New Hall—See 1451	107 12
1860	" 17	Jones & Porter	College General Account—insurance	137 50
1861	" 21	M. Miles	Salary	150 00
1862	" 23	S. S. Rockwell	Boarding Hall—See 1399	218 62
1863	" 23	S. S. Rockwell	New Hall—See 1451	150 00
1864	" 23	S. S. Rockwell	Boarding Hall—See 1453	200 00
1865	" 23	House & Sanford	College General Account—insurance	125 00
1866	" 25	S. S. Rockwell	Boarding Hall—See 1399	132 09
1867	" 25	S. S. Rockwell	New Hall—See 1451	186 00
1868	" 29	R. Haigh, Jr.	Office of Assistant Secretary:	
			Furniture $19 37	
			Repairs 14 63	34 00
1869	" 29	R. Haigh, Jr.	Office of President and Secretary:	
			Postage, and 1,000 stamped envelopes $90 20	
			Printing address of Hon. Geo. Willard 70 00	
			Current expenses 8 25	168 45

WARRANT STATEMENT—Continued.

Number.	Date.	To Whom Drawn.	Object.	Amount.
	1870.			
1370	Mar. 29	H. G. Wells	State Board Expenses	$10 25
1371	April 8	S. S. Rockwell	Boarding Hall—See 1399	363 15
1372	" 8	S. S. Rockwell	New Hall—See 1451	110 00
1373	" 9	C. C. Stowe	Farm House:	
			Construction $40 48	
			Furniture 28 39	
			Current expenses 174 19	243 0[illegible]
1374	" 29	Sanford Howard	Secretary's Office, Lansing:	
			Current expenses	56 00
1375	" 29	A. J. Cook	Apiary $3 34	
			Museum 4 50	
			Farm Department:	
			Alsike clover seed 3 80	11 64
1376	May 14	S. S. Rockwell	Boarding Hall:	
			Current expenses $442 41	
			New Hall 20 13	462 54
1377	" 14	M. Miles	Farm Department:	
			Implements $75 91	
			Implements repaired 56 36	
			Tile 174 33	
			Seed 11 65	
			Plaster 12 00	
			Current expenses 304 21 $634 46	
			Farm House:	
			Furniture 18 00	652 46
1378	" 31	M. Miles	Salary	300 00
1379	" 31	R. C. Kedzie	Salary	375 00
1380	" 31	Geo. T. Fairchild	Salary	375 00
1381	" 31	A. J. Cook	Salary	375 00
1382	" 31	Will W. Tracy	Salary	370 00
1383	" 31	S. S. Rockwell	Salary	150 00
1384	" 31	C. C. Stowe	Salary	125 00
1385	" 31	E. H. Hume	Salary	90 00
1386	" 31	R. C. Kedzie	Services of Assistant Chemist	174 00

WARRANT STATEMENT—CONTINUED.

NUMBER.	DATE.	TO WHOM DRAWN.	OBJECT.	AMOUNT.
	1870.			
1887	May 31	T. C. Abbot	Salary	$562 50
1888	" 31	Sanford Howard	Salary	125 00
1889	" 31	R. Haigh, Jr.	Salary	250 00
1890	" 31	S. S. Rockwell	Salary	125 00
1891	" 31	R. Haigh, Jr.	Office of Pres. and Sec'y—ac't covers:	
			Warrant 1805 $200 00	
			Warrant 1814 200 00	
			Warrant 1823 100 00	
			Warrant 1832 200 00	
			Warrant 1891 134 48 $834 48	
			Expenditures:	
			Advertising opening of Spring term, 1870 $834 48	134 48
1892	" 31	A. J. Cook	Apiary—five bee-hives, @ $3	15 00
1893	" 31	S. S. Rockwell	Boarding Hall—account covers:	
			Warrant 1851 $100 00	
			Warrant 1898 300 00	
			Warrant 1893 156 07	
			Cash receipts 223 68 $779 95	
			Expenditures:	
			Furniture $ 32 65	
			Repairs 10 00	
			Current expenses 737 10 $779 75	156 07
1894	" 31	Will W. Tracy	Horticultural Department—See 1895	93 50
1895	" 31	Will W. Tracy	Horticultural Department—ac't covers:	
			Warrant 1394 $93 50	
			Warrant 1395 51 15 $144 65	
			Expenditures:	
			Greenhouse $90 40	
			Seed 9 38	
			Current expenses 44 87 $144 65	51 15
1896	" 31	Will W. Tracy	Horticultural Department—See 1460	175 00
1897	June 13	R. C. Kedzie	Chemical Laboratory	227 35
1898	" 14	S. S. Rockwell	Boarding Hall—See 1893	300 00

WARRANT STATEMENT—Continued.

Number.	Date.	To Whom Drawn.	Object.	Amount.
	1870.			
1399	June 14	S. S. Rockwell	Boarding Hall—account covers:	
			Warrant 1362 .... $218 62	
			Warrant 1366 .... 132 09	
			Warrant 1371 .... 263 15	
			Warrant 1399 .... 584 82	
			Cash receipts 111 25 $1,309 93	
			Expenditures:	
			Repairs $350 71	
			Furniture 19 20	
			Current expenses 940 02 $1,309 93	$584 82
1400	" 24	J. J. Fernand	Services as French teacher	33 00
1401	" 24	A. J. Cook	Salary	125 00
1402	" 29	Will W. Tracy	Salary	85 00
1403	" 30	S. S. Rockwell	Boarding Hall:	
			Current expenses	281 64
1404	" 30	M. Miles	Farm Department:	
			Tile $218 75	
			Lumber 29 00	
			Lime 44 25	
			Stock 54 62	
			Implements 19 79	
			Current expenses 55 32 $421 73	
			Farm House:	
			Construction $103 20	524 93
1405	" 30	M. Miles	Salary	150 00
1406	" 30	C. C. Stowe	Farm House:	
			Current expenses	228 22
1407	July 27	Geo. T. Fairchild	Salary	100 00
1408	" 27	W. J. Beal	Salary	50 00
1409	Aug. 2	Will W. Tracy	Horticultural Department—See 1410	133 05
1410	" 2	Will W. Tracy	Horticultural Department—ac't covers:	
			Warrant 1409 .... $133 05	
			Warrant 1410 .... 90 31 $223 36	

WARRANT STATEMENT—CONTINUED.

NUMBER.	DATE.	TO WHOM DRAWN.	OBJECT.	AMOUNT.
	1870.			
			Expenditures:	
			Seeds and Plants......$125 69	
			Hardware 22 12	
			Implements 9 30	
			Current expenses...... 66 25 $223 36	$90 31
1411	Aug. 2	Will W. Tracy.....	College General Account:	
			Wood..................$60 00	
			Chairs 21 00	
			Crayons................. 1 25 $82 25	
			Museum 1 25	83 50
1412	" 2	M. Miles...........	Salary..............................	150 00
1413	" 5	T. C. Abbot........	College General Account:	
			Repairs on house.................$ 9 50	
			French class.................. 16 00	
			Express 3 55	
			Implements 1 63	
			Current expenses.............. 27 55	58 23
1414	" 26	S. S. Rockwell.....	Boarding Hall:	
			Current expenses..................	2,677 88
1415	" 26	A. J. Cook.........	Museum:	
			New shelves...............$2 87	
			Current expenses.......... 6 40 $9 27	
			Apiary:	
			Boxes and hives...........$6 50	
			Comb 1 00 $7 50	16 77
1416	" 26	Sanford Howard...	Secretary's Office, Lansing—See 1417...	5 87
1417	" 26	Sanford Howard...	Secretary's Office, Lansing—ac't covers:	
			Warrant 1284..........$50 00	
			Warrant 1416.......... 5 87	
			Warrant 1417.......... 18 62 $74 49	
			Expenditures:	
			Current expenses.......... $74 49	18 62

WARRANT STATEMENT—Continued.

Number.	Date.	To Whom Drawn.	Object.		Amount.
	1870.				
1418	Aug. 26	R. Haigh, Jr.	Office of President and Secretary:		
			Advertising $18 85		
			Current expenses 20 42	$38 77	
			College General Account	5 08	
			Farm Department—current exp.	16 46	
			Boarding Hall—current expenses	20 47	$80 78
1419	" 30	S. S. Rockwell	Boarding Hall—See 1444		1,000 00
1420	Sept. 2	Sanford Howard	Salary		83 34
1421	" 2	Sanford Howard	Secretary's Office, Lansing:		
			Current expenses		14 05
1422	" 5	Will W. Tracy	Salary		215 00
1423	" 5	T. C. Abbot	Salary	$562 50	
			French class	56 00	618 50
1424	" 5	M. Miles	Salary		150 00
1425	" 5	R. C. Kedzie	Salary		375 00
1426	" 5	Geo. T. Fairchild	Salary		275 00
1427	" 5	A. J. Cook	Salary		250 00
1428	" 5	Will W. Tracy	Salary		140 00
1429	" 5	R. Haigh, Jr.	Salary		150 00
1430	" 5	S. S. Rockwell	Salary		150 00
1431	" 5	C. C. Stowe	Salary		125 00
1432	" 5	E. H. Hume	Salary		90 00
1433	" 5	R. C. Kedzie	Services of Assistant Chemist		87 50
1434	" 5	W. J. Beal	Salary		350 00
1435	" 5	Sanford Howard	Salary		41 66
1436	" 8	T. C. Abbot	State Board Expenses:		
			H. G. Wells	$23 40	
			J. W. Childs	35 60	
			A. C. Prutzman	33 50	
			S. O. Knapp	52 45	
			Exchange	30	145 25
1437	" 8	S. S. Rockwell	New Hall—See 1451		55 00
1438	" 8	E. R. Merrifield	College General Account—Insurance		27 08

WARRANT STATEMENT—Continued.

Number.	Date.	To Whom Drawn.	Object.	Amount.
	1870.			
1439	Sept. 19	M. Miles	Farm Department:	
			Implements $48 26	
			Implements repaired 7 61	
			Stock 12 00	
			Current expenses 335 21 $402 98	
			College General Account:	
			Expenses at Fair 25 00	$427 90
1440	" 19	Will W. Tracy	Horticultural Department—See 1460	120 84
1441	" 21	C. C. Stowe	Farm House:	
			Current expenses	220 64
1442	" 21	C. C. Stowe	Farm House:	
			Current expenses	75 00
1443	" 21	D. L. Case	Secretary's Office, Lansing:	
			Rent for one year, to July 31, 1870	150 00
1444	" 21	S. S. Rockwell	Boarding Hall—account covers:	
			Warrant 1419 $1,000 00	
			Warrant 1444 338 58 $1,338 58	
			Expenditures:	
			Current expenses $1,338 58	338 58
1445	Oct. 1	W. S. George & Co.	Office of President & Sec'y:	
			Account books, blank bills, receipts, regulations, etc. $44 25	
			Farm Department—Day-book 6 00	50 25
1446	" 7	S. S. Rockwell	Boarding Hall—See 1453	112 56
1447	" 7	T. C. Abbot	College General Account $12 00	
			Office of President and Secretary:	
			Current expenses 12 85	24 85
1448	" 13	S. S. Rockwell	Boarding Hall—See 1453	100 00
1449	" 24	Davis & Larned	New Hall—furniture	100 00
1450	" 31	Jones & Porter	College General Account—insurance	114 00

WARRANT STATEMENT—Continued.

Number.	Date.	To Whom Drawn.	Object.	Amount.
	1870.			
1451	Nov. 21	S. S. Rockwell......	New Hall—acc't extends from Dec. 1, 1869, to Nov. 21, 1870, and covers:	
			Warrant 1193......$2,400 00	
			Warrant 1293...... 907 00	
			Warrant 1307...... 1,200 00	
			Warrant 1313...... 600 00	
			Warrant 1331...... 600 00	
			Warrant 1350...... 150 00	
			Warrant 1359...... 107 12	
			Warrant 1363...... 150 00	
			Warrant 1367...... 186 00	
			Warrant 1372...... 110 00	
			Warrant 1437...... 55 00	
			Warrant 1451...... 105 69	
			Receipts........... 8 60	
			Cash on hand...... 304 62 $6,884 03	
			Expenditures:	
			Extra construction of new hall......$1,364 65	
			Cistern 654 84	
			Steam-fitting 2,081 83	
			Gas-fitting......... 312 74	
			Water supply...... 90 80	
			Tank 142 09	
			Range 708 23	
			Furniture.......... 1,529 35 $6,884 03	$105 69
1452	" 29	A. J. Cook.........	Museum:	
			Current expenses................ $1 40	
			Apiary—Italian bees.............. 21 50	22 90
1453	" 29	S. S. Rockwell......	Boarding Hall—account covers:	
			Warrant 1364..........$200 00	
			Warrant 1446.......... 112 56	
			Warrant 1448.......... 100 00	
			Warrant 1453.......... 240 00	
			Receipts.............. 9 74 $662 30	

WARRANT STATEMENT—Continued.

Number.	Date.	To Whom Drawn.	Object.	Amount.
	1870.			
			Expenditures:	
			Current expenses $662 30	$240 00
1454	Nov. 29	M. Miles	Farm Department—See 1461	67 93
1455	" 29	Sanford Howard ...	Secretary's Office, Lansing:	
			Current expenses	23 83
1456	" 29	R. Haigh, Jr.	Office of President and Secretary:	
			Current expenses	8 50
1457	" 29	R. Haigh, Jr.	Office of President and Secretary:	
			Services as Assistant Secretary, 1869	15 00
1458	" 29	T. C. Abbot	College General Account:	
			Current expenses	10 00
1459	" 30	Will W. Tracy	Horticultural Department—See 1460	27 60
1460	" 30	Will W. Tracy	Horticultural Department—acc't covers:	
			Warrant 1329 .... $25 00	
			Warrant 1396 .... 175 00	
			Warrant 1440 .... 120 34	
			Warrant 1459 .... 27 60	
			Warrant 1460 .... 17 00 $364 94	
			Expenditures:	
			Greenhouse $22 77	
			Implements repaired .. 29 53	
			Current expenses 312 64 $364 94	17 00
1461	" 30	M. Miles	Farm Department—account covers:	
			Warrant 1339 ....$200 00	
			Warrant 1454 .... 67 93	
			Warrant 1461 .... 45 00 $312 93	
			Expenditures:	
			Current expenses $259 32	
			Stock 48 00	
			Implements repaired. 5 60 $312 93	45 00
			Total	$40,754 53

SUMMARY OF WARRANT ACCOUNT.

1870.	
Expenses of Board	$ 299 15
Salaries	10,360 43
Secretary Howard, expenses of office, traveling, etc.	318 37
Office of President and Secretary at College. (Amount includes furniture for office of Assistant Secretary, office books, advertising, etc.)	1,332 61
Farm Department. (Amount includes experiments, new implements, etc.)	3,085 96
Horticultural Department. (Amount includes expenses and care of Greenhouse)	1,170 30
Boarding Hall. (Amount includes wages paid to students for labor)	7,708 87
Farm House. (Amount includes board of men for the garden)	1,254 93
Chemical Laboratory	227 35
Library and text books	35 00
Museum	23 42
Apiary	47 34
New Dormitory (on contract)	5,900 00
New Dormitory, extra construction, steam fitting, fixtures, and furniture	6,965 73
New Farm House, construction	1,241 34
Expenditures unclassified in this summary	783 73
	$40,754 53

TREASURER'S REPORT.

Joseph Mills, Treasurer, in Account with State Board of Agriculture.

DR.

1869.			
Dec.	1.	To balance on hand	$4,079 48
	4.	To cash of S. Howard, Secretary	350 00
	28.	" State Treasurer	1,600 00
1870.			
Jan.	5.	To cash of S. Howard, Secretary	900 00
	18.	" State Treasurer	5,000 00
	21.	" " "	1,000 00
Feb.	8.	" " "	2,400 00
	19.	" " "	7,000 00
Mar.	26.	" S. Howard, Secretary	2,500 00
May	18.	" " "	1,100 00
	26.	" State Treasurer	5,000 00
June	3.	" S. Howard, Secretary	2,000 00
Sept.	1.	" State Treasurer	3,000 00
	6.	" " "	1,800 00
	8.	" S. Howard, Secretary	3,633 40
	10.	" " "	150 00
Oct.	7.	" " "	112 56
Nov.	3.	" " "	125 00
	18.	" State Treasurer	979 89
	30.	" S. Howard, Secretary	809 94
		Total	$43,540 27

1870.			
Dec.	1.	To balance from old account	$1,016 29

1870.
Nov. 30. Cr.

By paid warrant No.	1,267	$41 67
" "	1,274	224 82
" "	1,275	677 47
" "	1,276	147 35
" "	1,277	150 00
" "	1,278	18 43
" "	1,279	22 40
" "	1,280	308 65
" "	1,281	156 91
" "	1,282	21 83
" "	1,283	44 92
" "	1,284	50 00
" "	1,285	300 00
" "	1,286	187 50
" "	1,287	1,000 00
" "	1,288	35 00
" "	1,289	60 00
" "	1,290	50 00
" "	1,291	50 00
" "	1,292	30 00
" "	1,293	907 00
" "	1,294	375 00
" "	1,295	500 00
" "	1,296	75 00
" "	1,297	836 34
" "	1,298	93 43
" "	1,299	41 66
" "	1,300	1,000 00
" "	1,301	187 50
" "	1,302	29 15
" "	1,303	52 00
" "	1,304	13 76
" "	1,305	200 00
" "	1,306	18 81
" "	1,307	1,200 00
" "	1,308	75 00
" "	1,309	44 95
" "	1,310	500 00
" "	1,311	200 00
" "	1,312	65 00

By paid warrant No.	1,313	$600	00
" "	1,314	200	00
" "	1,315	50	00
" "	1,316	465	00
" "	1,317	200	00
" "	1,318	1,000	00
" "	1,319	6	06
" "	1,320	70	00
" "	1,321	400	00
" "	1,322	495	79
" "	1,323	100	00
" "	1,324	263	67
" "	1,325	187	50
" "	1,326	119	37
" "	1,327	37	68
" "	1,328	60	00
" "	1,329	25	00
" "	1,330	41	66
" "	1,331	600	00
" "	1,332	200	00
" "	1,333	1,000	00
" "	1,334	2,662	79
" "	1,335	35	00
" "	1,336	63	70
" "	1,377	500	00
" "	1,338	31	50
" "	1,339	200	00
" "	1,340	150	00
" "	1,341	250	00
" "	1,342	250	00
" "	1,343	125	00
" "	1,344	245	00
" "	1,345	41	68
" "	1,346	150	00
" "	1,347	125	00
" "	1,348	90	00
" "	1,349	47	33
" "	1,350	150	00
" "	1,351	100	00
" "	1,352	49	70
" "	1,353	133	67
" "	1,354	170	07
" "	1,355	91	31

5

By paid warrant No. 1,356	$533 77
" " 1,357	15 00
" " 1,358	95 43
" " 1,359	107 12
" " 1,360	137 50
" " 1,361	150 00
" " 1,362	218 62
" " 1,363	150 00
" " 1,364	200 00
" " 1,365	125 00
" " 1,366	132 09
" " 1,367	186 00
" " 1,368	34 00
" " 1,369	168 45
" " 1,370	10 25
" " 1,371	263 15
" " 1,372	110 00
" " 1,373	243 01
" " 1,374	56 00
" " 1,375	11 64
" " 1,376	462 54
" " 1,377	652 46
" " 1,378	300 00
" " 1,379	375 00
" " 1,380	375 00
" " 1,381	375 00
" " 1,382	370 00
" " 1,383	150 00
" " 1,384	125 00
" " 1,385	90 00
" " 1,386	174 00
" " 1,387	562 50
" " 1,388	125 00
" " 1,389	250 00
" " 1,390	125 00
" " 1,391	134 48
" " 1,392	15 00
" " 1,393	156 07
" " 1,394	93 50
" " 1,395	51 15
" " 1,396	175 00
" " 1,397	227 35
" " 1,398	300 00

By paid warrant No.	1,399	$584	82
" "	1,400	33	00
" "	1,401	125	00
" "	1,402	35	00
" "	1,403	281	64
" "	1,404	524	93
" "	1,405	150	00
" "	1,406	228	22
" "	1,407	100	00
" "	1,408	50	00
" "	1,409	133	05
" "	1,410	90	31
" "	1,411	83	50
" "	1,412	150	00
" "	1,413	58	23
" "	1,414	2,677	88
" "	1,415	16	77
" "	1,416	5	87
" "	1,417	18	62
" "	1,418	80	78
" "	1,419	1,000	00
" "	1,420	83	34
" "	1,421	14	05
" "	1,422	215	00
" "	1,423	618	50
" "	1,424	150	00
" "	1,425	375	00
" "	1,426	275	00
" "	1,427	250	00
" "	1,428	140	00
" "	1,429	150	00
" "	1,430	150	00
" "	1,431	125	00
" "	1,432	90	00
" "	1,433	87	50
" "	1,434	350	00
" "	1,435	41	66
" "	1,436	145	25
" "	1,437	55	00
" "	1,438	27	30
" "	1,439	427	98
" "	1,440	120	34
" "	1,441	220	64

By paid warrant No. 1,442	$75 00
" " 1,443	150 00
" " 1,444	338 58
" " 1,445	50 25
" " 1,446	112 56
" " 1,447	24 85
" " 1,448	100 00
" " 1,449	100 00
" " 1,450	114 00
" " 1,451	105 69
" " 1,452	22 90
" " 1,453	240 00
" " 1,454	67 93
" " 1,455	23 83
" " 1,456	3 50
" " 1,457	15 00
" " 1,458	10 00
" " 1,459	27 60
" " 1,460	17 00
By balance to new account	1,016 29
Total	$43,540 27

SUMMARY.

Joseph Mills, Treasurer.

DR.

1869.		
Dec. 1.	To balance on hand	$4,079 48
1870.		
Nov. 30.	To cash of Secretary Howard and State Treasurer at sundry times	39,460 79
	Total	$43,540 27

CR.

1870.		
Nov. 30.	By paid warrants 1267, and 1274 to 1282, inclusive, issued previous to December 1st, 1869, and so not included in warrant account of 1870	$1,769 52
" "	By paid warrants 1283 to 1460, inclusive	40,754 45
" "	Balance to new account	1,016 20
		$43,540 27

DR.

1870.		
Dec. 1.	To balance from old account	$1,016 29

NOTE.—The warrants drawn up to December 1, 1870, are all paid, except No. 1461, for $45 00.

SUMMARY OF CASH ACCOUNT

Of the Superintendent of the Farm Department, for the year ending December 1, 1870.

RECEIPTS.

Warrants	$4,553 14	
Farm produce	1,645 57	
Labor	105 68	
Wood sold	204 81	
Tile sold	2 40	
Cement sold	2 75	
Freight returned	10 00	
		$6,524 35

EXPENDITURES.

Paid Secretary, farm receipts, 1870	$1,645 57	
" " wood receipts	204 81	
" " labor receipts	105 68	
" " tile receipts	2 40	
" " cement receipts	2 75	
" " freight returned	10 00	
		$1,971 21
Paid for construction of Farm House		1,322 08
" current expenses at Farm House		73 98
" furniture for Farm House		181 99
" barn machinery		542 13
" labor		728 00
" tile		393 08
" feed		543 90
" on account of stock		109 40
" for implements on shop account		33 10
" " farm account		142 53
" " wood account		68 76
" " draining account		46 33
" for hardware		12 22
" seeds		22 77
" plaster		12 00
" wells and pumps		54 42
" lumber		85 98
" for office expenses		23 28

Paid for blacksmithing	$48 24	
" College, sundry items	35 46	
" on account of repairs of buildings	25 35	
" " barns (current expenses)	9 09	
" " experiments	21 60	
" for flour barrels	13 50	
" harness repairs	3 39	
		$6,524 35

STATEMENT OF RECEIPTS AND EXPENDITURES

Of College Farm, for the year ending December 1, 1870.

RECEIPTS.

Cash sales of produce	$1,645 57	
86 tons of hay, @ $10	860 00	
1085 lbs. of wool	335 21	
1549 bushels of corn (ears) @ 30c	464 70	
180 " wheat, @ $1.25	225 00	
550 " oats, @ 40c	220 00	
1940 " roots, @ 10c	194 00	
50 tons of cornstalks, @ $3 (estimated)	150 00	
10 " wheat straw, @ $2.50	25 00	
13½ " oat straw, @ $4	54 00	
Produce sold to Boarding Hall	538 32	
" " Farm House	311 09	
" " J. N. Smith	130 35	
Increase in value of stock, as per inventory	693 00	
		$5,846 24

EXPENDITURES.

Labor of men, @ $1.40 per day	$902 06	
" teams, @ $1.50 per day	441 60	
" students	771 77	
Account of 1869 with crops of 1870	255 47	
Seeds purchased	22 77	
Plaster	12 00	
Repairs of implements	41 34	
Produce on hand Dec. 1st, 1869	1,946 59	
Profit to balance	1,452 64	
		$5,846 24

STATEMENT OF RECEIPTS AND EXPENDITURES

Of the Farm Department, for the year ending December 1, 1870.

RECEIPTS.

Warrants	$4,553 14	
Farm produce	1,645 75	
Labor	105 68	
Wood	204 81	
Sundry cash items	15 15	
Amount of student's labor paid by College	2,134 99	
Amount of student's labor in recess paid by College	174 56	
Board of men at Farm House, 133⅓ weeks, @ $4	533 33	
Produce on hand December 1st, 1869	2,697 60	
Account of last year with crops of 1870	255 47	
Amount of accounts unpaid	570 48	
Sundry bills paid by College	9 80	
		$12,900 58

EXPENDITURES.

Secretary of College:		
Farm receipts of 1870	$1,645 57	
Wood " "	204 81	
Labor " "	105 68	
Sundry " "	15 15	
		$1,971 21
College:		
Labor	$35 71	
Expenses of Fairs	53 48	
Experiments	315 97	
Construction, new hall	9 35	
General account	10 46	
		424 97
Boarding Hall:		
Wood	$128 46	
Labor	24 08	
Farm produce	538 32	
Keeping horse	110 00	
		800 86
New Farm House:		
Cash (construction account)	$1,322 08	
Labor " "	89 63	
Furniture	181 99	
		$1,593 70

Permanent Improvements—Buildings:		
Cash (general account)	$25 35	
President's house	8 33	
Steps to old Boarding Hall	15 43	
Sheep barn	7 26	
Cattle barn	187 83	
Shops and tool house	24 15	
Clearing	181 58	
Removing stones from fields	34 24	
Draining labor	438 78	
Draining tile	412 36	
Farm roads and bridges	26 16	
Fences	5 77	
Grading around buildings and lawn	273 27	
Hauling gravel for roads on lawn,	131 22	
Wells and pumps	54 42	
Lumber	247 08	
Hardware	54 51	
		$2,127 74
Farm House (current expenses):		
Wood	$7 50	
Labor	29 18	
Farm produce	311 09	
Cash	73 98	
		421 75
Horticultural Department:		
Labor on hay	$26 11	
Labor	2 90	
		29 01
Stock:		
Paid on account of stock purchased and freights	109 40	
		109 40
Implements:		
Barn machinery	$542 13	
On shop account	35 85	
On farm account	144 89	
On wood account	68 76	
On draining account	50 66	
		842 29
Miscellaneous:		
Toll	$ 56	
Office and tool record expenses	71 91	
Labor on wood on hand	20 00	
		$92 47

Current Farm Expenses:			
Labor of men (including students' labor during recess)	$902 06		
Labor of teams	441 60		
Labor of students	771 77		
Account of 1869 with crops of 1870	255 47		
Seeds purchased	22 77		
Plaster purchased	12 00		
Repairs of implements	41 34		
Produce on hand Dec. 1st, 1869	1,946 51		
		$4,393 52	
Account with crops of 1871:			
Labor in fields Nos. 2 and 6	$72 66		
15 bushels seed-wheat, @ 1.40	21 00		
		93 66	
			$12,900 58

NOTE—The account of receipts and expenditures of the Farm Department sets down as paid into the hands of the Secretary, $1,971.21, while the Secretary gives credit for $2,316.04, and $529.39, and $120.83; total, $2,966.26. The difference, $995.05, appears in the Farm Department account of 1869, page 13, as part of the $2,399.82 on both receipt and expenditure sides of the account.

SUMMARY OF CASH ACCOUNT

Of the Superintendent of the Horticultural Department, for the year ending December 1, 1870.

RECEIPTS.

Warrants	$939 32	
Sale of apples	57 60	
" garden vegetables	32 34	
" plants from greenhouse	124 35	
Board of horses	44 00	
Labor of team	48 75	
		$1,249 36

EXPENDITURES.

Keeping horses	$404 49	
Labor	166 71	
Plants and seeds	214 02	
Implements and repairs	36 50	
Greenhouse current expenses	67 12	
Moving entrance gate	17 57	
Unsettled account of 1869 (team)	25 05	
Manure	13 25	
Museum	5 60	
Current expenses	15 03	
Balance due College	21 00	
Receipts paid Secretary	260 02	
		$1,246 36

INDEBTEDNESS.

December 1, 1870.

To officers, on account of salaries	$3,162 50
Assistant Chemist	87 50
W. S. George and Co.	150 00
S. R. Greene	61 25
S. S. Rockwell	125 00
Board of Agriculture (estimated)	120 00
On account of Farm Department	240 00
" Garden, Greenhouse, etc. (estimated)	100 00
" Boarding Hall	3,338 87
" Farm House	154 58
Total	$7,539 70
The Treasurer reports balance to new account	$1,016 29
	$6,523 41

This indebtedness was occasioned by the expenditures on the part of the College in supplying the new hall with fixtures and furniture, and the sums due as above were left unpaid in order that all accounts due to parties remote from Lansing might be settled.

ESTIMATES FOR 1871.

CURRENT EXPENSES.

Salaries of President, professors, superintendent of the Horticultural Department, foreman, steward, assistant chemist, secretary and assistant	$15,000 00
Expenses of the State Board of Agriculture. (The members receive no compensation for their services.)	300 00
Expenditures of Secretary of the Board	400 00
Printing of catalogues, binding, postage, blanks, blank books, etc.	500 00
Labor of students	3,000 00
Farm Department, permanent improvements, help, implements, etc.	1,500 00
Horticultural Department, Greenhouse	800 00
Insurance, library, museum, apiary, chemicals, etc.	1,000 00
	$22,500 00

INCOME ESTIMATED.

From Agricultural College interest fund (the U. S. grant), receipts last year were $2,779.89; receipts for next two years, estimated by Governor Baldwin to average	$ 4,000 00
Receipts from sales of swamp lands (estimated)	1,500 00
Appropriation asked for	17,000 00
	$22,500 00

ESTIMATES FOR NEW LABORATORY.

Estimate (and bid) of Messrs. Edwards & Cooper, contractors of the New Hall.

Brick building, including lecture room	$8,300 00
Ten work tables	300 00
Ten bowls, traps	100 00
Fifteen evaporating hoods	120 00
Tiles and drains	40 00
Iron gas ventilating shaft, 16 in. drain	96 00

Water-tanks and pipes	$400 00
Gas fixtures	75 00
Wardrobe stalls, etc	50 00
Pneumatic cistern	75 00
Seating lecture room for 60 persons	220 00
Steam bath	250 00
Cases for apparatus	300 00
	$10,326 00
Appropriation asked for	$10,000 00

ESTIMATES FOR 1872.

Same as per current expenses of 1871	$17,000 00

NOT MADE TO BE CUT DOWN.

The estimates are submitted to the good sense of the Legislature. They are put as low as it was known how to make them. Certainly there was no thought of putting in a margin to be cut off. The Legislature has not cut down the appropriation for current expenses below what the Board of Agriculture has asked, since 1863, when the reduction was $1,000. Nothing not absolutely necessary has been put in the estimates, and desired improvements are put off to the time when the endowment fund can be used for them.

THE LABORATORY.

Accurate meteorological records have been kept since April, 1863, which have been published in full in the reports of the Board. They have been used monthly in the reports of the Washington Department of Agriculture; have been referred to as important for their practical value by the park commissioners of Chicago, etc. These, and the Professor's observations on the meteorology of Michigan, and address in reports of 1865 and 1866, are referred to in the annual address of the president of the New York Agricultural Society of 1868, as worth study, and as "of great practical value." The Professor's report on the destruction of forest trees, in 1866, was largely quoted in reports in New England and in the West, sometimes by page upon page, as of great importance. In May of 1870 he published his investigations into the magnetic properties of the magnetic wells, showing that there was *no* magnetism in the water, being, I believe, the first to announce that fact. His experiments in agricultural chemistry, analysis of products useful to agriculturists, lectures upon the land dunes, etc., are only introduced here to show that a new laboratory would fall into the hands of one honored in the State and abroad, and who would use it for the public good, by instructing students in analysis and the taking of observations.

Let it be remembered that the laboratory cannot be built from the interest of the Congressional land grant, the law of the United States providing that: "No portion of said fund, nor the interest thereon, shall be applied, directly or indirectly, under any pretense whatever, to the purchase, erection, preservation, or repair of any building or buildings."—*Chapter CXXX, United States Laws, 1862.*

Other States are providing chemical laboratories for their agricultural colleges. Maine has just completed one, modeled after that of Brown University, which cost, for the building only, $12,000 in 1862, prices having advanced but little. The Vermont Agricultural College has a new laboratory, presenting, their reports say, "the amplest facilities." The Massachusetts Agricultural College laboratory is a part of a large building. This part cost $10,000. The agricultural colleges of Rhode Island and New York are large enough for their classes. Pennsylvania Agricultural College advertises "three large chemical laboratories," all furnished.

Colby University, in Maine, appropriates this year $25,000

for laboratory and museum; and of the $255,000 which the Industrial University of Illinois asks of the Legislature this winter, $50,000 is for laboratory and apparatus. The laboratory, as planned for the Michigan Agricultural College, has tables for forty, a lecture-room for ninety, furnace, and other rooms.

EXPERIMENTS.

The report for 1869 gives nineteen pages (page 52) on pig feeding, continued from report of 1868, pages 73 to 97. Report of 1869 gives experiments in the application of manures. plats being left unmanured, manured before plowing, and on surface after plowing. It was continued from report 1868, pages 99 to 108. Report 1869, experiments with special manures, pages 81 to 86; with corn, pages 86 to 92; with special fertilizers, pages 93 to 101; nine new kinds of oats tested, pages 102 to 104; total pages, 64. Report for 1868 gives, besides those mentioned, experiments in sheep feeding, pages 47 to 71; corn in hills and drills, page 117, etc.; application of muck, ashes, and lime to clover, page 129; experiments on temperature of soils as affected by chemical constitution, pages 137 to 149; total pages of experiments in report 1868 is 107. There are 46 pages of experiments in the preceding volumes, making a total of 217 pages, with a good variety of subjects. The American Agriculturist for February, 1871, just out, details some of these experiments in discussing the profit of feeding grain to cattle and sheep. The able writer would not have taken half a column of them unless it tended to prove something. Joseph Harris, also, in his new work on the pig (1870) gives five pages of what he calls "valuable experiments in feeding young pigs." Mr. Harris once visited the College, and examined into the methods of conducting the experiments. Having experimented with Messrs. Lawes and Gilbert, of England, and being a successful farmer, he should know as to the value of the experiments. Mr. Harris declined the appointment of Professor of Agriculture in Cornell University. These are only some of the most recent of testimonials, given in very strong terms, by competent judges, as to the value of the experiments.

STUDENTS AND GRADUATES.

For account of these, see Senate Journal for January 19, 1871, page 126. I will only add that most parts of the State have been represented at different times in the students. Berrien has had 14 different students; Benzie, 2; Calhoun, 34; Genesee, 24; Hillsdale, 9; Ionia, 44; Jackson, 23; Kalamazoo, 29; Kent, 16; Lenawee, 33; Livingston, 23; Monroe, 9; Montcalm, 6; Newaygo, 3; Oakland, 42; Ottawa, 10; Shiawassee, 24; St. Clair, 6; St. Joseph, 17; Van Buren, 8; Washtenaw, 12; Wayne, 41; Tuscola, 3, etc., etc.

An institution of learning is successful when it turns from its doors a reasonable number of educated and efficient men. If it is a professional school, its graduates will be sought in a good degree in the profession in which the school gives instruction. The Agricultural College probably would not find so large a proportion of its graduates farmers, as schools of law and medicine would find of their graduates in their respective professions, for the Agricultural College is not a merely professional school. It does not give lectures for two seasons to its students, and then confer upon them a degree, but it keeps them through a four years' course of very varied, although largely professional study.

Young men aspiring to success as lawyers, clergymen, and physicians, know that an education *is the only way to the positions they seek.* Young men desiring to be farmers, know, on the other hand, that an education is not essential to the usual amount of success as farmers. The desire to possess an education peculiarly adapted to a farmer's life is *to be created.* "Of course," says the London Times, "a man may be a good farmer, and not able to read and write." If a farmer does not need an education because of his work, he needs it because he is a man and a citizen; and in these days of agricultural papers and general reading, every intelligent farmer feels indignant that such a sentence as that of the London Times can be written of farmers of any country.

This College sends half its graduates back to work upon farms. Is not its tendency, then, to do something to bring to an end the divorce that is thought to exist between agricultural theories and agricultural practice? It has been too much the case heretofore that men who have theorized have not had practice, and that men who worked were unacquainted with scientific principles relating to agriculture. Mutual confidence between theorists and practical men must be established by the experiments of the one and the education of the other class. In this College agriculture is taught as an art, founded not on any theories, but on observation and experiment.

Finally, the Michigan Agricultural College has an excellent reputation throughout the nation, with those who interest themselves in scientific and agricultural education. The testimony to this effect is ample and from excellent sources. If sustained, it should be made to do honor to the State. I know the smallness of a tax is no proof that the tax can be afforded, yet it is nevertheless true that if the institution is doing good, and succeeding in the work of educating farmers, a work which many say is impossible to be done, then the State can afford it all it asks, which amounts to one-half of a tenth of a mill on a dollar, for each of the two years 1871 and 1872, taking Governor Baldwin's estimate of the aggregate assessed valuation of the real and personal estate of the State. The tax is $1 on $10,000.

Shall it not be permitted to keep its present high rank as compared with the similar schools of other States, and be enabled to improve from year to year?

DONATIONS TO THE COLLEGE—1870.

From the DEPARTMENT OF AGRICULTURE, Washington, D. C.:
Seeds and grains; also Monthly Reports of Department.

From U. S. BUREAU OF STATISTICS *(E. Young)*:
Commerce and Navigation, No. 9, Series 1869–70;
" " " June, 1870;
" " " for the year ending June, 1870.

From SMITHSONIAN INSTITUTE:
Smithsonian Miscel. Coll., Vols. VIII. and IX.; also, the Smithsonian Contributions to Knowledge, Vol. XVI.

From COM. B. F. SANDS, Sup't of U. S. Astronomical Observatory:
Meteorological Observations, 1867.

From PROF. BENJ. PIERCE, Sup't U. S. Coast Survey:
Report of U. S. Coast Survey, 1867.

From the ESSEX INSTITUTE:
Bulletin of Essex Institute, Vol. II., Nos. 3, 4, 5.

From HON. ZACH. CHANDLER of Detroit:
An order for plants from public conservatories; Geological Survey of Colorado and New Mexico; Navy Register for 1870; Report of Paris Exposition, Bibliography U. S. S., 1867; Report of Paris Exposition, Weights, Measures, and Coins, U. S. S., 1867; Report of Special Com., Revenue, Commerce, and Navigation, 1869; Report of Education, Paris Exposition, by J. W. Hoyt, Commissioner, Washington, D. C.

From HON. J. M. HOWARD of Detroit:
An order for plants from public conservatories.

From CLEVELAND ABBE, Director of Cincinnati Observatory:
Annual Report of Cincinnati Observatory for year ending June, 1870.

From GEO. H. COOK, State Geologist:
Geology of New Jersey.

From PROF. S. W. JOHNSON, of Sheffield Scientific School:
Green Sand Marl; Green Sand Marl with Shells; Russell & Coe's Superphosphate; South Carolina Phosphate; Sombrero Phosphate; Redondo Phosphate; Phosphate of Iron; Coster Pumice; Quinnipeac Co. Fish Guano.

From CHARLES E. BARNES of Battle Creek:
Slab of Fossiliferous Sandstone.

From C. P. BROWNING:
Fossil Wood.

From S. M. TRACY:
Minerals from Wisconsin and Vermont, consisting of Fossils, Crystals, and Metallic Ores.

From E. J. COOK:
An Indian Arrow Head.

From Israel Harris:
Quartz Crystals; Iron Ore; Heavy Spar; Calcite; Arrow Head; Marble and Clay.
From W. S. George & Co.:
Manufacturer and Builder, Vol. I., bound.
From the Manufacturers:
The Bay State Hay-Rake; Mower Knife Grinder.
From Publishers:
Lansing Republican.
Bay City Journal.
Flint Globe.
Grand Haven Herald.
Grand Haven Union.
Grand Traverse Herald.
Traverse Bay Eagle.
Michigan Argus.
Hillsdale Standard.
Monroe Commercial.
Sturgis Journal.
Peninsular Herald.
Western Rural. (The acknowledgment of the Rural for 1869 was omitted by mistake.)
Book-Buyer.
Journal of the Farm.
Prairie Farmer.
Romeo Weekly Observer.
Western Farmer.
Wolverine Citizen.
Bee Keeper's Journal.
Rural New Yorker.
American Bee Journal.
Communist.
American Missionary.
From College Officers:
Detroit Advertiser and Tribune.
Detroit Post.
Detroit Free Press.
Detroit Commercial Advertiser.
Michigan Teacher.
Nation.
Advance.
Michigan Farmer.
Springfield Republican.

SANFORD HOWARD,

LATE SECRETARY OF THE STATE BOARD OF AGRICULTURE.

Sanford Howard, the subject of this memoir, was born in Easton, Bristol County, Massachusetts, August 7, 1805. He was the sixth descendant from John Howard, who came from England in 1651 and subsequently settled in West Bristol.

His father, Roland Howard, was a somewhat noted farmer, who thought, justly, that the education of his children would be very much promoted by home training and a judicious selection of books. He accumulated what in those days was considered quite a library,—many, if not most of his books, relating to agricultural subjects. To these, and his father's judicious instructions, Mr. Howard was mainly indebted for the tastes which gave character to the pursuits of his maturer years.

Living in a country neighborhood, his advantages for education were limited to three or four months in a year at the district school, but being of a studious turn and quick to learn, he supplemented his studies with such reading as assisted him in laying the foundation for future usefulness.

When quite a boy, he evinced a decided love for natural history, especially that relating to domestic animals.

In early life he became acquainted with Col. Samuel Jaques and the Hon. John Welles, two of the most noted breeders of their times. The judgment of the first named gentleman, in the breeding and raising of all animals belonging to the farm, was considered nearly infallible. Though much older than Mr. Howard, an attachment grew up between the two men that was cemented and intensified by the spirit of inquiry and love of the favorite pursuit which animated the minds of both. This friendship was only severed by the death of Col. Jaques, in 1859. To this intimacy the world is indebted, in a measure, for much of the information disseminated through Mr. Howard's pen during the last thirty years of his life.

When about seventeen years of age, he was placed in a dry goods and grocery store, where he remained about two years, when finding him still disinclined to any pursuit but farming, his father consented to his return home. From this time he remained with his father in his favorite occupation till 1830, when he married Miss Matilda Williams, the daughter of a farmer in his native town, and removed to Hallowell, Maine, where he took charge of the celebrated Vaughan farm. He took with him short-horn stock from the herds of Col. Jaques and Hon. John Welles, at that time the best in the country, having descended from imported stock. There was stock already on the farm, imported by Charles Vaughan, Esq., considered at that time the best in Maine. He also introduced upon the Vaughan farm the Dishley or Bakewell sheep, and the Bedford and Mackay breeds of swine; the latter purchased direct from Capt. Mackay, the founder of the variety.

In the Maine Farmer of March, 1871, Mr. Boardman, the editor, justly remarks: "In introducing these breeds of stock, and in the example of improved farm culture and breeding which he exhibited to the farmers of Maine, he rendered a service to the agriculture of the State, the influence of which has come down to the present time, and will not soon be forgotten. While residing at Hallowell he was a frequent contributor to the columns of the Maine Farmer, then just established, and his friendship for the paper and its several editors continued through life; the present writer having carried on a somewhat lengthy correspondence with him. He rendered great aid in our report on the agriculture of Kennebec, and his reminiscences of early agricultural operations and early farmers in this county were of much interest, and could have been furnished by no other pen."

Having seen, in Massachusetts, the benefits of agricultural societies to a farming community, Mr. Howard became anxious that Kennebec County should enjoy like advantages; and it is not claiming too much to say that to Dr. Ezekiel Holmes, the able editor of the Maine Farmer,—then in its infancy,—and to Mr. Howard, the county is indebted for the establishment of the Kennebec County Agricultural Society. This was the pioneer agricultural society of the State. On the occasion of its first or second exhibition, Mr. Howard was called upon to deliver the address, which he reluctantly consented to do: not that he feared he should not have truths to tell them, but feared that he could not give them an acceptable setting, as he had then never spoken in public excepting before a country lyceum. The society had the address printed in one of the early numbers of the Maine Farmer, where it stands as the production of a young farmer desirous that others should realize the nobility of the calling.

Up to the time of his death Mr. Howard retained a lively interest in the farming prosperity of Maine, and was frequently consulted on the disputed pedigrees of horses and cattle, his evidence being generally conclusive on the subject.

In 1837 Mr. Howard removed with his family to Zanesville, Ohio, where he became engaged in farming, and also for some years conducted an agricultural department in the Zanesville Gazette. Here, too, he was chiefly instrumental in establishing the Muskingum County Agricultural Society. Such was the confidence reposed in his judgment, that he was sent to Massachusetts and New York to purchase stock which continues to take the lead in that part of the State.

In 1844 he received and accepted a position as associate editor of the Albany (N. Y.) Cultivator, with Mr. Luther

Tucker, who has grown venerable in the work of disseminating useful knowledge through his agricultural publications.

His residence in Albany, and connection with a leading agricultural journal, led him to form acquaintances among the most prominent stock breeders and farmers throughout the State. That being the headquarters of the principal men of the New York State Agricultural Society, he formed many personal friendships that ended only with his life.

January 14th, 1852, Mr. Howard removed to Boston, Mass., to take charge of the agricultural department of the Boston Cultivator, which position he maintained with benefit to its readers and satisfaction to its proprietor during twelve years. While there, his character became widely known. He always attended, and often presided, at the weekly farmers' meetings, always held at the State House during the sessions of the Legislature. These weekly discussions, in which he generally took a part, with his editorial labors, caused an intelligent class of farmers to place a high estimate on his judgment and honesty.

In 1857 the Massachusetts Society for the Promotion of Agriculture decided to make an importation of stock for the improvement of their domestic animals. Mr. Howard was employed for the purpose of visiting England, Scotland, Ireland, and France. Furnished with some explicit directions, and large discretionary powers, he fulfilled his mission so satisfactorily in the purchase of cattle, horses, and sheep, that he was solicited by individuals, the next year, to go over the same ground for the same purpose. At this time he purchased a herd of Ayrshire cattle for H. H. Peters, Esq., then of Southborough, Mass.,—a herd that has become too well known and too widely scattered to need any comments here. Mr. Peters, in giving Mr. Howard suggestions as to the purchase of a herd of Ayrshires, says: "I do not mean to have them understood as instructions," and adds that if he will do as well by him at this time as on his previous visit to England, he "can ask for nothing more" and "desire nothing better."

During these two visits of Mr. Howard, he was furnished with so abundant letters from individuals, societies, and men in official stations, to societies and persons in Great Britain, as to secure for him all the attention requisite to the examination of the stock, and methods of agriculture in the islands. He attended the fair of the Royal Agricultural Society at Chester, the Ayrshire County show, and that of Yorkshire, noted for its Booth short-horns.

His chief attention was given to stock of every kind; but he also gave a critical examination to the implements upon trial at the shows, and the improvements by drainage as carried on upon some of the larger estates of England and Scotland. On the occasion of his second visit, Mr. Howard was commissioned by Lord Berwick in person to purchase two Morgan or Blackhawk mares for the improvement of the trotting qualities of his stud.

The editor of the Middlebury (Vt.) Register, in speaking of him, says: "Especially as a judge and improver of farm stock did Mr. Howard excel. Probably, in a general knowledge of all kinds and breeds, no man in this country was his equal. Vermont has always been under obligations to Mr. Howard for the earnest manner in which he has advocated the claims of her Morgan horses upon the public, and not a little of the popularity of Vermont horses abroad is due to his interest in them."

The purchase of Ayrshire cattle led him to the birthplace of the poet Burns, for whom he entertained a great admiration. One of his first purchases of books consisted of an early edition of the life and writings of Burns. With this love for the poet, and as great a familiarity with the incidents of Burns' life as if he had been born in Scotland, no wonder every object connected with him became a sacred memento. On the occasion of his first visit he was introduced to Mrs.

Begg, the sister of Burns, and visited at her house. He found her a hale old lady of eighty-six years, with a memory that enabled her to repeat accurately and with spirit the longest poems written by her brother. This was to Mr. Howard a delight he often dwelt upon. "Auld Kirk Alloway," the scene of the revels described in "Tam O'Shanter," was an object of peculiar interest. The remains of the elder Burns, father of the poet, for whom he entertained a great veneration, repose in the "kirk-yard."

Mr. Howard's importations were not confined to any one kind of stock. So much reliance was placed upon his judgment that he was employed in the selection and introduction of Kerry cattle, Shetland ponies, horses, sheep, and fowls. He was permitted to make some selections from the game fowls of the Earl of Derby, although they were not for sale. He also received some specimens of the best breeds of game fowls from Lord Berwick.

Facilities were freely furnished him in France for the inspection of whatever he desired to examine. In his letters from that country, he mentions the Percheron horse as the best for an omnibus horse that he had seen. He says he rode in a car drawn by three of these horses, from Versailles to Paris,—fourteen miles,—on what was there called the "American railroad," at a speed, considering the load, that was surprising: Time, an hour and a half, including stoppages.

On Mr. Howard's return from Europe he continued to edit the Boston Cultivator until he removed to Michigan. His judgment on all agricultural subjects was highly esteemed, and his papers have been widely copied.

In February, 1864, Mr. Howard was elected Secretary of the Michigan State Board of Agriculture, and in May he removed from Boston to Lansing, Michigan, and entered upon the duties of his office.

His removal from Massachusetts was the occasion of a dinner and presentation to Mr. Howard. The doings of this occasion, as reported specially for the Ploughman, are here given in full, as exhibiting the hold which Mr. Howard had, as well on the friendship as on the respect of the agriculturists and others with whom he had been associated:

DINNER AND PRESENTATION TO SANFORD HOWARD, ESQ.

On Saturday afternoon last the members of the Massachusetts Agricultural Club and others gave a farewell entertainment at the Parker House, to Sanford Howard, Esq., editor of the Boston Cultivator, and well known in this State for his skill in agricultural matters and his efforts for their improvement. Mr. Howard is a veteran agriculturist and editor. His first connection with the press commenced in 1843, when he associated himself with the Albany Cultivator up to 1852, when he entered on an engagement with the Boston Cultivator. The sound practical character of his writings has done much to advance the cause of enlightened agriculture among us, under safe scientific principles. In his public lectures and addresses Mr. Howard has also done much for the spread of usefnl and profitable knowledge among farmers in New England, who will, while they regret his retirement from among them, be pleased to learn that he goes to occupy a situation of great responsibility in connection with his favorite pursuits and studies. He has accepted the office of Secretary of the Michigan Board of Agriculture and of the Agricultural College of that State,—the latter institution being under the control of the Board. His duties will place him among the Faculty, and will probably embrace those of making himself acquainted with the agricultural resources of the State, of making annual reports concerning the progress of improvement, of advising in regard to the farm of six hundred acres attached to the College, to institute, supervise, and report upon agricultural experiments made thereon, and generally to direct matters associated with the rural economy of the institution. His abilities are equal to any requirements that may be made on him. He is an excellent farmer and judge of horses and cattle, and has several times visited Europe for the Massachusetts Society for the Improvement of Agriculture, and for other parties, to purchase horses and cattle, which he has most judiciously done, greatly to the benefit of individuals and the agricultural interest of the country. Mr. Howard will leave for his new situation about a fortnight hence.

Cheever Newhall, Esq., President of the Massachusetts Agricultural Club, presided on the occasion of the dinner, which was in Parker's best style. At the close of the physical entertainment the chairman announced the special object of the assembling of the party, which was to present Sanford Howard, Esq., with a token of the respect of his agricultural friends in this neighborhood, and called on Dr. George B. Loring, of Salem, to present to Mr. Howard a massive silver pitcher which bore the following inscription:

"Presented to SANFORD HOWARD, ESQ., by members of the Massachusetts Agricultural Club, and other friends, as a token of their appreciation of his services for the improvement of agriculture, and their respect for his character. April 30, 1864."

DR. LORING'S PRESENTATION SPEECH.

Dr. Loring, in fulfilling a duty for which he is so eminently qualified, said:

GENTLEMEN—I beg leave to express the obligations I am under for the compliment you have paid me, in electing me as your messenger of pleasant recognition to one of our friends. There are few duties more agreeable than this. He who carries with him his own generous sentiments alone, is always welcome. But he who, in every word he utters, gives expression to the kind impulses of appreciative friendship, as well as to the more sober judgment of the intelligent and impartial stranger, bears with him a double burthen of honor and pleasure. I am not called on to speak for myself alone. The gentlemen who have assembled here to pay a tribute of respect to one of the most useful members of the agricultural fraternity, have long been known to me as the guides and teachers in the great art which we all admire. The light which they have shed upon the path of those who would cultivate the earth for the necessary wants of man, and of those who would adorn and embellish it, has guided many of us in the service of nature, which always brings its reward. It is not easy to express the respect and admiration I feel for him who points out to us some manifestation of the never-ending bounties of the earth. He who adorns the landscape with a fresh tinge of color from some rare and unknown foliage,—he who develops some new and still more beautiful variety of fruit,—he who paints a flower with choicer colors, and clothes it with brighter graces,—he who transforms the rudeness of untamed earth into the luxuriant field, and with his own hand arranges the beauties of the garden and the grandeur of the forest, gives to social life that warming and invigorating influence without which more brilliant attainments would fall upon stony and ungenial hearts. If you would realize the value of him, lay the axe at the root of the trees which he has planted; upturn the soft and swelling lawn which he has shaped; strip your homes of all the beauty which his skill has gathered around them; send the cultivated landscape, with its waving grain and luxuriance of herbage, back to its original wild and rugged features, and leave mankind, without the refinement which all these bestow, to struggle in the forum and the market place. Could anything be more sullen and frigid than a community deprived of these embellishments? Could wealth and honors and power fill the vacant spot, or impart a ray of warmth to this cold and heartless scene?

In behalf of the gentlemen here assembled, whose leisure hours have been spent in efforts to beautify the dwelling-places of men, I desire to present to you, Mr. Howard, a token of their respect and esteem. We all know how faithfully, and with how little ostentation, you have devoted your life to the study of agriculture, and to the work of imparting your knowledge to others; and I trust we all appreciate the services of one who successfully performs his duty as an agricultural editor.

The farmer's newspaper is, in our country, almost the sole guide of the farmer's labor. It has thus far performed the part of college and teacher. It constitutes a large portion of the literature of that profession which all men love, and upon which all men depend, directly or indirectly, for their subsistence. There is in its pages a common ground where all conflict ends; a platform upon which all can stand; a creed which all can believe. And who does not know the inward peace and satisfaction with which the unhappy voyager across the stormy surface of a partisan press finds repose in these columns, which remind him of the calm and steady and luxuriant promises of nature,—of growing crops, and of animals devoted to the "service of man?" And more: Who does not know that whatever progress has been made in agriculture has received its stimulus and direction from these same columns? By suggestion, by investigation, by *records of experiments,* by statements of successes, has the agricultural newspaper press of our day kept the agricultural mind stimulated and informed. When larger and more ambitious designs accomplish in a more imposing manner what the agricultural editor is quietly doing every week, we shall be sure that something positive is done in the way of agricultural education.

As you pass then, sir, from the humble sphere of usefulness to the more conspicuous, and are called to fill a high position in the agricultural institutions of a great producing State, you have but to carry with you the knowledge which you have dispensed so freely here to all who would read, and to supply still the universal demand for agricultural information. In doing this, be assured you carry with you the best wishes of all those who have always found in you a sound and intelligent agricultural adviser, and a wise and discriminating sympathizer. Our choicest dairy herds bear the marks of the best blood, selected by you from the Highlands of Scotland. The best families of New England horses have found in you an unwavering defender. The improvements in cultivation, in machinery, in drainage, have not escaped your vigilance. And we congratulate ourselves that, however far from us may be your future field of action, your counsels and your information must still remain.

In begging you to accept this token of our high estimation of your character and services, we would add our warm wishes for your usefulness and prosperity and happiness.

Following the remarks of Dr. Loring, Mr. Howard spoke in substance as follows:

Mr. President and Gentlemen—I feel that I shall hardly be able to reply in fitting terms to this token of your favor. I can only say that it will always be regarded by me, and by those with whom I am connected by family ties, with feelings of gratitude and pride.

Of my labors for the improvement of agriculture, to which allusion has been so flatteringly made, it does not become me to say much. I was in my youth impressed with the truthfulness of the maxim that Agriculture is the nursing mother of nations; and from my earliest recollections I have entertained an ardent desire to investigate the principles involved in vegetation, and in the organization and characteristics of animals. The passion grew with my growth, and strengthened with my strength. For the past twenty years, circumstances have placed me in a position in which it has been my duty to collect and disseminate information relating to rural affairs. In regard to the results of my labors in this field, my friends can judge better than myself; but I may be permitted to say that I have been actuated by honest motives, and that, though I may have committed mistakes, I have never advocated or countenanced anything that I did not believe would be really and intrinsically useful to the public.

The improvements which have been made in agriculture

since my connection with the press have been numerous and striking. I have not the vanity to assume that my own feeble efforts have availed much in the origination of these improvements. I have watched with great interest the progress which has been made in the substitution of machinery for manual labor, by which a vast change has been effected in some of the most important farm processes; and it is a matter of laudable pride to us Americans that these improvements emanated from our country, and that through the agency of our artisans the husbandry of every civilized nation in the world has been advanced. In reference to ourselves as a nation, the importance of these improvements can scarcely be estimated. It is through them alone that we have been able to fill our armies, and to obtain from our soil the means of supporting our people.

But I can only allude in general terms to these things. The association under whose auspices we are assembled has performed its part in the rural improvements of the age. Particular conditions or circumstances have rendered it expedient that particular branches of agriculture should here receive special attention. And who can estimate the advantages which have resulted, or may yet result, from the light brought out by the members of this association, collectively or individually, on the cultivation of some of the most valuable fruits? I intend no empty compliment when I say that I have been proud to hear the names of gentlemen enrolled in your association mentioned in terms of profound respect in foreign countries. It has been my fortune to witness examples of high cultivation in the Old World; and yet, I am happy to say, on lands owned and cultivated by members of the Massachusetts Agricultural Club, I have seen examples which would compare favorably with any I have met with elsewhere.

Among the "other friends" to whom I am this day placed under lasting obligations, I recognize leading improvers and cultivators in various departments of agriculture.

Gentlemen, the destinies which control me seem to require that I shall take up my residence in a distant locality. What there awaits me cannot be foreseen. But

> "Who that in distant lands has chanced to roam,
> Ne'er thrilled with pleasure at the name of *home?*"

That endearing name I first learned to love in the Commonwealth of Massachusetts. May God bless the dear old State!

> "I love her rocks and rills,
> Her woods and templed hills!"

Neither these, nor the friends of my childhood, nor those of later years, will be forgotten while the power of memory remains. And when in coming time the scenes of the past shall arise—as arise they will—before my mental vision, none will be more fondly cherished, or reviewed with greater pleasure than this, in which, by your kindness, I am made so prominent a participator.

Speeches—all of them incidentally and highly complimentary to the private and professional character of Mr. Howard, and also conveying very valuable agricultural information, were made by C. M. Hovey, Esq., President of the Massachusetts Horticultural Society, Albert Fearing, Esq., President of the Hingham Agricultural Society, Charles L. Flint, Esq., Secretary of the Board of Agriculture, Hon. J. S. Cabot of Salem, Dr. Eben Wight of Dedham, H. H. Peters, Esq., of Southboro, Prof. J. D. Runkle of the U. S. Nautical Almanac, Josiah Stickney, Esq., of Watertown, J. A. Billings, Esq., of West Roxbury, Aaron D. Weld, Esq., of West Roxbury, John R. Brewer, Esq., Mr. John C. Moore of the Boston Journal, and others.

Intermingled in the addresses made on the occasion were remarks commendatory of the character of Col. Marshal P. Wilder, and sympathizing with him in his physical affliction, which has, for a time, deprived the Agricultural Society (not only of the locality, but of the country at large), of one of its most energetic and useful men and brightest ornaments.

The proceedings were extremely pleasant in every detail, and justly complimentary to Mr. Howard, who leaves Boston with the best wishes of the best men of the agricultural community of New England.

Mr. Howard's labors in Michigan were more in the interests of the agriculture of the State at large than with the State Agricultural College. His office was not at the College, but in the city, some three and a half miles away. He was, however, a member of the Faculty, and took his turn in the general lectures delivered before the College as a whole, and his addresses were always highly welcomed by his audience. He was also in a position to give his advice in all matters pertaining to the management of the College, with the entire Faculty of which he livedon terms of personal friendship.

But the chief duties of his place led him to form acquaintance with the State, its resources, productions, and capabilities; to become acquainted with its chief agriculturists, and the efforts they were making for the improvement of their lands, crops, and herds. In the exercise of these duties he made a large and varied acquaintance with agriculturists, fruit growers, and stock raisers of the State, and many of the improvements carried on by these men were suggested, or furthered, or encouraged by him. He superintended the publication of the annual report of the Secretary of the Board of Agriculture, and six bound volumes of these reports, filled with matter selected and edited, or entirely written by him, attest his diligence, his painstaking, and his worth in his department of labor.

Mr. Howard's judgment was often sought in regard to the merits of agricultural implements and machinery. His eye was critical to detect unsound principles, and his fear of recommending or encouraging a worthless article made him cautious. Nor was he ever known to allow his judgment to be warped by anything approaching a bribe. Of him it might justly be said, he was

> "A friend to truth, of soul sincere,
> In action faithful, and in honor clear."

He served on the committees employed in every trial of implements by the New York Agricultural Society. In the last and most important one, held at Utica, in October, 1870, he acted as chairman. Making the report of that trial, where over two hundred implements and machines were tested, was one of his latest labors.

As early as 1832 or 1833, Mr. Howard, with two or three other intelligent farmers, was invited to examine the threshing-machine just made by the Pitts brothers of Winthrop, Maine. The machine was in a barn, where it had been built privately. What the opinion was on the value of the machine it is needless to say, as that, with all the subsequent improvements, is now well known throughout the civilized world. Mr. Howard often remarked that those men were deserving of statues from their countrymen; and it was with him a subject of much pride that he was in a degree identified with the introduction of such a machine to the agricultural world.

In the spring of 1871, Mr. Howard was stricken with partial paralysis, affecting his right side. He went from his office, where the fatal disease had found him at his work, to his home, where, after an eight days' illness, during which he was conscious and in possession of his mind to the last, he died on the 9th of March.

Mr. Howard left a widow, one son, and three daughters, to mourn his loss.

The newspapers throughout the entire country paid their tribute of respect to the memory of the deceased. A few extracts from the extended obituaries may serve in this place as a record of the esteem in which he was held.

The Country Gentleman says: "In the Cultivator for the month of January, 1844, it was announced that Mr. Howard had accepted a position in the editorial duties of this office. He had then been conducting an agricultural department for some years in the Zanesville (Ohio) Gazette, near which place he was engaged in farming; and it was the high character of his writings for that paper, together with occasional communications for our own columns, which led to the engagement. In announcing it, he was spoken of as 'a practical farmer from his youth, a close and accurate observer, intimately acquainted with the husbandry of different sections of our country, a good judge of farm stock, and familiar with all the best breeds among us.' Indeed, it was the personal knowledge of agricultural affairs, derived from hard-earned experience in actual farming in the States of Massachusetts, Maine, and Ohio, at a time when the labors of the farm were much more severe and its rewards smaller than now, which formed the sound and substantial basis on which Mr. Howard's subsequent judgment and writings were founded. * * He wrote nothing without full consideration, and the patient expenditure of whatever time the proper investigation of the subject might require."

The Boston Cultivator spoke of him as a "public loss," "to be honorably mentioned among the pioneers of the revival of agricultural literature in this country," and as having "left a name that will not soon be forgotten."

The Buffalo Daily Courier, in an editorial written by one intimately acquainted with Mr. Howard, says: "He was a cordial hater of shams of every description, and labored faithfully to elevate the tone, and give more reliable character to the agricultural journals of the country. Socially he was one of the best and purest of men. His temperament and habits of thought were not such as to make him a leader of masses of men; but he had deserved influence with those who knew him intimately, and possessed the fullest confidence of all who met him in the home circle. He was a kind-hearted, generous, and just man, a true friend, and a liberal, catholic, and Christian gentleman."

The Michigan Farmer, after speaking of the value of his labors, says: "His sincere, unassuming character made him to be highly esteemed. Mild and quiet in manner, and studious in habit, while he helped others and became interested in their labors, he was not selfish, and derived but little benefit from their labors beyond the satisfaction of knowing that he had aided in the increase of a correct knowledge of the practice and principles of agriculture, and that he had done something to render it a nobler pursuit than it was when he first entered upon its study."

The announcement of Mr. Howard's death was followed by the voluntary offer of the use of the Hall of Representatives, by the House then in session, for the funeral services, which occurred on the 11th of March. The Legislature of the State paid further tribute to the memory of the deceased.

Copy from Lansing Republican, Senate resolutions and House resolutions:

RESOLUTIONS OF RESPECT.

The following resolution in relation to the death of Hon. Sanford Howard, offered by Senator Begole, was unanimously adopted in the Senate last Saturday:

Resolved, That, having learned with deep regret the death of Hon. Sanford Howard, Secretary of the State Board of Agriculture, so suddenly stricken down while in the midst of life and apparent health, we extend to his mourning and bereaved family our sympathy in this hour of deep affliction, and that as a tribute of respect to the memory of one who has so long and ably served one of the great public interests of our State, both at home and abroad, who has stood first and foremost as one of the most eloquent and successful advocates of the great agricultural interests of the country, who for years has done so much to develop and bring it to its present advanced and prosperous condition, and whose loss cannot be considered less than a public calamity, this body will adjourn this afternoon and attend the funeral at 2 o'clock, P. M.

On the same day the House adopted the following resolutions, offered by Mr. Walton:

Whereas, This House has learned, with profound sorrow, of the decease of Hon. Sanford Howard, Secretary of the State Board of Agriculture of this State, and long and intimately connected with said Board and the State Agricultural College, and thus closely identified with the agricultural interests of the State; therefore

Resolved, That the members of this House, as a body, tender to the family of the deceased their heartfelt sympathies in this their great affliction.

Resolved, That in the death of Mr. Howard the agricultural interests of this State have lost a faithful, efficient, and peculiarly intellectual friend and co-laborer in the great cause of progressive agriculture, and the State Board of Agriculture, and the State Agricultural College an able and invaluable officer.

Resolved, That this preamble and resolutions be entered in full on the journal of this House, and the Clerk be directed to transmit a copy thereof to the family of the deceased.

Resolved, That in respect for the memory of the deceased, this House do now adjourn.

The funeral services were conducted by President Abbot, of the State Agricultural College. He delivered a discourse to a large audience. Speaker Woodman, Senator Begole, Representatives Hart and Cameron, and Professors Miles and Fairchild of the Agricultural College, acted as pall-bearers.

"The discourse was a fitting eulogium of the moral and mental qualities of the deceased. He was alluded to as a man of great purity of soul and purpose, and one who possessed a high-toned religious faith and trust. His aim had been to raise the standard of agriculture to the highest attainable point, and he only sought to disseminate truth in the great science of agriculture.

"Thus passes away another true and able man. With a long life of mental labor and contact with humanity, he kept the principles of truth and purity ever dear, and his moral nature uncontaminated or debased by aught that was wrong. He rests, and his place will not easily be filled. His teachings remain to elevate and ennoble the cause of agriculture, to which he gave his manhood. Like other brain-workers, he has gone in the prime of mental vigor, and all is well."

The following resolutions on the death of Mr. Howard were passed by the State Board of Agricuture:

Whereas, The Secretary of this Board, the Hon. Sanford Howard, was removed from us by death on the ninth of March, therefore

Resolved, That we recognize in our late Secretary a man of eminent abilities and acquirements, and national reputation in matters relating to agriculture in all its branches; and that we cherish a grateful memory of the many private virtues that endeared him to us and made him universally respected.

Resolved, That we tender to his afflicted widow and children our sincere sympathy with them in their deep sorrow.

NOTE.—The closing words of the first paragraph in this memoir should read "West Bridgewater," instead of "West Bristol." This error was discovered too late to be rectified in the proper place.

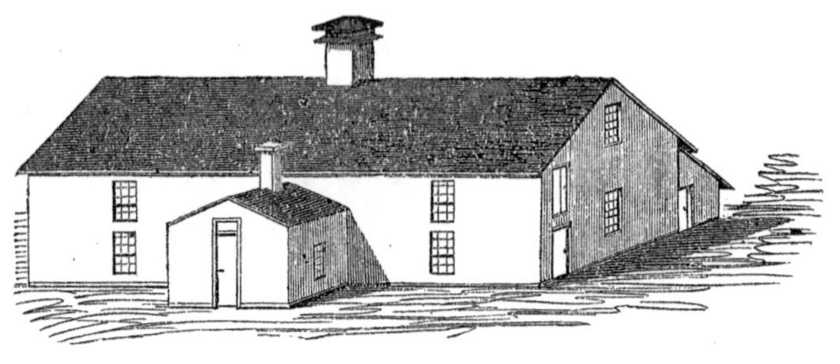

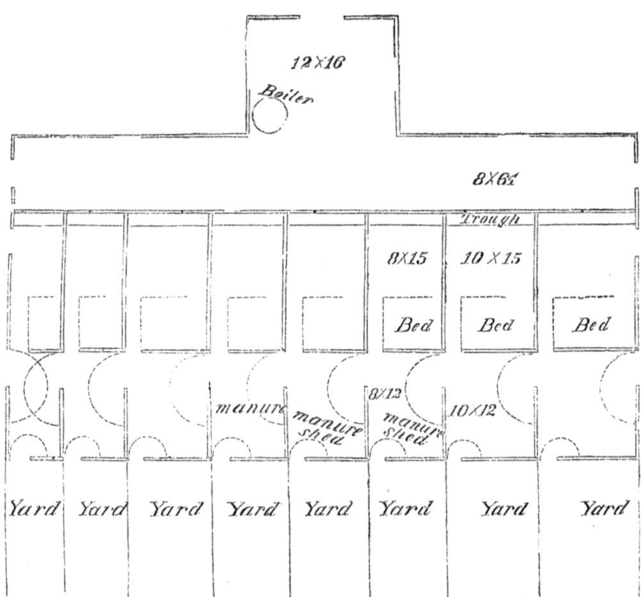

PIGGERY OF THE MICHIGAN STATE

AGRICULTURAL COLLEGE.

NOTE.—The plan has been somewhat changed since this cut was made.

EXPERIMENTS.

To the President of the State Agricultural College:

The following report of experiments made in the Farm Department, for the year 1870, is respectfully presented.

M. MILES, *Superintendent.*

EXPERIMENTS IN PIG-FEEDING.

The experiments in feeding, made in the years 1868 and 1869, have been repeated the present year for the purpose of verification, and the correction of any errors that may have arisen from peculiarities of the season.

The method of feeding and weighing described in the Report of 1869 (see Report of Secretary of State Board of Agriculture for 1869, pp. 53, 54, 55), has been strictly followed in the present experiments, with satisfactory results.

The plan of placing a single animal in each pen has many advantages, and we have thus far been unable to discover any unfavorable influences arising from this mode of treatment.

As in the experiment of 1869, the animals selected were of various ages and of different strains of blood, for the reason that uniformity in these particulars could not be secured. The experiment was commenced August 30th, and continued for sixteen weeks. Eleven pigs were put in the pens at the beginning of the experiment, as follows:

Pen No. 1, pure-bred Essex, age seven and one-half months.
" " 3, " " " " " " " "
" " 2, " " Suffolk, " one year and seven weeks.
" " 4, " " " " " " " "

Pen No. 5, pure-bred Suffolk, age one year and ten weeks.
" " 6, " " " " " " " "
" " 8, " " " " " " " "
" " 7, "Grades or Natives," age four months (estimated).
" " 9, " " " " " "
" " 10, " " " " " "
" " 12, " " " " " "

Pigs Nos. 1 and 3 were from the same litter.

Pigs Nos. 2 and 4 were from the same litter.

Pigs Nos. 5, 6, and 8 were from the same litter.

Pigs Nos. 7, 9, 10, and 12 were from the same litter.

Pigs Nos. 5, 6, and 8 were the same as Nos. 7, 9, and 11 of the experiment of 1869, that were turned out the seventeenth week on account of the "oily secretion from the skin," and their "stunted" appearance. (See Report of Secretary of State Board of Agriculture, 1869, p. 71).

The feed of these pigs, after they were removed from the experimental pens, in 1869, to the commencement of the experiment of 1870, was ordinary slops from the house, and a small allowance of corn during the winter, and slops and grass during the summer.

Pig No. 13 of the experiment of 1869 was disabled by an accident, and killed in the spring.

On the 18th of October a pure-bred Essex sow, four years and seven weeks old, was put in pen No. 13. This sow was No. 3 of the pigs under experiment in 1866 that were fed with milk, by means of the "nursing-bottle," from their birth. (See Report of Secretary of State Board of Agriculture for 1866, p. 59.)

The "grade or native" pigs, Nos. 7, 9, 10, and 12, were part of a lot picked up in the country to consume the slops from the boarding-hall that were not needed by the pure-bred swine on the farm, as they were in high condition on grass alone. The "*blood*" of these "*natives*" was not known, as their ancestry could not be traced but one generation by their owner, and that on the side of the dam. They answered the purpose admirably for which they were purchased, but the return for feed consumed was apparently very small. They were, however, better than the average "native" swine of this vicinity.

As in the previous experiments, the pigs were weighed at the beginning of each week, before feeding, and their ration of unbolted corn meal (their exclusive feed) was weighed out at the time of feeding. Care was also taken to give each pig all the meal it would readily consume, the amount of each feed being regulated by the appetite of the animal, so that no waste could take place by leaving a portion of the ration in the trough.

The weight of each pig at the beginning of the experiment and at the close of each week, together with the total gain and total gain per cent is given in Table No. 1.

The weights in this and the following tables are given in pounds and decimals of a pound.

TABLE No. 1.

WEIGHT OF EACH PIG AT THE BEGINNING OF THE EXPERIMENT, AND AT THE CLOSE OF EACH WEEK, TOGETHER WITH THE TOTAL GAIN, AND TOTAL GAIN PER CENT.

DATE.	WEEK OF EXPER'MT.	PEN 1.	PEN 3.	PEN 5.	PEN 7.	PEN 9.	PEN 13.	PEN 2.	PEN 4.	PEN 6.	PEN 8.	PEN 10.	PEN 12.
August 30.	*Beginning of Exper'nt*	155,00	157.50	236.50	66.00	57.00	----------	210.00	179.00	176.00	160.50	75.00	71.00
Sept.... 6.	1st..........	169.00	173.50	245.00	66.50	56.50	----------	213.50	181.00	186.00	171.50	81.00	75.00
" ....13.	2d..........	178.00	180.00	255.00	74.00	61.50	----------	223.50	192.00	188.00	176.50	88.00	81.00
" ....20.	3d..........	180.50	186.00	260.00	76.50	63.00	----------	227.00	195.50	191.00	180.00	92.00	86.00
" ....27.	4th..........	196.00	201.00	273.00	82.50	70.50	----------	237.50	209.50	203.50	190.50	104.00	97.50
Oct..... 4.	5th..........	208.50	215,50	282.00	89.00	78.00	----------	245.50	221.50	214.50	203.00	112.00	103.00
" ....11.	6th..........	222.00	229.50	297.00	96.00	79.50	----------	257.50	230.50	227.00	210.00	121.50	107.00
" ....18.	7th..........	232.00	236.00	306.00	104.00	84.00	438.00	262.50	235.50	236.00	224.00	128.50	113.50
" ....25.	8th..........	250.00	244.00	332.00	111.00	89.00	434.50	279.50	247.50	247.00	236.00	138.50	122.00
Nov.... 1.	9th..........	256.50	262.00	337.00	109,00	93.00	475.50	287.50	256.50	257.00	243.00	144.00	130.50
" .... 8.	10th..........	272.00	275.00	342.00	125.00	97.00	493.00	294.00	266.00	266.50	252.50	152.00	135.50
" ....15.	11th..........	283.00	284.00	352.00	125.00	101.50	512.00	300.00	272.50	274.00	258.00	157.00	128.00
" ....22.	12th..........	286.50	289.50	359.00	130.00	101.00	516.50	308.50	281.00	280.00	263.00	161.00	138.50

Nov.....29.	13th........	300.00	297.00	356.00	134.00	104.00	528.00	317.00	290.00	287.00	268.00	164.00	141.00
Dec..... 6.	14th........	314.00	306.00	374.00	135.00	105.00	530.00	322.00	298.00	294.50	275.00	168.00	137.00
" .....13.	15th........	313.50	310.00	383.50	136.50	106.00	538.00	331.00	305.00	302.50	282.00	169.00	137.00
" .....20.	16th........	316.00	318.00	404.00	134.50	105.50	526.00	339.00	312.50	303.00	286.50	165.00	142.50
Total gain............		161,00	160.50	167.50	68.50	48.50	88.00	129.00	133.50	127.00	126.00	90.00	71.50
Total gain per cent..		103.80	101.90	70.80	103.78	85,08	20.09	61.42	74.58	72.16	78.50	120.00	100.70

The amount of meal consumed in each pen for each week, the total amount of meal consumed in each pen, and the amount of feed required on the average to produce one pound of increase of live weight, will be found in Table No. 2.

TABLE No. 2.

AMOUNT OF MEAL CONSUMED IN EACH PEN FOR EACH WEEK, THE TOTAL MEAL CONSUMED, AND THE AVERAGE AMOUNT OF MEAL CONSUMED TO PRODUCE ONE POUND OF INCREASE.

WEEK.	PEN 1.	PEN 3.	PEN 5.	PEN 7.	PEN 9.	PEN 13.	PEN 2.	PEN 4.	PEN 6.	PEN 8.	PEN 10.	PEN 12.
1st week	32.75	36.25	36.75	20.00	19.50	----------	23.75	26.50	32.00	29.75	25.00	27.00
2d week	41.75	41.75	41.75	24.00	22.25	----------	29.00	30.25	30.25	30.25	30.25	28.50
3d week	46.50	45.00	46.00	23.25	20.25	----------	33.00	33.00	33.00	33.00	29.75	29.25
4th week	42.50	42.50	42.50	24.00	24.00	----------	34.00	34.00	34.00	34.00	31.25	30.25
5th week	42.75	42.75	42.75	26.75	23.75	----------	37.50	37.50	37.50	37.50	32.50	28.75
6th week	48.00	48.00	48.00	30.75	21.00	----------	43.75	43.50	43.50	43.50	35.00	30.50
7th week	52.50	52.50	52.50	28.25	28.25	----------	43.75	43.75	43.75	43.75	35.00	28.25
8th week	52.50	52.50	52.50	31.25	21.50	52.50	43.75	43.75	43.75	43.75	35.00	30.50
9th week	51.00	52.50	51.00	32.50	23.25	84.00	42.50	42.50	42.50	42.50	34.00	34.00
10th week	52.00	52.00	52.00	33.50	23.25	88.00	43.50	43.25	43.75	43.75	40.00	35.00
11th week	52.50	52.50	52.50	30.00	21.25	87.50	43.75	43.75	43.75	43.75	37.50	33.00
12th week	51.00	51.00	51.00	24.00	19.00	68.50	42.50	42.50	42.50	32.50	31.25	27.00
13th week	52.50	52.50	52.50	28.00	21.00	87.50	43.75	43.75	43.75	44.00	25.00	30.00
14th week	52.50	52.50	52.50	27.00	18.75	65.00	43.75	43.75	43.75	43.75	30.00	26.00
15th week	49.00	58.50	70.00	22.00	19.50	58.00	45.75	45.75	45.75	41.25	26.00	19.00

16th week	42.75	51.25	70.00	16.50	15.00	50.75	48.75	43.75	48.75	48.75	15.75	17.75
Total	762.25	784.00	813.75	420.75	336.50	641.75	637.75	641.25	647.25	630.75	493.25	454.75
Average am't of meal to 1 ℔. increase	4.73	4.88	4.85	6.14	6.93	7.29	4.94	4.80	5.09	5.01	5.48	6.36

In Table No. 3 is given the average weight of each pig for each week of the experiment.

TABLE No. 3.

AVERAGE WEIGHT OF EACH PIG FOR EACH WEEK OF THE EXPERIMENT.

WEEK.	PEN 1.	PEN 3.	PEN 5.	PEN 7.	PEN 9.	PEN 13.	PEN 2.	PEN 4.	PEN 6.	PEN 8.	PEN 10.	PEN 12.
1st week	162.00	165.50	240.75	66.25	56.75	----------	211.75	180.00	181.00	166.00	78.00	73.00
2d week	173.50	176.75	250.00	70.25	59.00	----------	218.50	186.50	187.00	174.00	84.50	78.00
3d week	179.25	183.00	257.50	75.25	62.25	----------	225.25	193.75	189.50	178.25	90.00	83.50
4th week	188.25	193.50	266.50	79.50	66.75	----------	232.25	202.50	197.25	185.25	98.00	91.75
5th week	202.25	208.25	277.50	85.75	74.25	----------	241.50	215.50	209.00	196.75	108.00	100.25
6th week	215.25	222.50	289.50	92.50	78.75	----------	251.50	226.00	220.75	206.50	116.75	105.00
7th week	227.00	232.75	301.50	100.00	81.75	----------	260.00	233.00	231.50	217.00	125.00	110.25
8th week	241.00	240.00	319.00	107.50	86.50	436.25	271.00	241.50	241.50	230.00	133.50	117.75
9th week	253.25	253.00	334.50	110.00	91.00	455.00	283.50	251.75	252.00	239.50	141.25	126.25
10th week	264.25	268.50	339.50	117.00	95.00	484.25	290.75	261.25	261.75	247.75	148.00	133.00
11th week	277.50	279.50	347.00	125.00	99.25	502.50	297.00	269.25	270.25	255.25	154.50	131.75
12th week	284.75	286.75	355.50	127.50	101.25	514.25	304.25	276.75	277.00	260.50	159.00	133.25
13th week	293.25	293.25	357.50	132.00	102.50	522.25	312.75	285.50	283.50	265.50	162.50	139.75
14th week	307.00	301.50	365.00	134.50	104.50	529.00	319.50	294.00	290.75	271.50	166.00	139.00
15th week	313.75	308.00	378.75	135.75	105.50	534.00	326.50	301.50	298.50	278.50	168.50	137.00
16th week	314.75	314.00	393.75	135.50	105.75	532.00	335.00	308.75	302.75	284.25	167.00	139.75

For the purpose of showing the relation of feed consumed to the weight of the animal, Table No. 4 has been made, which gives the amount of feed consumed per week in each pen for each 100 lbs. live weight.

TABLE No. 4.

MEAL CONSUMED PER WEEK FOR EACH 100 POUNDS OF LIVE WEIGHT.

WEEK.	PEN 1.	PEN 3.	PEN 5.	PEN 7.	PEN 9.	PEN 13.	PEN 2.	PEN 4.	PEN 6.	PEN 8.	PEN 10.	PEN 12.
1st week	20.21	21.90	15.26	30.18	34.36	--------	11.21	14.72	17.68	17.92	32.05	37.00
2d week	24.06	23.62	16.70	34.16	37.71	--------	13.23	16.22	16.17	17.38	35.79	36.54
3d week	25.94	24.59	17.86	30.89	32.53	--------	14.65	17.03	17.41	18.51	33.05	35.03
4th week	22.57	21.96	15.94	30.19	35.94	--------	14.63	16.79	17.23	18.35	31.88	32.97
5th week	21.13	20.50	15.40	31.20	31.98	--------	15.52	17.40	17.94	19.06	30.09	28.67
6th week	22.29	21.57	16.58	33.24	26.66	--------	17.39	19.24	19.70	21.06	29.98	29.04
7th week	28.12	22.55	17.41	28.25	28.44	--------	16.82	18.77	18.89	20.16	28.00	25.62
8th week	21.78	21.87	16.45	29.06	24.85	12.03	15.77	18.11	18.11	19.02	26.22	25.90
9th week	20.13	20.75	15.24	29.55	25.54	18.46	14.99	16.88	16.85	17.74	24.61	26.93
10th week	19.67	19.36	15.31	28.63	24.47	18.19	14.96	16.55	16.71	17.65	27.02	26.31
11th week	18.09	18.78	15.12	24.00	21.41	17.41	14.73	16.24	16.18	17.14	24.27	25.14
12th week	17.91	17.78	14.34	18.82	18.76	13.32	13.93	15.35	15.34	12.47	19.65	20.26
13th week	17.90	17.90	14.68	21.21	20.48	16.75	13.98	15.32	15.43	16.57	15.38	21.46
14th week	17.10	17.41	14.38	20.07	17.94	12.28	13.69	14.88	15.04	16.11	18.07	18.70
15th week	15.61	18.99	18.48	16.20	18.48	10.86	14.01	15.17	15.32	14.81	15.43	13.87
16th week	13.58	16.33	17.77	12.17	14.18	9.53	13.06	14.16	14.45	15.39	9.43	12.74

In Table No. 5 is given the gain in live weight made in each pen for each week of the experiment. A loss in weight is indicated by the minus sign.

TABLE No. 5.

GAIN PER WEEK OF EACH PEN.

WEEK.	PEN 1.	PEN 3.	PEN 5.	PEN 7.	PEN 9.	PEN 13.	PEN 2.	PEN 4.	PEN 6.	PEN 8.	PEN 10.	PEN 12.
1st week	14.00	16.00	8.50	0.50	—0.50	----------	3.50	2.00	10.00	11.00	6.00	4.00
2d week	9.00	6.50	10.00	7.50	5,00	----------	10.00	11.00	2.00	5.00	7.00	6.00
3d week	2.50	6.00	5.00	2.50	1.50	----------	3.50	3.50	3.00	3.50	4.00	5.00
4th week	15.50	15.00	13.00	5.00	6.50	----------	10.50	14.00	12.50	10.50	12.00	11.50
5th week	12.50	14.50	9.00	6.50	7.50	----------	8.00	12.00	11.00	12.50	8.00	5.50
6th week	13.50	14.00	15.00	7.00	1.50	----------	12.00	9.00	12.50	7.00	9.50	4.00
7th week	10.00	6.50	9.00	8.00	5.50	----------	5.00	5.00	9.00	14.00	7.00	6.50
8th week	13.00	8.00	26.00	7.00	5.00	—3.50	17.00	12.00	11.00	12.00	10.00	8.50
9th week	6.50	18.00	5.00	—2.00	4.00	41.00	8.00	9.00	10.00	7.00	5.50	8.50
10th week	16.00	13.00	5.00	16,00	4.00	17.50	6.50	9.50	9.50	9.50	8.00	5.00
11th week	11.00	9.00	10.00	0.00	4.50	19.00	6.00	6.50	7.50	5.50	5.00	—7.50
12th week	3.50	5.50	7.00	5.00	—0.50	4.50	8.50	8.50	6.00	5.00	4.00	10.50
13th week	13.50	7,50	—3.00	4.00	3.00	11.50	8.50	9.00	7.00	5.00	3.00	2.50
14th week	14.00	9.00	18.00	1.00	1.00	2.00	5.00	8.00	7.50	7.00	4.00	—4.00
15th week	—.50	4.00	9.50	1.50	1.00	8.00	9.00	7.00	8.00	7.00	1.00	0.00
16th week	2.50	8.00	20.50	—2.00	—0.50	—12.00	8.00	7.50	0.50	4.50	—4.00	5.50

The amount of feed required to produce one pound of increase in live weight for each week, is given in Table No. 6. "No gain" is entered in the columns where losses in weight occurred.

TABLE No. 6.

FEED REQUIRED TO PRODUCE ONE POUND OF INCREASE FOR EACH WEEK.

WEEK.	PEN 1.	PEN 3.	PEN 5.	PEN 7.	PEN 9.	Pen 13.	PEN 2.	PEN 4.	PEN 6.	PEN 8.	PEN 10.	PEN 12.
1st week	2.34	2.27	4.32	40.00	No gain	----------	6.79	1.33	3.20	2.70	4.17	6.75
2nd week	4.64	6.42	4.17	3.20	4.45	----------	2.90	2.75	15.13	6.05	4.32	4.75
3rd week	18.60	7.50	9.20	9.30	13.50	----------	9.43	9.43	11.00	9.43	7.44	5.85
4th week	2.74	2.83	3.27	4.80	3.70	----------	3.24	2.43	2.72	3.24	3.60	2.63
5th week	3.42	2.94	4.75	4.11	3.17	----------	4.69	3.13	3.41	3.00	4.06	5.23
6th week	3.56	3.43	3.20	4.39	14.00	----------	3.65	4.83	3.48	6.21	3.68	7.62
7th week	5.25	8.08	5.83	3.53	4.22	----------	8.75	8.75	4.86	3.13	5.00	4.35
8th week	2.92	6.56	2.02	4.46	4.30	No gain	2.57	3.65	3.98	3.65	3.50	3.59
9th week	7.85	2.92	10.20	No gain	5.81	2.05	5.31	4.72	4.25	6.07	6.18	4.00
10th week	3.25	4.00	10.40	2.09	5.81	4.75	6.70	4.55	4.60	4.60	5.00	7.00
11th week	4.77	5.83	5.25	No gain	4.72	4.61	7.29	6.73	5.82	7.95	7.50	No gain
12th week	17.43	9.27	7.29	4.80	No gain	15,22	5.00	5.00	7.08	6.50	7.81	2.57
13th week	3.89	7.00	No gain	7.00	7.00	7.61	5.15	4.86	6.25	8.80	8.33	12.00
14th week	3,75	5.83	2.92	27.00	18.75	32.50	8.75	5.47	5.82	6.25	7.50	No gain
15th week	No gain	14.62	7.37	1.47	19.50	7.25	5.08	6.54	5.72	5.89	26.00	No gain
16th week	17.10	6.41	3.41	No gain	No gain	No gain	5.47	5.83	87.50	9.72	No gain	3.23

From the weekly fluctuations in the weight of animals and in the feed consumed, as shown in the preceding tables, the results of the experiment are more readily seen when tabulated in longer periods.

In the following table (No. 7) the results have been tabulated in periods of four weeks, and the pigs have been grouped so that the average results are given for those of the same age and description.

The influence of age and breed and size may be seen approximately, although it is impossible to fully determine the influence of each of these conditions:

TABLE No. 7.

SHOWING THE AVERAGE RESULTS IN PERIODS OF FOUR WEEKS EACH, GROUPING THE PIGS ACCORDING TO AGES.

	PERIODS.	Total increase.	Total meal consum'd.	Meal consumed for each 100 lbs. of live weight per week.	Meal consumed to produce 1 lb. of increase of live weight.	Average gain per ct. per week.	Average weight for each period.
Essex—Age 7½ m. Pens 1 and 3.	FIRST PERIOD: 1st, 2d, 3d, and 4th weeks	84.50	329.00	23.19	3.89	6.76	354.75
	SECOND PERIOD: 5th, 6th, 7th, and 8th weeks	97.00	391.50	21.98	4.04	6.11	445.25
	THIRD PERIOD: 9th, 10th, 11th, and 12th weeks	82.00	413.00	19.30	5.04	4.15	535.00
	FOURTH PERIOD: 13th, 14th, 15th, and 16th weeks	58.00	411.50	17.00	7.09	2.52	605.00
	Average and totals	321.50	1545.00	20.40	4.81	6.43	473.25
Suffolk—Age 1 year and 7 weeks. Pens 2 and 4.	FIRST PERIOD: 1st, 2d, 3d, and 4th weeks	58.00	243.50	14.56	4.20	3.73	418.00
	SECOND PERIOD: 5th, 6th, 7th, and 8th weeks	80.00	337.25	17.31	4.22	4.48	487.00
	THIRD PERIOD: 9th, 10th, 11th, and 12th weeks	62.50	344.25	15.42	5.51	2.78	558.25
	FOURTH PERIOD: 13th, 14th, 15th, and 16th weeks	62.00	354.00	14.26	5.71	2.63	620.50
	Average and totals	262.50	1279.00	15.37	4.87	4.22	520.25

TABLE No. 7—CONTINUED.

	PERIODS.	Total increase.	Total meal consum'd.	Meal consumed for each 100 lbs. of live weight per week.	Meal consumed to produce 1 lb. of increase of live weight.	Average gain per ct. per week.	Average weight for each period.
Suffolk—Age 1 year and 10 weeks. Pens 5, 6, and 8.	FIRST PERIOD: 1st, 2d, 3d, and 4th weeks	94.00	423.25	17.07	4.50	4.10	620.00
	SECOND PERIOD: 5th, 6th, 7th, and 8th weeks	148.00	532.75	17.97	3.60	5.55	741.00
	THIRD PERIOD: 9th, 10th, 11th, and 12th weeks	87.00	541.50	16.74	6.22	2.67	808.50
	FOURTH PERIOD: 13th, 14th, 15th, and 16th weeks	91.50	594.75	15.69	6.50	2.54	947.75
	Average and totals	420.50	2092.25	16.17	4.98	4.59	783.25
"Natives"—Age 4 months. Pens 7, 9, 10, and 12.	FIRST PERIOD: 1st, 2d, 3d, and 4th weeks	85.50	408.50	32.76	4.78	7.95	311.75
	SECOND PERIOD: 5th, 6th, 7th, and 8th weeks	106.00	462.00	28.34	4.36	7.48	407.50
	THIRD PERIOD: 9th, 10th, 11th, and 12th weeks	70.00	478.50	24.14	6.84	3.80	495.50
	FOURTH PERIOD: 13th, 14th, 15th, and 16th weeks	17.00	357.25	16.57	21.01	0.80	539.00
	Average and totals	278.50	1706.25	26.12	6.13	6.47	408.25
Essex — Age 4 y'rs and 7 w'ks. Pen 13.	FIRST PERIOD: 8th week	— 3.50	52.50	12.03	no g'n	no g'n	436.25
	SECOND PERIOD: 9th, 10th, 11th, and 12th weeks	82.00	328.00	17.27	4.00	4.72	475.50
	THIRD PERIOD: 13th, 14th, 15th, and 16th weeks	9.50	261.25	12.53	27.50	0.46	521.25
	Average and totals	88.00	641.75	14.79	7.29	2.23	482.00

By leaving out the first week, in which there was a loss of 3.50, and the last week of the third period, in which there was a loss of twelve pounds, the table would stand as follows:

PERIODS.	Total increase.	Total meal consum'd.	Meal consumed for each 100 lbs. of live weight per week.	Meal consumed to produce 1 lb. of increase of live weight.	Average gain per ct. per week.	Average weight for each period.
SECOND PERIOD: 9th, 10th, 11th, and 12th weeks	82.00	328.00	17.27	4.00	4.72	475.50
THIRD PERIOD: 13th, 14th, and 15th weeks	18.00	210.50	13.31	11.69	1.16	527.25
Average and totals	100.00	538.50	15.82	5.39	3.29	486.25

In comparing the results given in the preceding table, it should be remembered that age undoubtedly has an important influence on the amount of feed consumed in proportion to the live weight of the animal,—and also on the amount of feed required to produce a given increase.

The Suffolk pigs in pens 2 and 4, and in pens 5, 6, and 8, were nearly of the same age, and a close agreement will be noticed in the amount of feed consumed in proportion to live weight and in the feed required to produce one pound of increase.

The Essex pigs in pens 1 and 3, consumed on the average 20.40 lbs. of meal for each 100 lbs. of live weight, which is considerably *more* than the amount consumed by the Suffolks in pens 2 and 4, and pens 5, 6, and 8.

The amount of feed required to produce a pound of increase of live weight is less with the Essex than it is with the Suffolks. This must not be taken as an indication of the superior fattening qualities of the Essex breed, as the variation in the particulars referred to is readily accounted for by the difference in age.

The Essex pigs being younger, *should* consume less in proportion to live weight, and give a better return for the feed consumed.

The pure-bred pigs (Essex and Suffolk), in pens 1 and 3, 2 and 4, and 5, 6, and 8, required less feed to produce a pound of increase than the "grade or native" pigs in pens 7, 9, 10, and 12, notwithstanding the larger amount of feed consumed by the latter in proportion to weight.

This difference in results is undoubtedly owing to the superiority of the pure-breds in feeding qualities.

If the feeding quality of the pigs had been the same in all the pens, the "natives," being younger, should have consumed (as they did), more in proportion to weight; but they should also have given a better return for the feed consumed. In other words, the pure-breds gave the best return for feed consumed, notwithstanding the difference in age which gave the "natives" a decided advantage.

The "native" pigs (pens 7, 9, 10, and 12), did not fatten, but made their increase in weight in bone and muscle.

At the termination of the experiment, as they were only in fair store condition, they were turned out of the pens and kept for experiment when they were a year older.

The Essex sow in pen 13, furnishes a good illustration of the influence of age and condition (ripeness), on the profits of feeding. This sow always kept in high condition, even on grass alone. She had a litter of pigs when two years old, and failed to breed afterwards. She had been running in the pasture where the feed was very short during the spring and summer until put up for experiment. Her form was almost faultless, and she was thought to be a remarkable feeder. She was very fat, weighing 438 lbs. when put in the experimental pen; in fact she was almost as fat as she could be. It was often remarked by farmers who visited the experimental pens, that this sow must be the most profitable animal to feed in the lot. The result, however, showed that she gave the smallest return for feed consumed, while she ate less than the others in proportion to live weight. This result, although directly the reverse of what was expected by experienced feeders, who were not familiar with the exact determinations of weight and measure, was not unlooked for.

Our experiments thus far show that animals in high condition that are fully matured do not, other things being equal, give as good a return for feed consumed, as younger animals that are only in fair store condition.

The pigs in pens 1, 2, 3, 4, 5, 6, 8 and 13, were killed the 20th of December, after being weighed as usual in the morning.

The loose inside fat, the intestines and stomach and contents, and the *thoracic viscera* with the liver and tongue (commonly called the "pluck,") were each separately weighed.

In table No. 8, will be found the weights of the parts above mentioned, together with the shrinkage per cent for each animal.

TABLE NO. 8.

FASTED LIVE WEIGHT, DRESSED WEIGHT, SHRINKAGE PER CENT, PLUCK, INTESTINES, AND CONTENTS, INSIDE FAT.

PENS.	Fasted Live Weight.	Dressed Weight.	Shrinkage per cent.	Stomach, Intestines, and contents.	Liver, Heart, Lungs, and Tongue, etc.	Inside Fat.
No. 1	316.00	269.00	14.87	21.25	7.00	5.50
" 3	318.00	266.50	16.19	19.50	8.50	6.25
" 2	339.00	290.00	14.45	17.75	8.00	8.00
" 4	312.50	264.00	15.52	17.00	8.75	7.75
" 5	404.00	341.00	15.59	21.00	10.75	11.25
" 6	303.00	259.50	14.36	21.25	8.50	8.75
" 8	286.50	242.50	15.36	14.75	7.75	7.50
" 13	526.00	450.50	14.54	25.50	15.00	11.75

The pigs were killed in the following order: 1, 2, 3, 4, 6, 5, 13 and 8; the first at 8½ o'clock A. M., and the last at 3 o'clock P. M.

This difference in the time of killing would probably make some difference in the shrinkage, and also in weights of stomach, intestines, and contents.

The live weight given in the table was from the morning weighing before the first pig was killed.

The weekly return given in increase of live weight does not seem to depend entirely upon the amount of feed consumed during the week. This is shown very clearly in the experiments of 1868, '69, and '70.

It was found impossible to predict with any degree of accuracy whether a particular animal was making a satisfactory gain or not.

When there were no apparent disturbing influences, and the pigs were eating a uniform amount per day, the weekly gain frequently fluctuated in an unaccountable manner.

When a loss in weight occurred one week with a given amount of feed, the next week, with the same amount of feed, the gain in live weight would often be found unusually large.

A few cases from the present experiments will illustrate this remarkable peculiarity.

In Pen 1, the feed consumed the 7th, 8th, 9th, and 10th weeks, respectively, was 52.50, 52.50, 51.00, 52.00, while the gain for the same periods was 10.00, 18.00, 6.50, and 16.00.

In Pen No. 3, the feed consumed the 6th, 7th, 8th, and 9th weeks, respectively, was, 48.00, 52.50, 52.50, 52.50, and the gain for the same periods was, 14.00, 6.50, 8.00, and 18.00.

In pen No. 5, the feed consumed the 4th, 5th, 6th, 7th, and 8th weeks, respectively, was, 42.50, 42.75, 48.00, 52.50, 52.50, and the gain for the same periods was, 13.00, 9.00, 15.00, 9.00, 26.00.

In pen No. 7, the feed consumed the 8th, 9th, and 10th weeks, respectively, was, 31.25, 32.50, 33.50, and the gain for the same periods was, 7.00, 2.00 loss, 16.00.

Pen No. 2 consumed, the 6th, 7th and 8th weeks respectively, 43.75, 43.75, 43,75, and gave in gain for the same periods 12.00, 5.00, 17.00.

Pen No. 4 consumed, the 5th, 6th, 7th and 8th weeks respectively, 37.50, 43.75, 43.75, 43.75, and gave in return for the same periods 12.00, 9.00, 5.00, 12.00 of increase.

The weights given above are pounds and decimals of a pound, as in the tables.

Many more examples, quite as striking, might be given, but those are sufficient to illustrate the fact that the rate of increase with a given amount of feed is represented by an undulating line instead of one uniformly ascending.

EXPERIMENT IN THE APPLICATION OF MANURES.

This experiment was commenced is 1868, to test the different modes of applying manures.

In December, 1867, six loads of manure were applied to each of the strips marked G and C.

On the 21st of April, 1868, the same quantity and quality of manure were applied to the strip marked E.

The land was then all plowed from the 5th to the 7th of May.

The 9th of May the same quantity and quality of manure were applied to the strips marked A and I.

The entire surface was harrowed May 13th, and on the 21st of May the whole was planted to corn. (For detailed results of this part of the experiment, see Report of Secretary of State Board of Agriculture for 1868, pp. 99 to 108.)

In 1869 a crop of Swedes turnips was grown on the plats, and the results for the year were published in the Annual Report. (See Report of Secretary of State Board of Agriculture for 1869, p. 73 to 80).

In 1870, the plats were all plowed the 7th of May, and on the 9th they were all harrowed and sowed to oats at the rate of three bushels per acre. No manure had been applied since 1868, as described in the report for that year.

The plats numbered 1, 2, and 3 of each strip were cut on the 11th of August, and the plats numbered 4 in each strip were cut on the 12th of August. On the 20th of August the plats were all carefully weighed and hauled to the barn and threshed. The same precautions were taken as in 1868 and

1869 to give the plats the same treatment throughout the year.

The strips indicated by the letters are 2 by 16 rods, and the subdivisions indicated by figures are therefore 2 by 4 rods.

Plats marked N were not manured.

Plats marked X were manured December 6, 1867, before plowing.

Plats marked Y were manured April 21, 1868, before plowing.

Plats marked Z were manured May 9, 1868, on the surface, after plowing.

In the following plan of the experiment, the yield of grain and of straw is marked on each plat:

PLAN OF EXPERIMENT WITH OATS IN FIELD NO. 2, WITH THE YIELD IN POUNDS OF GRAIN AND OF STRAW MARKED ON EACH PLAT.

TWO RODS. NORTH.

FOUR RODS.

N. K. 1. Grain. 62.25 Straw. 57.75	Z. I. 1. Grain. 79.00 Straw. 83.00	N. H. 1. Grain. 65.25 Straw. 76.75	X. G. 1. Grain. 79.25 Straw. 90.75	N. F. 1. Grain. 69.00 Straw. 71.00	Y. E. 1. Grain. 84.00 Straw. 88.00	N. D. 1. Grain. 76.50 Straw. 83.50	X. C. 1. Grain. 83.75 Straw. 92.25	N. B. 1. Grain. 63.50 Straw. 86.50	Z. A. 1. Grain. 71.25 Straw. 84.75
N. K. 2. Grain. 81.00 Straw. 77.00	Z. I. 2. Grain. 78.50 Straw. 95.50	N. H. 2. Grain. 67.25 Straw. 62.75	X. G. 2. Grain. 81.00 Straw. 75.00	N. F. 2. Grain. 74.75 Straw. 75.25	Y. E. 2. Grain. 83.75 Straw. 86.25	N. D. 2. Grain. 67.75 Straw. 64.25	X. C. 2. Grain. 81.00 Straw. 71.00	N. B. 2. Grain. 67.00 Straw. 67.00	Z. A. 2. Grain. 84.75 Straw. 89.25
N. K. 3. Grain. 63.00 Straw. 77.00	Z. I. 3. Grain. 89.50 Straw. 87.50	N. H. 3. Grain. 68.75 Straw. 71.25	X. G. 3. Grain. 80.25 Straw. 85.75	N. F. 3. Grain. 70.75 Straw. 65.25	Y. E. 3. Grain. 78.75 Straw. 87.25	N. D. 3. Grain. 66.00 Straw. 64.00	X. C. 3. Grain. 94.00 Straw. 64.00	N. B. 3. Grain. 59.00 Straw. 53.00	Z. A. 3. Grain. 82.25 Straw. 73.75
N. K. 4. Grain. 50.50 Straw. 61.50	Z. I. 4. Grain. 61.50 Straw. 76.50	N. H. 4. Grain. 43.75 Straw. 34.25	X. G. 4. Grain. 72.00 Straw. 78.00	N. F. 4. Grain. 66.00 Straw. 58.00	Y. E. 4. Grain. 89.50 Straw. 78.50	N. D. 4. Grain. 71.75 Straw. 86.25	X. C. 4. Grain. 81.00 Straw. 77.00	N. B. 4. Grain. 68.50 Straw. 65.50	Z. A. 4. Grain. 85.75 Straw. 82.25

SOUTH.

In Table No. 1 the yield *per acre* is given, of the grain in pounds and in bushels, and of the straw in pounds and in decimals of a ton.

TABLE No. 1.

YIELD PER ACRE.

LETTER AND NUMBER OF PLOT.	GRAIN.		STRAW.	
	Pounds.	Bushels.	Pounds.	Tons.
Z A 1	1425.00	44.53	1695.00	0.84
" 2	1695.00	52.97	1785.00	0.89
" 3	1645.00	51.40	1475.00	0.74
" 4	1715.00	53.59	1645.00	0.82
Average	1620.00	50.62	1650.00	0.82
N B 1	1270.00	39.69	1710.00	0.86
" 2	1340.00	41.87	1340.00	0.67
" 3	1180.00	36.88	1060.00	0.53
" 4	1370.00	42.81	1310.00	0.65
Average	1290.00	40.31	1355.00	0.68
H C 1	1675.00	52.34	1845.00	0.92
" 2	1620.00	50.62	1420.00	0.71
" 3	1880.00	58.75	1280.00	0.64
" 4	1620.00	50.62	1540.00	0.77
Average	1699.00	53.06	1521.00	0.77
N D 1	1530.00	47.81	1670.00	0.83
" 2	1355.00	42.34	1285.00	0.64
" 3	1320.00	41.25	1280.00	0.64
" 4	1435.00	44.53	1725.00	0.86
Average	1410.00	44.06	1170.00	0.59

TABLE No. 1—Continued.

LETTER AND NUMBER OF PLOT.	GRAIN.		STRAW.	
	Pounds.	Bushels.	Pounds.	Tons.
Y E 1	1680.00	52.50	1760.00	0.88
" 2	1675.00	52.34	1725.00	0.86
" 3	1575.00	49.21	1745.00	0.87
" 4	1790.00	55.94	1570.00	0.78
Average	1680.00	52.50	1700.00	0.82
N F 1	1380.00	43.12	1420.00	0.71
" 2	1495.00	46.72	1505.00	0.75
" 3	1415.00	44.22	1305.00	0.65
" 4	1320.00	41.25	1160.00	0.58
Average	1402.00	43.81	1347.50	0.72
X G 1	1585.00	49.53	1815.00	0.91
" 2	1620.00	50.62	1500.00	0.75
" 3	1605.00	50.12	1715.00	0.86
" 4	1440.00	45.00	1560.00	0.78
Average	1562.00	48.81	1647.00	0.82
N H 1	1305.00	40.78	1535.00	0.76
" 2	1345.00	42.03	1255.00	0.63
" 3	1375.00	42.97	1425.00	0.71
" 4	875.00	27.34	685.00	0.34
Average	1225.00	38.28	1225.00	0.61
Z I 1	1580.00	49.37	1660.00	0.83
" 2	1570.00	49.06	1910.00	0.95
" 3	1610.00	50,31	1750.00	0.87
" 4	1230.00	38.44	1530.00	0.76
Average	1498.00	46.81	1712.00	0.86

TABLE No. 1—Continued.

LETTER AND NUMBER OF PLOT.	GRAIN.		STRAW.	
	Pounds.	Bushels.	Pounds.	Tons.
N K 1	1245.00	38.90	1155.00	0.58
" 2	1620.00	50.62	1540.00	0.77
" 3	1260.00	39.38	1540.00	0.77
" 4	1010.00	31.56	1230.00	0.61
Average	1284.00	40.12	1366.00	0.68

From the arrangement of the plats, it will be seen by reference to the plan of the experiment, the results of four distinct experiments are obtained.

The plats numbered 1 in each letter are the duplicates of those numbered 2, and so on for the numbers 3 and 4. These are given separately in Table No. 2, together with the averages and the gain of the manured plats when compared with the unmanured:

TABLE No. 2.

RESULTS PER ACRE, IN FOUR SERIES, AS NUMBERED

UNMANURED.			MANUR'D MAY 9, '68, ON THE SURFACE, AFTER PLOWING.			MANURED DEC. 6, 1867, AND PLOWED UNDER IN THE SPRING.			MANURED APRIL 28, 1868, AND PLOW'D UNDER.		
Plat.	Bushels of grain.	Pounds of straw.	Plat.	Bushels of grain.	Pounds of straw.	Plat.	Bushels of Grain.	Pounds of Straw.	Plat.	Bushels of Grain.	Pounds of Straw.
N. K. 1	38.90	1155.00	Z. I. 1	49.37	1660.00	X.G. 1	49.53	1815.00	Y.E. 1	52.50	1760
N. H. 1	40.78	1535.00	Z.A. 1	44.53	1695.00	X.C. 1	52.34	1845.00	------	------	----
N. F. 1	43.12	1420.00	------	------	--------	------	------	--------	------	------	----
N. D. 1	47.81	1670.00	------	------	--------	------	------	--------	------	------	----
N. B. 1	39.69	1710.00	------	------	--------	------	------	--------	------	------	----
Ave.	42.06	1498.00	------	46.95	1677.50	------	50.93	1830.00	------	52.50	1760
Gain	------	--------	------	4.89	179.50	------	8.87	331.00	------	10.44	262
N. K. 2	50.62	1540.00	Z. I. 2	49.06	1910.00	X.G. 2	50.62	1500.00	Y.E. 2	52.34	1725
N. H. 2	42.03	1255.00	Z.A. 2	52.97	1785.00	X.C. 2	50.62	1420.00	------	------	----
N. F. 2	46.72	1505.00	------	------	--------	------	------	--------	------	------	----
N. D. 2	42.34	1285.00	------	------	--------	------	------	--------	------	------	----
N. B. 2	41.87	1340.00	------	------	--------	------	------	--------	------	------	----
Ave.	44.72	1385.00	------	51.01	1847.50	------	50.62	1460.00	------	52.34	1725
Gain	------	--------	------	6.29	462.50	------	5.90	75.00	------	7.62	340
N. K. 3	39.38	1540.00	Z. I. 3	50.31	1750.00	X.G. 3	50.12	1715.00	Y.E. 3	49.21	1745
N. H. 3	42.97	1425.00	Z.A. 3	51.40	1475.00	X.C. 3	58.75	1280.00	------	------	----
N. F. 3	44.22	1305.00	------	------	--------	------	------	--------	------	------	----
N. D. 3	41.25	1280.00	------	------	--------	------	------	--------	------	------	----
N. B. 3	36.88	1060.00	------	------	--------	------	------	--------	------	------	----
Ave.	40.94	1322.00	------	50.86	1612.50	------	54.44	1497.50	------	49.21	1745
Gain	------	--------	------	9.92	290.50	------	13.50	175.50	------	8.27	423

TABLE NO. 2—CONTINUED.

UNMANURED.			MANUR'D MAY 9, '68, ON THE SURFACE, AFTER PLOWING.			MANURED DEC. 6, 1867, AND PLOWED UNDER IN THE SPRING.			MANURED APRIL 28, 1868, AND PLOW'D UNDER.		
Plat.	Bushels of Grain.	Pounds of Straw.	Plat.	Bushels of Grain.	Pounds of Straw.	Plat.	Bushels of Grain.	Pounds of Straw.	Plat.	Bushels of Grain.	Pounds of Straw.
N. K. 4	31.56	1230.00	Z. I. 4	38.44	1530.00	X.G. 4	45.00	1560.00	Y.E. 4	55.94	1570
N. H. 4	27.34	685.00	Z.A. 4	53.59	1645.00	X.C. 4	50.62	1540.00	......	......	....
N. F. 4	41.25	1160.00	......	......	....	......	......	....	......	......	....
N. D. 4	44.53	1725.00	......	......	....	......	......	....	......	......	....
N. B. 4	42.81	1310.00	......	......	....	......	......	....	......	......	....
Ave.	37.50	1222.00	......	46.02	1587.50	......	47.81	1550.00	......	55.94	1570
Gain	......	....	......	8.52	365.50	......	10.31	328.00	......	18.44	348
Total Ave.	41.30	1356.75	......	48.71	1681.25	......	50.95	1584.00	......	52.50	1700
Ave. Gain	......	....	......	7.41	324.50	......	9.65	227.60	......	11.20	343

It will be seen that the plats manured in the spring, before plowing, have given the best return this season.

In 1869, the average yield of the plats manured in the spring, before plowing, and of the plats manured in the preceding December, gave nearly the same results, the yield being considerably larger than on the plats manured on the surface after plowing.

In 1868, the plats manured on the surface after plowing gave decidedly the best results, and the plats manured in December gave a better return than those manured in April, before plowing.

This experiment will be continued for the purpose of ascertaining how long the influence of the manure can be perceived. Thus far, the experiment seems to show that it makes but little difference, in the end, as to the manner in which the manure is applied. The surface application appears to give quicker results, while that plowed under seems to act slowly, giving the best return in subsequent crops.

The season of 1870 was remarkable for the excessive rainfall during the growing months. The oat crop was seriously injured, the grain being unusually light, particularly in those cases where the crop was not materially diminished in quantity. A remarkable difference was noticed in the weight of a measured bushel of grain from the different plats. The heaviest grain weighed 31 pounds per bushel, and the lightest 26.5 pounds. Adjacent plats, treated precisely alike, showed a difference of two pounds in the weight of a measured bushel, the difference in yield being at the rate of but 2.18 bushels per acre, the lightest grain being from the plat giving the largest yield. This difference in the weight of the grain was observed when weighing the first plats threshed, and it was thought to arise from an error in the measuring or in the cleaning of the grain. It was soon ascertained by repeated trials that the error in measuring, with the precautions taken, did not exceed an ounce and a half per bushel. The fanning mill used for cleaning the grain had a self-feeding hopper (presented to the College by Messrs. Farnum & Mosier of Lansing), which regulated the supply of grain with great accuracy, and particular care was taken to run the mill at a uniform rate when cleaning the different samples. After a series of trials, it was found that the error arising from cleaning the grain was so slight that it could not materially affect the results. This variation in the weight of grain is seen alike on the manured and on the unmanured plats,—adjacent plats, treated alike, often showing a greater difference than adjacent plats under different treatment. The variation in the yield of adjacent plats similarly treated, which has been mentioned in our former reports of experiments, in connection with the variation in the weight of grain produced under the same treatment, furnishes an interesting subject for investigation that may, perhaps, lead to some valuable practical results.

EXPERIMENTS WITH SPECIAL MANURES.

In 1868, 32 plats, each two rods square, were staked out in Field No. 3, for experiments with special fertilizers. On the A plats, common salt, at the rate of 300 pounds per acre, was applied. Berry's superphosphate of lime, at the rate of 300 pounds per acre, was applied to the C plats. Baugh's superphosphate of lime was applied, at the rate of 300 pounds per acre, to the E plats. Baugh's Chicago Blood Manure, at the rate of 300 pounds per acre, was applied to the G plats. All the other plats received no dressing of special fertilizers. The plats indicated by letters were separated by a narrow strip that was kept cultivated. Those indicated by figures were separated by a wire, so that a division of plats could be readily made in harvesting.

In 1868 a root crop was grown on the plats. (See Report of Secretary of State Board of Agriculture for 1868, pp. 109–16.) In 1869 an oat crop was grown on the plats. (See Report of Secretary of State Board of Agriculture for 1869, pp. 81–86.)

The present year a crop of winter wheat was grown on the plats, no fertilizers having been applied since 1868.

The plats were plowed Sept. 11th, 1869, and on the 23d they were thoroughly harrowed, and wheat was drilled in at the rate of 1¼ bushels per acre. The wheat was harvested July 20th, 1870, and set up in stooks. The plats numbered 1 and 2 were hauled to the barn and threshed July 22d, and the plats numbered 3 and 4 were hauled and threshed July 23d. The crop on each plat was weighed before threshing, the grain was afterwards weighed and deducted from the total yield of each plat, and the difference was recorded as straw. The same precautions were taken in the separation of plats as described in the experiments on the same plats in 1868 and 1869.

The following plan of the experiment shows the kinds and quantities of manures applied, and the yield of grain and of straw is marked on each plat.

PLAN OF EXPERIMENT WITH SPECIAL FERTILIZERS IN FIELD NO. 3.

Nothing.	Baugh's Chicago Blood Manure, 300 lbs. per acre applied in 1868.	Nothing.	Baugh's Super-phos. of Lime, 300 lbs. per acre applied in 1868.	Nothing.	Berry's Super-phos. of Lime, 300 lbs. per acre applied in 1868.	Nothing.	Common Salt, 300 lbs per acre, applied in 1868.
H. 1. Grain, 27.50. Straw, 40.50.	G. 1. Grain, 26.50. Straw, 39.50.	F. 1. Grain, 26.00. Straw. 36.00.	E. 1. Grain, 32.25. Straw, 45.75.	D. 1. Grain, 38.00. Straw, 54.00.	C. 1. Grain, 46.00. Straw, 82.00.	B. 1. Grain, 44.75. Straw, 79.25.	A. 1. Grain, 30.25. Straw, 37.75.
H. 2. Grain, 32.50. Straw, 41.50.	G. 2. Grain, 28.25. Straw, 39.75.	F. 2. Grain, 28.25. Straw, 33.75.	E. 2. Grain, 28.00. Straw, 34.00.	D. 2. Grain, 35.50. Straw, 52.50.	C. 2. Grain, 47.50. Straw, 68.50.	B. 2. Grain, 47.50. Straw, 82.50.	A. 2. Grain, 31.00. Straw, 65.00.
H. 3. Grain, 39.00. Straw, 55.00.	G. 3. Grain, 28.50. Straw, 35.50.	F. 3. Grain, 31.00. Straw, 43.00.	E. 3. Grain, 29.00. Straw, 35.00.	D. 3. Grain, 31.00. Straw, 49.00.	C. 3. Grain, 48.00. Straw, 64.00.	B. 3. Grain, 49.00. Straw, 79.00.	A. 3. Grain, 31.00. Straw, 47.00.
H. 4. Grain, 47.50. Straw, 68.50.	G. 4. Grain, 34.25. Straw, 37.75.	F. 4. Grain, 36.00. Straw, 44.00.	E. 4. Grain, 34.50. Straw, 51,50.	D. 4. Grain, 36.00. Straw, 44.00.	C. 4. Grain, 40.50. Straw, 51.50.	B. 4. Grain, 53.00. Straw, 81.00.	A. 4. Grain, 56.50. Straw, 93.50.

In Table No. 1 will be found the results per acre of the grain, in pounds and in bushels, and of the straw, in tons and decimals of a ton, together with the average results for similar plats.

TABLE NO. 1.

Showing the Results per acre.

NUMBER OF PLAT.	Wheat.		Straw.	
	Pounds.	Bushels.	Pounds.	Tons.
A. 1	1210	20.16	1510	0.75
A. 2	1240	20.66	2600	1.30
A. 3	1240	20.66	1880	0.94
A. 4	2260	37.33	3740	1.87
Average	1739	24.70	2433	1.21
B. 1	1790	29.83	3170	1.58
B. 2	1900	31.66	3300	1.65
B. 3	1960	32.66	3160	1.58
B. 4	2120	35.33	3240	1.62
Average	1943	32.37	3218	1.60
C. 1	1840	30.66	3280	1.64
C. 2	1900	31.66	2740	1.37
C. 3	1920	32.00	2560	1.28
C. 4	1620	27.00	2060	1.03
Average	1820	30.33	2660	1.33
D. 1	1520	25.33	2160	1.08
D. 2	1420	23.66	2100	1.05
D. 3	1240	20.66	1960	0.98
D. 4	1440	24.00	1760	0.88
Average	1405	23.41	1995	0.99

TABLE NO. 1—CONTINUED.

NUMBER OF PLAT.	Wheat.		Straw.	
	Pounds.	Bushels.	Pounds.	Tons.
E. 1	1290	21.50	1880	0.91
E. 2	1120	18.66	1360	0.68
E. 3	1160	19.33	1400	0.70
E. 4	1380	23.00	2060	1.03
Average	1238	20.61	1663	0.83
F. 1	1040	17.33	1440	0.72
F. 2	1130	18.83	1350	0.67
F. 3	1240	20.66	1720	0.86
F. 4	1440	26.00	1760	0.88
Average	1213	20.71	1568	0.78
G. 1	1060	17.66	1580	0.79
G. 2	1130	18.83	1590	0.79
G. 3	1140	19.33	1420	0.71
G. 4	1370	22.83	1510	0.75
Average	1175	19.66	1525	0.76
H. 1	1100	18.33	1620	0.81
H. 2	1300	21.66	1660	0.83
H. 3	1560	26.00	2200	1.10
H. 4	1900	31.66	2740	1.37
Average	1465	24.41	2055	1.03

As in the experiment with oats in Field No. 2, a difference in the weight of a measured bushel of grain from the different plats was observed.

The variation was not, however, as marked, the heaviest bushel weighing 57¼ pounds, and the lightest 56 pounds.

EXPERIMENTS WITH MANURES.

In 1868 several plats were staked out on the lawn north of the Boarding Hall, and planted to corn. All the plats were unmanured, and the treatment was the same in each,—the object being to ascertain the natural variation of soils of a similar character, and to furnish a standard of comparison for future experiments with manures. (See Report of Secretary of State Board of Agriculture for 1868, p. 122–128.)

In 1869 manure of the same quality was applied to each of the F., D., and B. plats at the rate of forty loads (of forty cubic feet each) per acre, and the plats were again planted to corn. (For results of this part of the experiment, see Report of Secretary of State Board of Agriculture for 1869, pp. 87–91.

In 1870 the plats were plowed May 4th, and on the 5th, after harrowing, oats were sown at the rate of three bushels per acre.

The oats were harvested August 9th, and hauled to the barn and threshed August 18th.

The weight of the grain after threshing was deducted from the weight of the crop from each plat before threshing, and the difference was put down as straw.

On the following plan of the experiment, the weight of grain and of straw is marked on each plat:

PLAN OF EXPERIMENT.

Showing the weight of Grain and of Straw on each Plat.

PLANK ROAD.

TWO RODS. FOUR RODS.

NORTH.

F. Oats, 71.50. Straw, 102.50.	E. Oats, 22.50. Str'w, 33.50.	D. Oats, 22.00. Str'w, 58.00.	C. Oats, 18.00. Str'w, 30.00.		
F. 1. Oats, 77.75. Straw, 86.85.	E. 1. Oats, 40.50. Straw. 57.50.	D. 1. Oats, 65.25. Straw, 77.00.	C. 1. Oats, 38.00. Straw, 64.00.	B. 1. Oats, 60.50. Straw, 77.50.	A. 1. Oats, 40.75. Straw, 39.25.
F. 2. Oats, 82.00. Straw, 106.00.	E. 2. Oats, 38.50. Straw, 47.50.	D. 2. Oats, 44.50. Straw, 73.50.	C. 2. Oats, 38.50. Straw, 47.50.	B. 2. Oats, 45.00. Straw, 74.00.	A. 2. Oats, 41.25. Straw, 61.75.
F. 3. Oats, 68.25. Straw. 57.75.	E. 3. Oats, 36.50. Straw, 57.50.	D. 3. Oats, 48.50. Straw, 61.50.	C. 3. Oats, 44.50. Straw, 53.50.	B. 3. Oats, 60.50. Straw, 79.50.	A. 3. Oats, 57.00. Straw, 87.00.
F. 4. Oats, 66.50. Straw, 89.50.	E. 4. Oats, 46.50. Straw, 39.50.	D. 4. Oats, 53.75. Straw, 46.25.	C. 4. Oats, 47.50. Straw, 86.50.	B. 4. Oats, 58.00. Straw, 61.00.	A. 4. Oats, 36.50. Straw, 57.50.

SOUTH.

TABLE NO. 1.

Showing the Yield per Acre.

LETTER AND NUMBER OF PLAT.	GRAIN.		STRAW.	
	Pounds.	Bushels.	Pounds.	Tons.
A. 1	815	25.47	785	.393
A. 2	825	25.78	1235	.618
A. 3	1140	35.62	780	.390
A. 4	730	22.81	1130	.565
Average	878	27.25	982	.491
B. 1	1210	37.81	1550	.775
B. 2	900	28.12	1480	.740
B. 3	1210	37.81	1590	.795
B. 4	1060	33.12	1220	.610
Average	1095	34.22	1460	.730
C	720	22.50	1200	.600
C. 1	760	23.75	1280	.640
C. 2	770	24.06	950	.475
C. 3	890	27.81	1110	.555
C. 4	950	29.69	730	.365
Average	818	25.56	1054	.527
D	880	27.50	2320	1.160
D. 1	1305	40.78	1540	.770
D. 2	890	27.81	1470	.735
D. 3	970	30.31	1230	.615
D. 4	1075	33.59	925	.463
Average	1024	32.00	1497	.748

TABLE NO. 1.—CONTINUED.

LETTER AND NUMBER OF PLAT.	GRAIN.		STRAW.	
	Pounds.	Bushels.	Pounds.	Tons.
E	900	28.12	1340	.670
E. 1	810	25.31	1150	.575
E. 2	770	24.06	950	.475
E. 3	730	22.81	1150	.575
E. 4	930	29.06	790	.395
Average	828	25.87	1076	.538
F	1430	44.69	2050	1.025
F. 1	1555	48.59	1727	.864
F. 2	1640	51.25	2120	1.060
F. 3	1365	42 65	1155	.576
F. 4	1330	41.56	1790	.895
Average	1465	45.75	1768	.884

In Table No. 1, the yield per acre is given of the grain in pounds and in bushels, and of the straw in pounds and in tons and decimals of a ton.

The grain varied in weight on the different plats, the heaviest measured bushel weighing thirty pounds and the lightest twenty-six pounds.

EXPERIMENTS WITH SPECIAL FERTILIZERS.

In 1869 forty-nine plats of one-hundredth of an acre each (exclusive of the strip separating the plats) were staked out in Field No. 6, and sowed to turnips. For details of this part of the experiment see Report of Secretary of State Board of Agriculture for 1869, pp. 93 to 101.

On the 25th of April, 1870, the plats were plowed, and the following day, after harrowing, oats were sown at the rate of three bushels per acre.

The crop was harvested July 26th and put up in stooks, and on the 6th of August it was hauled to the barn and threshed.

In the following plan of the experiment (No. 1) the arrangement of the plats is shown, and the fertilizers applied in 1869 are marked on each plat.

On the plan of the Experiment No. 2, the yield of grain and of straw is marked on each plat in pounds and decimals of a pound:

PLAN OF EXPERIMENT No. 1.

J. 7. Nothing.	K. 7. Bone Bust.	L. 7. Nothing.	M. 7. Baugh's Superphosphate.	N. 7. Nothing.	O. 7. Berry's Superphosphate.	P. 7. Nothing.
J. 6. Bone Dust.	K. 6. Nothing.	L. 6. Baugh's Superphosphate.	M. 6. Nothing.	N. 6. Berry's Superphosphate.	O. 6. Nothing.	P. 6. Bone Dust.
J. 5. Nothing.	K. 5. Baugh's Superphosphate.	L. 5. Nothing.	M. 5. Berry's Superphosphate.	N. 5. Nothing.	O. 5. Bone Dust.	P. 5. Nothing.
J. 4. Baugh's Superphosphate.	K. 4. Nothing.	L. 4. Berry's Superphosphate.	M. 4. Nothing.	N. 4. Bone Dust.	O. 4. Nothing.	P. 4. Baugh's Superphosphate.
J. 3. Nothing.	K. 3. Berry's Superphosphate.	L. 3. Nothing.	M. 3. Bone Dust.	N. 3. Nothing.	O. 3. Baugh's Superphosphate.	P. 3. Nothing.
J. 2. Berry's Superphosphate.	K. 2. Nothing.	L. 2. Bone Dust.	M. 2. Nothing.	N. 2. Baugh's Superphosphate.	O. 2. Nothing.	P. 2. Berry's Superphosphate.
J. 1. Nothing.	K. 1. Bone Dust.	L. 1. Nothing.	M. 1. Baugh's Superphosphate.	N. 1. Nothing.	O. 1. Berry's Superphosphate.	P. 1. Nothing.

PLAN OF EXPERIMENT NO. 2.

J. 7. Grain, 11.84. Straw, 33.16.	K. 7. Grain, 12.50. Straw, 62.60.	L. 7. Grain, 12.00. Straw, 38.00.	M. 7. Grain, 10.81. Straw, 77.19.	N. 7. Grain, 12.09. Straw, 35.91.	O. 7. Grain, 10.93. Straw, 37.07.	P. 7. Grain, 12.84. Straw, 25.41.
J. 6. Grain, 14.37. Straw, 41.63.	K. 6. Grain, 14.81. Straw, 45.19.	L. 6. Grain, 14.18. Straw, 53.82.	M. 6. Grain, 11.12. Straw, 34.88.	N. 6. Grain, 12.43. Straw, 37.00.	O. 6. Grain, 12.25. Straw, 37.75.	P. 6. Grain, 12.93. Straw, 55.60.
J. 5. Grain, 14.97. Straw, 65.07.	K. 5. Grain, 15.28. Straw, 64.72.	L. 5. Grain, 15.68. Straw, 24.32.	M. 5. Grain, 12.81. Straw, 55.19.	N. 5. Grain, 13.00. Straw, 25.00.	O. 5. Grain, 11.93. Straw, 28.00.	P. 5. Grain, 17.50. Straw, 40.50.
J. 4. Grain, 13.55. Straw, 48.48.	K. 4. Grain, 16.62. Straw, 39.38.	L. 4. Grain, 16.87. Straw, 57.13.	M. 4. Grain, 13.75. Straw, 32.25.	N. 4. Grain, 13.37. Straw, 51.00.	O. 4. Grain, 14.03. Straw, 33.97.	P. 4. Grain, 11.43. Straw, 22.57.
J. 3. Grain, 13.25. Straw, 58.75.	K. 3. Grain, 15.25. Straw, 50.75.	L. 3. Grain, 18.81. Straw, 47.19.	M. 3. Grain, 16.31. Straw, 31.69.	N. 3. Grain, 13.25, Straw, 26.75.	O. 3. Grain, 14.18. Straw, 39.82.	P. 3. Grain, 14.18. Straw, 53.82.
J. 2. Grain, 15.00. Straw, 35.00.	K. 2. Grain, 16.25. Straw, 43.75.	L. 2. Grain, 18.91. Straw, 39.09.	M. 2. Grain, 18.93. Straw, 41.07.	N. 2. Grain, 13.37. Straw, 32.63.	O. 2. Grain, 12.36. Straw, 59.64.	P. 2. Grain, 16.91. Straw, 23.09.
J. 1. Grain, 15.53. Straw, 42.47.	K. 1. Grain, 17.58. Straw, 52.42.	L. 1. Grain, 17.72. Straw, 50.28.	M. 1. Grain, 20.56. Straw, 35.45.	N. 1. Grain, 14.40. Straw, 35.75.	O. 1. Grain, 13.50. Straw, 50.50.	P. 1. Grain. 14.25. Straw, 21.75.

In Table No. 1 is given the yield per acre of the *nothing* plats—the grain in pounds and in bushels, and the straw in pounds and in tons—together with the average of all the nothing plats.

In Table No. 2 is given the yield per acre of the plats to which the fertilizers were applied, bringing together the plats that had the same treatment.

The quantity of grain and of straw is given for each plat as in Table No. 1, together with the average for the plats receiving the same manure.

TABLE NO. 1.

Yield per acre of Nothing Plats.

LETTER AND NUMBER OF PLAT.	GRAIN.		STRAW.	
	Pounds.	Bushels.	Pounds.	Tons.
J. 1, Nothing	1553	48.53	4247	2.12
J. 3, Nothing	1325	41.41	5875	2.94
J. 5, Nothing	1497	46.78	6507	3.25
J. 7, Nothing	1184	37.00	3316	1.66
K. 2, Nothing	1625	50.78	4375	2.19
K. 4, Nothing	1662	51.93	3938	1.97
K. 6, Nothing	1481	46.28	4519	2.26
L. 1, Nothing	1772	55.37	5028	2.51
L. 3, Nothing	1881	58.78	4719	2.36
L. 5, Nothing	1568	49.00	2432	1.22
L. 7, Nothing	1200	37.50	3800	1.90
M. 2, Nothing	1893	59.15	4107	2.05
M. 4, Nothing	1375	42.97	3225	1.61
M. 6, Nothing	1112	34.75	3488	1.74
N. 1, Nothing	1440	45.00	3575	1.79
N. 3, Nothing	1325	41.41	2675	1.34
N. 5, Nothing	1300	40.62	2500	1.25
N. 7, Nothing	1209	37.81	3519	1.79

TABLE NO. 1—CONTINUED.

LETTER AND NUMBER OF PLAT.	GRAIN.		STRAW.	
	Pounds.	Bushels.	Pounds.	Tons.
O. 2, Nothing	1236	38.62	5964	2.98
O. 4, Nothing	1403	43.84	3397	1.69
O. 6, Nothing	1225	38.28	3775	1.89
P. 1, Nothing	1425	44.53	2175	1.09
P. 3, Nothing	1418	44.31	5382	2.69
P. 5, Nothing	1750	54.69	4050	2.02
P. 7, Nothing	1284	40.12	2541	1.27

TABLE NO. 2.

Yield per acre of Plats to which Special Fertilizers were applied.

	LETTER AND NUMBER OF PLAT.	GRAIN.		STRAW.	
		Pounds.	Bushels.	Pounds.	Tons.
Berry's Superphosphate applied in 1869.	J. 2	1500	46.87	3500	1.75
	K. 3	1525	47.62	5075	2.59
	L. 4	1687	52.71	5713	2.86
	M. 5	1281	40.03	5519	2.76
	N. 6	1243	38.84	3700	1.85
	O. 1	1350	42.19	5050	2.52
	O. 7	1093	34.16	3707	1.85
	P. 2	1691	52.84	2309	1.15
	Average	1421	44.40	4321	2.16

TABLE NO. 2—CONTINUED.

LETTER AND NUMBER OF PLAT.		Grain.		Straw.	
		Pounds.	Bushels.	Pounds.	Tons.
Baugh's Superphosphate applied in 1869.	J. 4	1355	42.34	4848	2.42
	K. 5	1528	47.75	6472	3.24
	L. 6	1418	44.31	5382	2.69
	M. 1	2056	64.25	3545	1,77
	M. 7	1081	33.78	7719	3.86
	N. 2	1337	41.78	3263	1.63
	O. 3	1418	44.31	3982	1.99
	P. 4	1143	35.71	2257	1.14
	Average	1417	44.28	4683	2.34
Berry's Bone Dust applied in 1869.	J. 6	1487	44.90	4163	2.08
	K. 1	1758	54.94	5242	2.62
	K. 7	1250	39.06	6250	3.13
	L. 2	1891	59.09	3909	1.95
	M. 3	1631	50.97	3169	1.58
	N. 4	1337	41.78	5100	2.55
	O. 5	1193	37.28	2800	1.80
	P. 6	1293	40.41	5560	2.78
	Average	1474	46.03	4524	2.26

The average yield per acre for the different fertilizers is as follows:

	Oats. Bushels.	Straw. Tons.
25 nothing plats	45.14	1.99
8 plats, Berry's Superphosphate	44.40	2.16
8 plats, Baugh's Superphosphate	44.28	2.34
8 plats, Berry's Bonedust	45.24	2.26

In 1869, Berry's Superphosphate gave a gain of 40.26 bushels of roots per acre; Baugh's Superphosphate gave a gain of 37.79 bushels; and Berry's Bone Dust a gain of 40.93 bushels, when compared with the average of the nothing plats.

The variation in the weight of the grain per measured bushel is greater in this experiment than in either of the preceding. The heaviest grain was at the rate of 38 lbs. per bushel, and the lightest grain 28 lbs. per bushel.

In this experiment, the variety of oats sown was the White Schonen, which gave an average weight of 37.5 lbs. per measured bushel, in 1869, and a yield per acre, by weight, of 63.3.

In concluding this Report on the experiments conducted in the Farm Department, I would again acknowledge the very valuable assistance rendered by Mr. C. C. Stowe, foreman of the farm, Mr. E. H. Hume, assistant foreman of the farm, and the students assigned to the department.

CHARACTERISTICS OF THE SEASON OF 1870.

That part of the year 1870 which comprised the period of the growth of vegetation, was in Central Michigan more remarkable for heat and moisture than for any other atmospheric feature. It is true, that taking the entire season, the degree of moisture varied much in the different sections of the State, but until the time of wheat harvest there seemed to be no lack in any quarter. The wheat crop was, indeed, very much injured in various places, from frequent and copious rains that fell just before and at the time harvesting was going on. In some instances the grain sprouted considerably while standing in shock, and in a few cases it sprouted before it was cut. The loss from this cause, and from wastage in handling, must have been considerable, while the heavy extra expense of cutting by hand labor, on account of the ground being in many instances too soft to use a reaping machine (as was the case to a great extent in 1869), increased the cost of the crop.

In the central and eastern portion of the State, the season continued rather wet from its opening to its close, and in connection with 1869, we had the coincidence of two wet seasons in succession. From the first of May to the last of November, 1870, there fell 27.047 inches of rain, against 22.149 inches in 1869—showing nearly five inches more for the former than the latter season. Yet 1869 was called a very wet season. But the summer of 1870 was here still more on the extreme of wetness. The rain-fall of June and July, 1869, was 10.164

inches, and the rain-fall for the same period of 1870, was 16.294 inches—or 6.130 more in the latter than in the former season. The average annual rain-fall for the locality, is 32 inches, so that in June and July, 1870, we had more than half the usual supply for the year.

The following table, compiled from the records kept by Professor Kedzie, at the State Agricultural College, shows the rain-fall, in inches and fractions of an inch, for the months of May to November, inclusive, for the year 1870, compared with the like period for the five preceding years:

MONTHS.	1865.	1866.	1867.	1868.	1869.	1870.
May	1.772	3.478	3.809	2.800	2.054	1.160
June	3.552	3.366	2.828	3.546	4.396	7.270
July	3.906	4.194	1.782	1.108	5.768	8.024
August	3.331	3.442	1.740	2.420	4.853	4.530
September	4.792	5.866	1.419	2.954	1.330	2.846
October	2.788	3.566	2.108	1.106	1.723	2.312
November	0.680	2.600	2.180	2.439	1.925	0.905
	20.822	28.452	15.866	16.371	22.149	27.047

The most remarkable feature presented by the table is the great amount of rain for the months of June and July, 1870, compared with the corresponding months of the series. In commenting on the characteristics of the season of 1869, in the Report for that year, p. 105, we said: "There has been nothing like drought in this locality during the whole season,—no time when grass, even in the dryest places, was not fresh and green." The remark might with propriety be so extended as to embrace the season of 1870; and it may be said that during the two seasons not a spire of grass, in this vicinity, indicated the want of rain.

But notwithstanding the greater rain-fall in the season of

1870, most crops have grown more luxuriantly than in 1869, and the ground has not been so constantly saturated with water. This is probably to be accounted for from the less degree of cloudiness, and the higher temperature, which prevailed during the latter season. The rains fell in more copious showers, but during the more frequent intervals of sunshine the moisture was to a greater extent evaporated. The mean temperature of the months from May to November, inclusive, for the year 1870, in comparison with the five preceding years, is shown by the following table, prepared from the records kept at the Agricultural College:

MONTHS.	1865.	1866.	1867.	1868.	1869.	1870.
May	57°.65	55°.04	51°.11	59°.08	55°.53	64°.82
June	70.76	63.06	71.61	68.48	64.45	70.87
July	65.06	77.72	71.06	77.19	70.85	74.40
August	65.84	62.06	69.78	79.37	70.50	70.11
September	67.66	55.08	55.06	58.77	63.45	63.66
October	46.05	49.05	50.06	45.19	40.84	52.45
November	38.63	37.94	40.44	36.77	32.05	38.40

The comparatively light rain-fall of September was favorable to wheat-growing, and probably about the usual breadth of land was sown to this grain. The early autumn was not only generally mild, but exempt from cold turns and frosts,—no frost to injure vegetation occurring in this vicinity till the second week in October, and even then it was not severe; the temperature of the month being higher, as will be seen by the above table, than any corresponding month of the series. Both October and November were very mild compared with the temperature of those months in 1869. There was no snow till after the 1st of December, but during that month there fell, according to Professor Kedzie's record, 18½ inches.

The remarkably mild weather, which continued till the close of the year, with the exception of a few cold days during the last week in December, was of great advantage to farmers in enabling them to secure all their crops in good condition, to finish up all necessary farm work, and make various improvements. Grazing stock obtained support in the fields to a considerable extent till December, except in localities where the growth of grass was checked by drought in autumn.

The necessity and advantage of drainage, impressed on farmers by the season of 1869, has been, in various districts, even more strongly enforced by the season of 1870. In some instances where the operation was commenced by the laying of tiles in the former year, the advantages have been strikingly manifest in the success of crops during the latter, although more time is required to bring the drains into full action.

The results of the season in reference to the general yield of crops, will be more particularly noticed under another head.

YIELD OF CROPS, 1870.

In response to inquiries sent to the various agricultural societies in the State, returns have been received in regard to the yield of crops per acre, as shown in the following table:

COUNTIES.	Wheat, bush.	Barley, bush.	Oats, bush.	Buckwheat, bush.	Corn, bush.	Potatoes, bush.	Clover Seed, bush.	Hay, tons.
Barry	15	25	30	15	40	150	2	1½
Calhoun	13	25	40	20	80*	60	1¼	1½
Cass	14	------	35	6	40†	15	2	1⅓
Eaton	16	15	50	20	60	150	1½	1⅓
Genesee	12½	30	35	12	30	100	2	1¼
Hillsdale	18	15	45	15	45	25	1¼	1⅓
Huron	15	20	48	32	38	150	------	1
Ingham	18	------	40	------	40	75	------	1⅓
Ionia	15	20	40	15	30	75	2½	1
Lapeer	15	20	35	10	40	100	2	1½
Livingston	14	20	37½	20	41	50	------	1⅓
Macomb	12	30	30	20	40	120	2	1
Shiawassee	16	20	40	20	45	125	2	1½
Tuscola	15	------	45	18	50	75	------	1⅓
Washtenaw	14	20	30	12	40	80	1½	1⅓

* Ears. † Shelled.

The small crop of potatoes in some of the counties seems to have been caused mostly by the potato-beetle.

ON THE REARING AND FEEDING OF LIVE STOCK.

At the last monthly meeting of the Ballineer Farmers' Club, Mr. Hooper, of Droomkeen, read the following paper:

In discussing the rearing and feeding of live stock, I will begin with the bovine or the ox tribe, as being, perhaps, the most important to the farmers of this neighborhood, and commonly distinguished from sheep, pigs, and horses, by the term cattle. Our first considration, then, shall be the rearing of the calf. The calf being dropped, this is to be borne in mind: From that moment until it arrives at maturity, it must be kept progressing, improving, growing; in no other way will it pay. The question is, How is this best to be done? Now, of course, the best food for a calf is that provided by nature—the pure sweet milk of the cow; but dairymen and dairy-farmers have proved by experience that an ordinary calf will not pay for new milk, or, at least, that the consumers of milk and butter in our towns will pay a better price for it; and we also find by experience that a calf will grow and be healthy on milk from which all the cream has been taken; and, in my opinion, no better plan can be adopted to rear calves profitably than to give them plenty of thick milk (a thing almost unknown in England, I believe, but decidedly better food for calves than skimmed milk which has not turned thick and sour); but this feeding should be continued longer than is usual, say for at least five months. The calf should not be allowed to run out too young. I am convinced that a great many of the calves belonging to some of the smaller farmers of my neighborhood, receive a check from which they never recover, by being exposed to the cold and wet before they are able to bear it. The calf should be kept in a loose box, with plenty of light and air, and a little fresh hay given him now and then, which he will soon learn to eat; then, at six or eight weeks old, he may be let out for a short time on a fine day, and the time increased by degrees as the weather permits. I believe some good authorities object to giving young calves hay, thinking it too binding for them; but I suppose this must be when the calves are reared on new milk; for I have always found the calves that eat most hay to be the healthiest and do the best. When there is abundance of milk the calves may be reared at very little cost, but they may also be reared very profitably even when there is a scarcity of milk. A Mr. Rush, in Gloucestershire (as I see from a report in the Cirencester Farmers' Club), rears from fifty to fifty-five calves annually, and he only keeps four cows. This he does by using hay-tea, with the addition of a little oilcake mixed with gruel. His calves cost him 1s. 6d. or 1s. 9d., a week each. I am myself rearing some calves now with a very little milk by the help of hay-tea and oilcake, and I find they cost me almost the same. The hay-tea is made by boiling some hay, or putting some in a tub and pouring boiling water on it: the former plan is the best. We allow each calf four pints of skimmed milk at a meal, and add enough hay-tea to make the quantity up to eight or ten pints, according to the size and appetite of the calf; into this we put a little oilcake, which has been previously made into paste with hot water, in the proportion of 2 lbs. of cake to five calves; and they are doing very well. I may here mention what appears to me to be a great advantage in rearing calves at this time of year: all the time they are living on milk and the little bite of hay before mentioned there is no grass; but when the grass comes in the months of April and May, they are ready to eat it; and the difference between them and the summer calves the following winter is remarkable. The latter never overtake them, and we could scarcely think the calves were of the same year. I think at the present price of stock, any good calf, dropped from November to May, should be reared.

There is not a ready market market in these parts for the prime veal that fetches so high a price in London and Dublin; but a paper on feeding cattle would not be complete without some mention on fattening calves. I have occasionally fattened an early calf, and they have paid me well for the new milk they drank; but I have found that I was quite at the mercy of the buyer, and always felt in taking a fat calf to Cork that I was running the risk of having almost to give him away. I may here mention that to fatten a calf most profitably, he should be dropped not later than January, as the price of veal becomes much lower after March. As regards the mode of feeding for veal, I cannot do better than to quote the account given by a Mr. Dowdall at the Ballymahon Farmers' Club, in the county of Westmeath. He lets the calf suck the cow morning and evening, and gives it a mid-day meal from the pail, which, after a while, has an egg beaten up with the milk. At a month old, he sops white bread with the milk, and at eight weeks old he has a well-finished calf worth £6. I feed mine three times a day from the pail, but I consider Mr. Dowdall's a better plan, and were I fattening another I should adopt it.

We now come to the weaning of the calf. I find my calves do best on a clover stubble; that is, on the stubble of a barley or oat field that is laid out with clover and grass seeds; and I do not find that any such light stock do any harm to the young clovers, provided they are not kept on them longer than the 1st of January. On whatever kind of pasture they are weaned, however, the milk should not be taken from them too suddenly; they should get a meal of milk in the middle of the day for a week or ten days after they are turned out to grass, and the quantity of this should be diminished by degrees. When the nights begin to get cold, generally some time in the month of October, they should be housed at night; but, in my opinion, should be allowed out by day the whole winter, and therefore should not be kept in too warm or close a house by night. I have kept them in close houses, and only let them out for a few hours on fine days; and I have kept them in altogether, day and night, the whole winter, some tied and some in loose boxes; but I find they thrive and grow best on the system I have adopted for the last three or four years, and that is to tie them up at night in a shed open to the south, and let them out the whole day in all weathers, excepting, of course, when the ground is covered with snow and they could get nothing to eat. If they could be put into a yard at night, with sheds around it, perhaps it would be better still, as some have an objection to tying up young calves; but if it has its disadvantages, it has also its advantages, one of which is, that they are much quieter when tied up the next winter, at which age I fatten off my young stock; whereas I find those I buy take some time to get accustomed to the chain and trough, and lose time accordingly. As to the feeding of calves, the first winter I find they do better, with the out-run I have spoken of, on hay alone, than on hay or straw or turnips. Some may think this unreasonable, but I can only say I have wintered calves for eight years *with* turnips, and for four years *without* them, and I have no intention of altering my present system. Mr. Burke, whom I quoted before, who rears calves with so little milk, winters them on pulped mangels, mixed with straw-chaff, oilcake, and crushed corn; and he says, reckoning 5 cwt. of straw chaff at 5s., 10 cwt. of pulped mangel at 5s., 1 cwt. of oilcake at 10s., and 4 cwt. of mixed crushed corn at 30s., he has one ton of food for 50s. equal to the best hay. But as our hay is seldom worth even as much as this in our own yards, I do not think we should gain much here by the adoption of his system, which is certainly, though, a great improvement on the ordinary one of whole or sliced turnips and hay or staw. I make no difference in the wintering calves intended for beef and those intended for the dairy. No matter what a calf is intended for, he should be well fed the first winter, or he will receive a check from which he will never recover. To proceed with my own system. I give my yearlings the best grass I have all the summer, and fatten them off the following winter; selling them when twenty-four, twenty-five, and twenty-six months old. I give them oilcake to the amount of 30s. a head (beginning with 1 lb. a day and increasing gradually to 3 lbs.), and hay or straw and turnips *ad libitum*, and the best fetch from £18 to £20 a head, and the smaller ones from £15 to £16,—that is, at the present and recent high price of beef. My cattle get three feeds of sliced turnips in the day, the first between five and six o'clock, the next at eleven (immediately before which they get their oilcake), and the third and last at five, or later as the days get longer. At each of the meals, if any animals have finished before the others, or show any desire for more, more is given till they are satisfied. The cleaning out of the stall and currying of the cattle keep them disturbed a good part of the time between the first two meals; but after the second meal they are left to rest till the third, and after that are left undisturbed for the night. The racks placed above the turnips are filled with hay or straw as often as may be required. At whatever age cattle are put into the stall, they should have some turnips given them on the grass for a week or two before they are put in, to accustom them to a change of food. The heifers that appear best adapted for the dairy, I sell as springers in October and November, when two-and-a-half years old. In my opinion this is young enough for even the largest and best heifers to calve; and smaller and weaker ones should not calve till they are nearly or fully three years old, according to their size and strength.

In regard to the feeding of dairy cows, every one knows that the better you feed them the better they will pay; and I am sure they would pay for a little linseed or rape-cake, in addition to the best food the farm affords. I "let" my dairy cows, and have not had any in my own hands for some years; but I am keeping a few now, chiefly for the purpose of rearing calves, which have of late years increased so much in value that I mean to try two pounds of rape-cake, daily, to each cow. The cost of this will be about twopence, and I feel sure the cow will give an increase of one pound per week, besides being kept in a healthy state, and the rest of her food doing her more good; for I find that cattle getting oilcake not only improve in condition faster, but their general health is better, and a change of food does not affect them as it does those not getting it. Many have used oilcake and other artificial foods extravagantly, and that has set many farmers against their use altogether; but I believe, used with caution, and as auxiliaries to other feeding, they will repay the outlay.

But to resume: As I said before, every one knows that the better a milch cow is fed the better she will pay; and there is little fear but that her owner will give her plenty to eat, if he can, as long as she gives him milk and butter. But some are apt to shorten the supplies of food during the time the cow is necessarily dry before calving. This is a great mistake, and tells against the owner in many ways. It weakens a cow at a time when she requires extra strength. This has such an injurious effect on her constitution that it lessens her supply of milk the ensuing season, and above, all, it injures the calf. There is no surer way of having strong, healthy calves, than to feed your cows well when they are dry.

I will now pass to the consideration of sheep; and, as I have just remarked with regard to the feeding of the cow, so I will say, the only way to insure strong, healthy lambs, is to feed your ewes well up to lambing time. I have found nothing so good for ewes, during the winter, as hay, given in racks on the pasture. If the pastures are very bare, and a sufficient quantity of hay cannot be spared, a few turnips will be necessary; but I do not like them for ewes if it can be helped, and if they must be given, would prefer Aberdeens to Swedish turnips. After lambing, unless a better pasture can be provided for them, they should get an increased supply of hay and a few mangels (not too many), and in any case half a pound of oats each, every morning. I believe no man ever took his oats to a better market than to give them to a flock of ewes with a good crop of lambs, for a month or six weeks after lambing, according to the season and growth of grass. The whole cost is nearly repaid in preventing the ewe from shedding her wool, as so many are apt to do after lambing, especially those that have twins; and the effect on the lambs is simply incredible, except to those who have tried it. If circumstances permit, those sheep that have twins should be kept separate from the rest of the flock, and have better keep. By this means you will have a more even lot of lambs at weaning time. The middle of February is early enough for the first lambs to be dropped; and a Mr. Davidson, in an address delivered at the Athy Farmers' Club, (a most valuable paper, by the way, on the management of sheep), says he does not like to begin lambing before March, and that any lambs dropped before April 20th are soon enough, and often beat the March lambs. At whatever time they are dropped, however, this is certain: that a late lamb that has received no check is much better than an early lamb that has been stunted.

Supposing the lambs, then, to have been dropped through the month of March, they will be ready to wean during the month of July. For this purpose, neither young grass, nor clover, nor old pasture is suitable,—either would scour them,—but a field two or three years in grass, that has been fed with sheep for the past twelve months, or the aftermath of an old grass field, will generally keep them healthy. If the lambs are intended for the butcher as hoggets, they should now get a liberal allowance of corn or cake, which they should have been taught to eat before being taken from the ewes. This will bring them in good condition to the turnips, a most important point, for if put on turnips weak, many of them will die, and many more not improve for some time. Before being put altogether on turnips, a few should be given them on the grass, to teach them to eat them, and to accustom them to a change of food. About the middle of October they should be put on turnips altogether; and then, no matter what their allowance of oilcake may be (and some will be required to get them fat by spring), they should get from half to three-quarters of a pound of oats daily. Until I began this system, I lost a large percentage of sheep on turnips. They should also have a constant supply of good oaten straw, or hay, in racks. I prefer the former, and if a little is trodden into the ground it is not of so much consequence. The turnips should be cut in pieces the size of a man's finger, by a machine made for the purpose, and given to them in troughs. By this treatment, lambs bred from the common sheep of this country, by a good ram, will come to eighteen or twenty pounds a quarter, at fourteen months old.

If fat lambs are desired to sell to the butcher at a few weeks old, I have found it a good plan to have them dropped earlier: say at the end of December or the beginning of January. When dropped, they should be kept in a sheltered field, or, better still, one with a shed in it, with their dams, and the latter should have a constant supply of good hay, a few sliced turnips, and a liberal allowance of corn and cake. By this means the lambs will be ready for the butcher in about eight weeks; and the ewes, if they were in good condition when they lambed, very soon after.

We have now come to the pig, and I will fully confess I have not found the feeding of pigs very profitable. A good breeding sow will pay, and a pig that lives on the refuse of the house, helped out with a stone or so of meal or bran in the week, is a profitable animal to keep; but as a part of the live stock of the farm, I have very little worth hearing to say about them. I will, therefore, read you an extract from a letter which appeared last year in the *Farmers' Gazette*, from Mr. William Joyce, of the Abbey Farm, Waterford, a most successful breeder and exhibitor of Berkshire pigs. He says: "There is no animal that will put on the same amount of flesh in a given time with the same amount of food as the highly bred pig. The Berkshire pig is a particularly thrifty, strong, hardy animal, and can bear any sort of weather; he does not want a sunshade in the summer to keep him from the sun, nor a great-coat in winter to keep him warm; and another great advantage is that no pig will feed on the grass like him. My breeding sows nearly live on grass all the summer. Even up to this time (November), they get only one feed of pulped roots in the day, besides what they get on the grass fields. I never cook any roots (except potatoes), but give everything in a raw state to cows, horses, and pigs. I have at present over 50 pigs, and every one that is able to eat gets pulped turnips. I find they do well on it, and the saving in labor and fuel is something considerable. People who come to see my pigs are greatly surprised to see them eating raw turnips, and to see them in such condition. I do not feed them on mangels, if I can avoid it, till sometime after Christmas. The way I use my pulped food for the pigs, is by putting the pulp into a large cask, and pouring over it a few bucketfulls of boiling porridge, made from Indian meal, leaving it a few hours to cool, and then it is fit for use."

To this I will only add that the remark I made, that all young annimals should be kept in an improving state, applies with equal, if not with greater force, to the pig. From the time of his birth to the time he is slaughtered, he should increase daily in size and weight. Every day that he does not do so he is a loss to his owner.

We now come to the rearing and feeding of the horse,—the animal of most interest to the general community, if not to the agriculturist. The farmer may not have all the appliances and arrangements of a stud farm, with its paddocks, open yards, loose boxes, etc.; still he may like to rear a foal or two every year; and, without entering deeply into the question whether it pays to breed farm horses, I think most will agree with me that it is better for a farmer to have horses to sell than horses to buy. Moreover, few farmers can afford to give up the whole time of their brood mares to breeding, nor do I see any reason why they should. Neither the mare nor the foal she is carrying will be any the worse for her regular work on the farm; provided she is well fed, and not put to extra hard work, nor to work to which she has been unaccustomed. Of course, the nearer the mare gets to her time of foaling, the lighter must be the work required of her; and a week or two before the time she should be left perfectly idle, especially if not in high condition. If the mare foals very early in the season, and the weather is bad, she and the foal should be put in a loose box at night, and allowed as much liberty by day as the weather will permit; but if she does not foal before the month of April, she may be kept in a sheltered field day and night, and, unless the pasture is very good, a feed of oats should be given her every morning. Now, we all know that the foal would be better if his dam had nothing to do but to suckle him till he is old enough to wean; but I am speaking of rearing horses on farms where the work of the farm is of primary and the rearing of foals of secondary importance. In general, then, the mare may be put to work a fortnight after foaling, and when she is at work, the foal should be shut up in a light, airy, loose box, and the mare taken in to suckle him at intervals of not more than two hours. Of course the mare must be highly fed when doing this double work, and should have as much green food as possible. In addition to the dam's milk, I always give the foals of my working mares a quart of cow's milk, brought to a natural heat,—a pint about an hour and a half after the mare has gone to work in the morning, and a pint in the afternoon. At first the young thing is very shy, and is frightened at being caught and having his nose held into the milk; but he soon learns to expect it, and will come and drink it out of the vessel when held to him. When the turnips are all sown, and there is little work for the horses, the foal can get two or three months of uninterrupted liberty with his dam. I find the most convenient time to wean is when the working horses are brought into the stable at night. By that time the foal is accustomed to the company of the other colts (or a donkey will answer the purpose, if there are no other colts on the farm), and will stay quietly with them without his dam. In about another month (according to the season) I bring him in also at night, putting him with another of the same age, or not more than a year older, into a loose box, and giving them as much chopped hay and furze as they will eat, and three or four pounds of crushed oats; and here let me remark, there is no time of his life at which a horse gives a better return for a few oats than the first winter. I turn them out the whole day in all weathers, except very hard frost or snow, and put them out altogether as soon as the weather and grass are good enough. The feeding the second winter is merely a repetition of the first.

And now, having reared the horse, let us consider his feeding after he has arrived at maturity; and, as we are speaking only of the rearing and feeding of farm-stock, I shall confine myself to the feeding of farm-horses. Now, a farmer not only requires to feed his horses well, so as to keep them up to their work, but also economically. To this end, it is indispensable, in my opinion, that he possesses two machines, viz., the furze-cutter and oat-bruiser; the price of the hay and oats saved will soon pay the cost of both. I find that my horses do as well on 10½ lbs. of crushed oats as on 14 lbs. of whole oats. They are fed in the following manner: From the time they are put to regular work after harvest, and stabled at night, they get 3½ lbs. crushed oats the first thing in the morning (at five o'clock in the autumn and spring months, when they go to work at half-past six; and at six or a little before it, in the dark winter mornings, when they cannot work till after seven), about an hour and-a-half before they go to work. At twelve o'clock they get another 3½ lbs. of oats and a pannier of chopped furze, and at night another 3½ lbs. of oats and two full panniers of chopped furze. On this feeding I have kept several horses in really good working condition all this winter; but I have some horses (not of my own rearing) which will not eat enough furze, and to them we give a little hay in addition. About the end of March the furze becomes less nutritious, and we substitute hay for it; and as the days are then long and the work hard, we give, in addition to the three feeds of oats, a small bucketful of boiled Swedes, mixed with a little meal or bran, to each horse at eight o'clock in the evening. I do not allow the cost of meal or bran to exceed 1s per week for each horse, so that the feed, including turnips and firing, costs something less than a feed of oats. If the turnips are finished before we can cut any grass or vetches, I give a feed of boiled barley instead. This costs as much as a feed of oats, *exclusive* of the firing. As soon as grass or vetches can be substituted for the hay, I find two feeds of oats sufficient without any boiled food. As soon as the turnips are all sown, I turn my horses out; and as there is then very little work to be done, and only one, or at most two, are required to be at work at a time till harvest, I give them no more oats. This run out to grass during the summer months I consider essential to the horse's health.

The cost of each horse per annum will thus be as follows: 10½ lbs. oats a day, reckoning the oats at the extreme price of 1s. a stone, will be 9d., *i. e.*, 5s. 3d. a week, or £6 16s. 6d. for twenty-six weeks. The cost of the furze is merely the rent of the land on which it grows, as there is no expense attending its cultivation ; the cost of preparing it is about 1s. a horse per week ; this gives 26s. to be added to the £6 16s. 6d., making the cost of feeding for twenty-six weeks, from October to March, £8 2s. 6d. We now come to eight weeks during which the horse is fed on hay, and has an extra feed of boiled food at night, to be reckoned at the same price as a feed of oats. This will give 1s. a day, then, for four feeds—*i. e.*, 7s. per week, or £2 16s. for the eight weeks. As for the hay, I must confess I never weighed it, but I think we may put it at one hundred weight and a quarter a week, say 4s.; this gives 32s. to be added to the £2 16s., making £4 8s. for eight weeks, from the beginning of April to the end of May. We have still some four weeks left, during which the horse requires two feeds of oats a day ; these at 6d. a day will amount to 14s., and reckoning the cut grass at 3s. per week, will make 12s., or £1 6s. for the four weeks' feeding. During the remaining fourteen weeks of the year the horse gets no corn except a few sheaves of oats in harvest-time, and perhaps a feed of corn now and then, if he has to go a journey, which I think need not be reckoned in the cost of his keep. Valuing his pasturage then at 2s. 6d. per week, we have the sum of £1 15s. for the remaining fourteen weeks' keep. We have then £8 2s. 6d. for twenty-six weeks, £4 8s. for eight weeks, £1 6s. for four weeks, and £1 15s. for fourteen weeks' pasturage, giving the cost of each horse for the year £15 11s. 6d. To this must be added something for the furze (a horse will eat about a quarter of an acre), and something for the wear and tear of the furze-cutter and oat-bruiser (I have had mine in constant use for ten years, and they are doing their work as well as ever). With whole oats the cost would be £4 more on each horse per annum.

I do not know that I could add any more on the rearing and feeding of live stock from my own experience, I will, therefore, conclude with an extract from an American paper, the *Country Gentleman:* "It is, perhaps, well for the cause of good management that all neglected treatment of animals should result in loss to the owner. If he has starved his cattle, sheep, and pigs for a year or more, he cannot atone for it by sudden attempts to push them to fatness. On the contrary, the only way is to see that the growth continues without cessation, summer and winter, from the earliest period of their existence till they are finally sold in the market. A single check given to this continued progress may arrest or retard it for months. Our own observations lead us to the conclusion that the whole profits resulting from rearing and fattening, when this continued progress is kept up by careful, regular, but not extravagant feeding, are at least triple the amount realized from early neglect and heavy feeding afterwards, and often the difference is many times greater than here stated."

There is nothing that should be more strongly impressed on the mind of the young farmer who makes the feeding of animals a prominent part of his business, than the importance of keeping up an unremitting growth throughout the whole course of their existence.

PASTURES.

FROM A LECTURE BEFORE THE CENTRAL FARMERS CLUB, BY PROF. COLEMAN.

MIXTURE OF GRASSES FOR PASTURES.

The following may be taken as an average, but different soils have different quantities:—5 lbs. rape, 5 lbs. cow-grass (perennial red clover), 5 lbs. white clover, 2 lbs. red clover, 2 lbs. Alsike clover, 2 lbs. meadow foxtail, 1 lb. crested dogs-tail, 2 quarts meadow-fescue, 8 quarts Italian rye-grass, 8 quarts Pacey's rye-grass, 8 quarts Stickney's rye-grass, 2 quarts cocksfoot (orchard grass), sown latter end of May and beginning of June.

PERIOD AT WHICH FIRST FED.

Generally in August. The field is always divided by hurdles into three or four divisions, and these are fed in rotation, not allowing them to get too bare. The stock in the first year is always sheep, which fatten with surprising rapidity. The rape affords shelter to the young seeds (young herbage), and the additional feed gives increased droppings on the land and increased fertility.

FUTURE MANAGEMENT AND MANURES, IF ANY.

The second year a part is generally cut for hay; mown early, about the 4th to 8th of June, and a luxuriant fog follows. The part pastured is fed by a mixed stock of horses, cattle, and sheep, and so on from year to year. If judiciously stocked and attended to, the pasture becomes in the second or third year like old, good pasture-land, and will continue so. Land managed as above twenty years ago is still fine pasture.

John Grey, in an address to the Hexham Farmers' Club, called attention to some experiments, made many years since for the Highland Society, and recorded in Morton's Cyclopedia. Fine soil was sifted and put into boxes in which grooves could be made to any desired depth. Grooves were made of ¼ in., ½ in., 1 in., 1½ in., and 2 inches, and grass-seeds sown in each. It was only in the ¼ in. depth that any quantity of the seeds came up; about half the seeds were lost in the half-inch groove; and at 1 in. hardly any penetrated. Mr. Grey adds: "What was to become of those small seeds if they were thrown upon clods, and a heavy roller passed over them afterwards? No doubt one-half of them would be lost sight of altogether; and if they wanted permanent pasture, they must be at the pains to make their mould as fine as possible, rolling the surface, and having sown the seeds, covering them as gently and lightly as possible with a light harrow."

IMPROVEMENT OF WORN-OUT AND NEGLECTED PASTURES.

The first point is to remove superfluous water. We cannot grow nutritious grasses so long as the soil is saturated with moisture. We may have, it is true, a considerable bulk of watery, bad herbage, and it is this fact that has led to an idea that grass-land does not want drainage so much as arable; indeed, in some instances it has been said that drainage has injured grass-land. This I deny as regards clay. Instances may occur in which drainage does no good,—where the subsoil is gravel, for instance, and where the moisture which found its way along the porous beds from higher land may have actually nourished, and only been in excess at rare intervals; for be it remembered that such water is not necessarily stagnant; it may fertilize and pass on, and so after a time escapes. Now close draining in such a case, by cutting off the water before it reaches the roots, may do harm, and I have heard of cases where this was so apparent that the drains were blocked up again. Draining alone will not *renovate* a worn-out pasture, as some people appear to imagine it should do. The coarse grasses which have taken possession of the ground, and driven out or dwarfed the better sorts, die out when the water goes; the soil contains but little available food; the atmosphere has not been able to act upon the minerals; and consequently, if drainage is not followed by liberal treatment, the produce, at any rate for a time, falls off, and the farmer who may have looked upon drainage as the only necessity, is disgusted to find he has less grass than before, or perhaps jumps to a too hasty conclusion that the drainage was a mistake. Constant rolling is beneficial, tending to produce an even, firm surface. Good grass only grows on a firm surface.

* * * * * * *

The cheapest way to improve grass land, is to fold sheep eating oil-cake or other artificial food. It of course takes time, but if slow it is very effectual. "The sheep's belly is the best dung-cart." This is the plan I most strongly recommend. The mechanical effects of the treading is considerable. I have seen wonderful improvement made in grass-land by sheep folding. At the same time applications of a bulky nature may be collected and applied without a very heavy outlay. In this way great improvement may be made in neglected grass-land—an improvement which will be found highly remunerative. Grass-land once put in good heart must be maintained by periodical dressings, by careful attention, rolling and chain-harrowing the surface, and by either collecting the droppings of cattle into a heap, and mixing, or by knocking them to pieces, and by keeping the out-falls of the drains cleaned out.

Farmers who have been brought up in a dry climate, on getting into a wet one, are too apt to continue the old system of farming. With regard to the remark of Mr. Thomas as to the period for laying down grass-land, he was in a position to speak practically. He had at that moment 300 acres of grass-land, which had been seeded down without a corn crop. Instead of directing his attention to corn, he had directed it to grass, on account of the character of the climate. He entirely differed from Mr. Thomas's friend, who sowed his grass-seed about June or July, a period of extreme heat, when the plants would not cover the ground. So far as his experience went, the best way of laying down land for permanent pasture was to prepare and clean it as well as possible for the root crop in the previous year, and in the next season, or as early as the frost would allow, to put in the grass-seed—say by the first of April. Rape-seed should be sown with the grass-seeds. If the seeds were sown thick enough, they would soon be up, and they would prove most verdant, and be ready to stock by midsummer. The land would thus be covered before the heat of the sun came, and be most valuable in the scarce, hot months of July to September, a period when they found the most difficulty in getting food for their animals. The young plants would throw up such an immense quantity of succulent feeding-grasses, that no doubt would remain in the minds of those who saw them, as to what period was best for sowing. * * He was convinced that the adoption of a better system of grass production, especially in suitable climates, would in many cases yield, for the outlay made, a return of not less than twenty per cent. Mr. Thomas made a remark about grass changing after draining. There was a principle involved in this: The aquatic grasses were killed, and Nature had to replant. Again, showing the effect of warmth, the grasses were fine on the southern slope of a hill, while those on the north were coarse.

Mr. T. Congreve maintained that grazing land required even a greater share of attention than ploughed land; and there were great numbers of men who paid as much attention to the grazing of their cattle and sheep as others did to the cultivation of wheat and barley. * * It was an old saying that any fool could make a farmer, but it certainly was not every fool that could make a grazier. The grazier had his stock to buy and to manage; and it was one thing to grow crops and another thing to buy and feed a lot of cattle and make money on them. The graziers had to keep their eye on the grass during the whole time of feeding. If they let it grow too long they were beat, and, on the other hand, if they kept it too short they were beat; and, therefore, grazing required as much attention as any other occupation.

Mr. J. A. Nockolds said, many years ago he was an advocate for the breaking up of old clay pastures; but he entirely agreed with Prof. Coleman that the renovation of old pastures was a far easier task than the making of new. There was another point to which he would refer. Valuers who visited a parish for the purpose of making an assessment, were in the habit of putting a spade in the ground to see what was the character of the subsoil.

Mr. Fisher Hobbs said, as to top-dressings, he had found that by draining, high manuring, and the occasional application of unfermented manure, say eight cubic yards per acre, immediately after the mowing, he benefited the grass more than he could by any other means.

THE WHEAT SPECIALTY.

In the central portion of the West, I have no hesitation in declaring that wheat-growing as a specialty should cease. It may be tolerated by the pioneer farmer, without capital, who adopts it as an inevitable necessity of his poverty, and as an expedient for exchanging a part of the intrinsic value of his farm for houses and implements, for horses and cattle, for fences, and for other improvements. The danger is, that having converted half his fixed investment into working capital, he may by force of habit continue to reduce its productive value, until his improvements are worn out, his surplus wasted, his fields barren, and the strength of his manhood gone, leaving him only the infirmities of age with which to renew the battle of life upon other and untried fields; or if he enriches himself while despoiling the land of its fatness, he entails hard labor and comparative poverty upon his successor. It is a practice unworthy of this age of progress, and of the wonderful capabilities of the soil which it impoverishes; it is one that engenders weeds, deteriorates seed, discourages system, and repudiates science. The vandals of cultivation should be kept on the outskirts of the domain of agriculture.

Exporters of breadstuffs, and political enthusiasts who neither know nor care for the interests of the farmers, sometimes prate of the great value of cereal exportations. It is, and ever must be, if persisted in as a settled policy and relied upon as a permanent source of prosperity, an unmitigated curse rather than a blessing. It disturbs the equilibrium of production, despoils the soil of its fertility, fattens a horde of go-betweens, and often gives the larger portion of the crop for the transportation of the remnant. No man in his senses, or unblinded by the glare of gold, can fail to see the wastefulness of the exportation to another continent of so bulky a product as wheat or flour. If, as is asserted, the price of the quantity consumed at home is fixed by the price of that which is sold abroad, the comparatively large exportation of last year might as well have been burned with the straw, so far as the grower was concerned, for the reduction in the home consumption far exceeded the value of the exports at the seaboard.

Very few, indeed, realize the comparative paucity of our exports of wheat and flour. In forty-three years, up to the close of the fiscal year 1868, the wheat exported was less than the quantity produced in the United States in 1869, and the wheat and flour together would aggregate little more than the crops of the past two years, namely, 236,942,887 bushels of wheat, or with flour included as wheat, 670,900,182 bushels.

Another consideration is found in this fact,—but one nationality, Great Britain and its dependencies, requires it; the purchases of all other nations are usually a mere bagatelle—not worth consideration. Thus we depend practically upon supplying the necessities of but one single people, and have in competition with us all the wheat-exporting nations of the earth; and the extent of those necessities varies with every change of season, and our share in the supply fluctuates with every caprice of trade.

A nation that lives by exporting wheat five thousand miles by land and sea must have a short existence. It is a folly which the practical sense of our farmers will never practice extensively, the vestiges of which will vanish gradually as population increases and advances westward. Let our agricultural exports increase in meat and cheese and wool, and in a hundred extended and valuable products of agriculture—not in wheat.

The burden of transportation is one that must ever bear heavily upon the farmers. Should ship canals, of sufficient size for the use of steam power, ever be opened to the sea-coast, or should a double-track freight railway, on which continuous trains could be run at lowest possible freights, ever be constructed, either by the United States or by a company under a charter drawn in the interest of the people, a burden of monopoly and excessive charges might be removed, but the expense would still be unendurable in the transportation of certain heavy products. The necessity would still exist for converting grass into butter and cheese, corn into beef and pork, roots and hay into mutton and wool; and better still, into woolens and worsted, and the elements of earth and air, through the alchemy of labor, into new and desirable forms of production required by the increasing wants of advancing civilization, of high value in proportion to bulk or weight, and of small expense for transportation in proportion to value.

The time has come when the farmers, in the exercise of their political prerogative, and in the strength of imperious numbers, should combine to protect themselves against the usurpations of railroad monopoly, and to thwart the schemes of gamblers in railroad stocks, who, with equal unscrupulousness, "water stock" and despoil the honest stockholder, and fleece the public with high fares and excessive freights. If farmers have no power to protect themselves, then they are indeed at the mercy of a set of reckless gamesters.—*Exts. from an address delivered by the Commissioner of Agriculture, at the Fair of the Ill. State Ag. Soc., at Decatur, on the 28th of September, 1870.*

SORGHUM AND ITS PRODUCTS.

After numerous attempts to produce sugar from the different varieties of the so-called Chinese sugar-cane, opinion in this country seems to have generally settled down to the fact that it is not an object to cultivate it here for any other purpose than the making of syrup. It is not unlikely that erroneous ideas have prevailed, more or less, in regard to the inducements which led to its cultivation in Europe. At one time we called to notice its cultivation in France, in competition with the beet, for the production of sugar. But it should be borne in mind that it was introduced into that country at a time when the grape crop was seriously injured by a fungoid disease; and some of the leading French cultivators of the sorghum stated verbally to the writer, that the production of alcohol which might be used in the manufacture of brandy, was the principal object in the cultivation of that plant. It is true that the beet was also cultivated to a considerable extent for the same purpose, and experiments with sorghum were made with the view of ascertaining whether it could be profitably used as a substitute. The result seemed to show that it could not; though the subsequent restoration of the vine and grape to comparative healthfulness, superseded the necessity of resorting, to as great an extent as previously, to other substances for spirit.

The excitement under which the cultivation of sorghum in this country was for some time carried on has now passed away, and it hence becomes important to learn its exact status as an economical product. In reference to this point, the statements of men who have given the subject close attention for many years, deserve attention. Mr. John Richard of Tecumseh writes that he considers the cultivation of sorghum for production of syrup, permanently established in the State, though he doubts whether it can be made profitable north of the three southern tiers of counties. He believes it to be the most profitable crop that can be grown, where it is properly managed, and says: "I raised the past season (1870), on a little less than an acre of ground, 200 gallons, which sold readily at seventy-five cents per gallon, and I consider the clear profit at least $75. No other crop on my farm paid so well." Mr. R. thinks the profit of sorghum depends very much on the care used in the manufacture of syrup, and that the lack of this in many instances causes the article to be of inferior quality, and consequently unprofitable. He speaks of the last crop as good—the yield of syrup large, and of fine quality, which will probably tend to increase the cultivation of the crop next season.

Mr. William Tafft of Plymouth, well known for the great interest he has taken in sorghum cultivation, etc., writes as follows:

"More attention was given to the cultivation of sorghum the past season, than for several previous years, and so far as my knowledge extends, the crop was excellent. The number of acres under cultivation in 1870 was about double that of the previous year. The yield to the acre was also about double, making four times as much syrup for the past, as the previous year, and of a better quality. The prospect for the coming season is, therefore, very good. The result is, however, very much owing to the seasons,—1869 being a very short, cold summer, and that of 1870 long and warm,—favorable to the growth and maturity of the sorghum plant.

"The price of ordinary farm products has also much to do with the sorghum crop. When wheat is worth two or three dollars per bushel, and other things in proportion, sorghum is neglected; but when prices are low the inquiry is, 'Where can I get some sorghum seed? I want to plant some this season; it is a good crop, after all.'

"The fact is now established that our soil and climate are favorable to the growth and even the *improvement* of the sorghum plant. We must have more and cheaper sugar (or its substitute), from some source, and our hopes center on the beet or the sorghum plant. My opinion is that both should receive the attention of the producers of the country. There are difficulties to be overcome in establishing the manufacture of sugar successfully in any country; but what has been done we may hope to do.

"There is not good syrup enough in the market to warrant the establishing of a refinery. The increase of production over a domestic want will be slow, and will depend very much on the price of other productions. I see a great improvement in the handling of the cane, a better understanding as to the requirements of the plant, and the manufacture of the syrup. The novelty has passed away; sorghum stands now on its merits; the increase in its production will be slow, but lasting and substantial."

QUESTIONS ABOUT DAIRYING.

Sanford Howard, Secretary of the Michigan State Board of Agriculture and the State Agricultural College, writes us as follows:

" How have the profits of cheese making in this country during the year 1870 compared with those of previous years?"

" What has been the range of prices in New York or at Little Falls, for the season?"

" At what age is American cheese sold for exportation?"

" Can you tell me the quantity exported last year, or tell me how I can obtain a correct statement in regard to it?"

" Do you think we can greatly increase the exportation of cheese from this country without lowering the price?"

I. Cheese dairying in the old dairy district of Central New York has been less remunerative during the past year than in 1869, and indeed less than for some years previous. This state of things is not confined wholly to New York, but there are causes affecting the dairy interest in this State which do not obtain in other sections.

But it must not be supposed that this less remunerative result of dairy farming comes from want of market, or lack of demand for dairy products, or from any over production of the same. The causes are of a different character, and may be briefly stated as follows: In the first place, drought, commencing early, and extending over a wide range of territory, so cut off pasturage that even on our best grazing lands the depreciation in the quality and quantity of milk commenced unusually early,—indeed, the herds at no time during the season of pasturage gave a full yield of milk. In some sections drought was excessively severe throughout the whole season, and the consequence was that, although the crop of the whole country, owing to the increase of dairy farmers, was not much, if any, less than in 1869, still, upon individual farms the decrease in quantity was sufficient to be very sensibly felt. Then the extremely hot weather of July and August, with an unusual trouble from flies in the curing-room, so injured the cheese product of these months that it could not be sold for the price of good cheese. Immense quantities of cheese throughout the country were not only "off flavor," but were also tainted with skippers, and a low price had to be taken.

Again, in some of the cheese-dairying counties of New York, abortion among cows prevails to an alarming extent. Of late years many are trying the experiment of keeping over, or "milking up" such stock, and in consequence, a less yield of cheese is obtained in these dairies on that account. Finally, the fall in gold, while it affected the price of cheese, did not, on the other hand, touch the price of labor in a corresponding ratio. The dairy farmer, with a less price in greenbacks for his product than in 1869, was compelled to pay for labor, for stock, and for other running expenses of his farm, about the same as during preceding years when the difference between gold and our paper currency was very much greater.

All these causes combined, it will be seen, have conspired to reduce the profits of dairy farming, notwithstanding there has been an unusually good demand for cheese for export and the gold prices in Liverpool and London have not fallen off. The facts here stated are important in considering the relative profits of the dairy for the year with those of previous years.

II. At the Little Falls market the range of prices for good factories during the year has been from 12c to 16c per pound. The best factories will not average much, if any, above 14½c. If the cheese from all the factories was averaged, the price would be less than the figures named. From a recent report by the Treasurer of the Willow Grove Factory,—one of the fancy factories of Oneida county, and which in former years has taken the lead in prices obtained for Oneida county cheese, —it appears that the average for the year has fallen a little short of 14½c. As the items of this report may be of interest in this connection we give them as printed in the Utica Herald as follows:

STATEMENT FROM WILLOW GROVE FACTORY.

The whole number of pounds of cheese sold was 363,302. The total value was $51,563 23, which was disposed of as follows: Paid to Alonzo Westcoat, for making and taking care of the cheese, 65 cents a hundred, amounting to $2,361 46; paid to the owners of the factory and machinery, for the use thereof, 40 cents a hundred, amounting to $1,453 21; paid to Major Miller 70 cents a hundred, amounting to $2,543 11. This constituted the furnishing fund, and was used by Mr. Miller in paying for wood, boxes, bandage, annotto, salt, rennets, government tax, stamps for checks, service and expense of salesman, the treasurer, etc. There remained of the furnishing fund, after paying all expenses, $166 07, which was paid to the patrons.

It will be seen that the actual cost for furnishing was $2,377 04, and $45,571 52 was paid to the patrons, making the average price a pound for cheese sold $.1424796, or very nearly 14¼ cents, and the patron's net receipts a pound were $.125438, or a little over 12½ cents.

HENRY BROADWELL,
Treasurer.

The "fancy factories" of Herkimer have done better than this, but as extreme prices are paid for "gilt-edged goods" at the Little Falls market, and, as is well known, often above the ruling prices at New York city, the average for "Herkimer fancies" would not be an average for "the fancies" of the State.

Farm dairies have averaged less than factories,—of the former we have seen "very good things" sold at Little Falls when the market was dull, at 8 to 10 cents, a price, of course, below the cost of production. On the other hand, some of the fancy farm dairies have run along pretty close with the factories. As an average, we should say that the farm dairies have sold at about 2c. below the factories.

III. The rule with factories and farm dairies is to make cheese that will mature early—cheese that will be ripe for export at thirty to sixty days from the press. Cheese that is not ready for market at this age is considered to be improperly made, or improperly managed during the curing process. We do not mean to say that cheese must necessarily be poor that is long in maturing, since under some of the English processes of cheese making an extremely fine product is so made. But the American cheese-maker seeks to get his product ready for market from month to month, and if it fails to ripen in the usual time, it results from over salting, or from being kept at too low a temperature while curing, or from some cause which is sought to be avoided as soon as discovered. The majority of New York factories are constructed with limited capacity for holding cheese for any considerable length of time during summer, and it is desirable to get the goods into market as soon as may be, since there is a saving from shrinkage, and a relief from labor in the care of cheese which, in a large accumulation, is considerable. We have seen cheese during favorable weather in summer, that was mellow and ripe, and in good condition for export, at twenty days old.

IV. The exports of cheese from New York city during the year 1870, according to the custom-house returns, were 61,716,400 pounds. Of this quantity, 57,663,000 pounds went to Great Britain. In 1869 the shipments from New York abroad were 55,232,900 pounds, showing an increase in 1870 over the previous year of nearly six and a half millions of pounds.

V. The exportation of an extra fine quality of cheese could be largely increased without lowering the price. This is the opinion of leading cheese-merchants in England, and from our observations abroad we are satisfied it is correct. But a very considerable increase in the export of ordinary or inferior sorts would have a tendency to lower prices. That these conclusions are well founded we have only to refer to the history of the trade, when it will be seen that excessive shipments at any one time of bad-flavored and inferior stuff has been followed by a sudden decline in prices, while large shipments of really fine goods like those of 1869 were worked off without any marked reduction, the market generally being sustained.

From statistics of cheese consumption among our own people, it would appear that our home demand keeps well up to the increase of production. There is, this year, no surplus in the country, and it may well be doubted whether the production of meaty, good-flavored cheese,—a really fine article,—will be in excess of the demand, at fair prices, for some years to come. The only fear that we entertain is the overproduction of ordinary and inferior sorts, as it is very difficult to coax people to pay a high price for poor stuff that gives no pleasure in the eating.

GENERAL FARM MANAGEMENT.

ESSAY BY MILTON J. GARD, TO THE VOLINIA FARMERS' CLUB.

In accordance with the programme, it has fallen upon me to furnish the Club with an essay. I have chosen for my subject, "General Farm Management." I am aware of the magnitude of the subject. It comprises many topics, each of which is sufficient in importance for a lengthy essay. It is also a subject that every cultivator of the soil is interested in. It is a subject on which depends the success or failure of the agriculturist. More depends upon the proper management of a farm than upon the quality of the soil and capital invested. It matters not how rich the land, how much capital is invested or labor expended, if there is a lack of proper management the result will be failure. We often see men on poor land gaining in wealth, when their neighbors, on better land, are growing poorer from year to year.

A farm is understood to be a portion of land properly improved with buildings and fences, and should afford a comfortable home for the farmer and his family, and supply them with all the necessaries and comforts of life, and accumulate something more for objects of charity and for support in old age. Farming may be classed under two heads, or divisions, —special and mixed. By special farming, we understand the farmer to be engaged in some special business, as grain-raising almost exclusively, or stock. By mixed farming, we are to conclude that the farmer is engaged in raising all, or nearly all, the grains adapted to his locality, in connection with all kinds of stock generally kept upon farms.

Now, as to which kind of farming is the most profitable, and to be recommended, depends on the knowledge and ability of the farmer, and the location and adaptability of the farm. It would be poor policy to undertake to adopt a course of mixed farming upon a farm that is only adapted to grass, and just as bad for a dairy farmer to undertake that special branch of farming on land that was too dry and poor for grass to grow. In order to be successful, we must know what our farms are best adapted to, and possess the knowledge requisite for the business we are to engage in, and of the markets we have to supply. My opinion is that where our lands are well adapted for a course of mixed farming, it is decidedly the best mode, for several reasons. Where there is sufficient amount of stock kept on the farm, all the products can be economized. All the coarse grain, hay, straw, and cornstalks can be made useful in wintering stock, and thereby converted into manure, and the fertility of the land kept up. But in order to adopt a course of mixed farming that will be profitable, it will be necessary that the farm be properly fenced, and suitable buildings constructed for the care of the animals. The farm should be subdivided in accordance with the system adopted in regard to the rotation of crops; and in order to keep up the fertility of the soil a regular rotation is necessary, always making as few fields or lots as will answer for the course adopted; for the smaller the fields the greater the expense of fencing, and hence the smaller the profits. The expenses are greater, in proportion to the capital invested, on a small than on a large farm. In order to pursue a proper rotation it will be necessary to have the farm divided into at least four fields, besides the orchard, yard, and garden. To fence forty acres into four fields will require 480 rods of fence, or twelve rods to the acre; and to fence eighty acres in the same number of fields will require 720 rods, or nine rods per acre, and the same proportion on larger farms. You see that the eighty-acre farm requires one-quarter less expense to the acre than the forty-acre farm. The same rule holds good in relation to the buildings, as it requires the same number of buildings on the small as on the large farm,—the difference will be in regard to size. The expense of repairs on buildings will be nearly the same We will suppose that a farm of forty acres is sufficient to support one farmer and his family, occupy the time and labor of the family, keep up repairs and expenses, and there is nothing gained above a living. Then suppose we add forty acres more to the farm. There will be but little additional expense for buildings; fences will cost one-fourth less per acre, and the same team and implements will perform the labor, and will last nearly as long as they would on forty acres. The farmer then will realize a profit above expenses.

There is no part of farm management requiring greater care and judgment than that of keeping up the fertility of the soil· In order to do this at the least expense, it is necessary to utilize everything raised upon the farm. All of the straw, cornstalks, hay, and coarse grain should be used upon the farm. To thresh and pile the straw in the fields, and let it there remain and rot, is more of a nuisance than a benefit. It not only encumbers the ground for several years in rotting, but it harbors grubs and various other troublesome insects, and furnishes a good breeding-place for them. Cornstalks left standing as they grew, are little better than a nuisance. The corn should, as soon as ripe, be cut up. The stalks, if properly taken care of, are excellent feed for cattle and sheep, and the refuse, if not of much value in itself as a manure, is a good absorbent of urine, and is a great help in utilizing manure. Wheat straw is also good feed for horses and cattle, and excellent bedding. All surplus straw not needed for feed and bedding should be stacked in the barn-yard where the stock can work at it all winter. It will cover the yard a good depth, and will afford all stock clean, comfortable quarters, and absorb a great amount of liquid that would otherwise go to waste. Thus, instead of being a nuisance, it will be of great benefit, and stop a leak where many dollars would ooze away.

Rotation of crops has much to do in keeping up the fertility of the soil, and is another part of farm management requiring thought and judgment. Each crop or kind of grain or grass requires different food, and when one kind of grain is growing and using up certain ingredients necessary to its growth, the land is recuperating an ingredient of another kind suitable for plants requiring different food. Land, by constant cropping with wheat, will in time exhaust the elements necessary for its healthy growth, and become, for that crop, poor; but probably if we plant corn, or seed the same to clover, we get large crops, and the land seems to be rich. Hence the necessity of a rotation of crops. I have adopted the following course, and I find it to work well with me, on my soil (prairie). First clover, followed by wheat, with but once plowing, in August; then corn, which is also seeded to wheat, and seeded to clover in the following spring. All the manure made is applied on the wheat stubble in the fall or winter, and spread as drawn, for the corn crop. My land is seeded and partly manured every fourth year, and for each seeding I get three grain crops,—two of wheat and one of corn. But so soon as wheat fails to bring such high prices as at present, I should change the course to two of grain to one of clover. I consider the best mode of renovating and keeping up the fertility of the soil is by the use of clover and plaster. There is no investment that pays me so well as plaster sown on clover at the rate of fifty pounds to the acre; and I believe that the fertility of our farms can be kept up by a judicious rotation of crops, and by carefully husbanding all the manure made on the farm, with no other foreign fertilizer.

Land seeded down for the purpose of renovating should not be pastured too closely, for by so doing the result will be partially, if not entirely, lost. In order to get the best results from the plaster, the clover should be allowed to get a good growth. Too closely pasturing our clover fields is very bad management, and causes a leak that wastes many dollars. I am of the opinion that *properly* pasturing the clover is an advantage in renovating. I have arrived at that conclusion from observation and experience. The question is often asked, "How can a part of the crop, as a fertilizer, be of more benefit than the whole?—as it is certain that a portion of it has been converted into the growth of stock and sold." The best answer that I can give is, that the part left on the field in excrements and urine is in a better condition for plant food than natural grass. Clover has a mechanical as well as manurial effect. The roots, penetrating the soil, make it porous and mellow.

Cultivation has very much to do in making farming profitable. We may have a well-arranged system of rotation of crops, and observe the most careful system of making and applying manure, and everything pertaining to the renovation of land, but unless the land is properly cultivated there will be failures. We must see to it that the land is put into proper condition, and the crop sown or planted in its proper season, and give it the necessary cultivation.

The labor employed on the farm may be regarded as so much capital invested; and, like all investments of capital, requires proper management. Unless labor is properly directed and expended, instead of being profitable it will cause a leak sufficient to drain away all the profits. It would be impossible for me to recommend just how much labor could be profitably employed on a farm, unless I could tell just how much there was to do. But I think I can safely recommend employing a sufficient amount of labor to keep up all necessary repairs,—to get the crops all in in good order, and in proper season. Less than that would be bad economy, and more would be a useless waste. Much may be done to economize labor by good management. On all farms there are many jobs that can be done on rainy or cold days, in the shop or barn. I am led to think, when I pass by a farm in harvest or haying time, and see the farmer, with one or two of his men, making or repairing hay-racks, rakes, or other tools,—work that should have been done on some of the rainy days that they were lounging around, or perhaps worse: spending their time in some village saloon,—that the labor on that farm is not well directed, and there are, very likely, somewhere, leaks that are draining away the profits. When I pass a neighbor farmer's cornfield that has been neglected until the weeds have such growth that the cultivator will not cover them, and I see him and his men hoeing where a timely use of the cultivator would have obviated the hoeing, I am led to think that the labor on that farm has not been well directed or economized.

STOCKING THE FARM.

The kinds of stock most profitable, and the number of each kind to be kept on the farm, per acre, or the amount of stock in proportion to the size of the farm, is probably one of the hardest questions to determine. There is no branch of agriculture requiring more thought and judgment than properly stocking the farm, and I think that many failures in farming may be attributed to an error in judgment in regard to this question. I should recommend keeping all kinds of stock usually kept on the farm. We should, at least, have enough of each kind to consume all the coarse products of the farm, such as straw, cornstalks, coarse grain, grass, and vegetables. I deem that amount of stock necessary to a profitable course of mixed farming. I think, in order to derive the greatest profit, a farm should be stocked to its full capacity for an ordinary season of plenty. Then, if drought, or failure of any crop should happen from any cause, we had better supply the deficiency, than to lack stock to consume the products of a plentiful year. If drought should shorten our pastures, as is frequently the case, we should be able to supply the deficiency by some forage crop. An acre or two of corn will supply a great deficiency in pasturage; and if our corn crop should fail, it would be a very exceptional case that the failure would be general so that we could not supply the deficiency. The corn belt is so extensive in this country, that a failure in any particular locality hardly affects the price of that article, or places it out of the reach of profitable feeding; but, if it should, we can always realize a paying price for store stock. One reason why we should have a variety of stock is, that it very often happens that one kind is too low in price to pay a profit, while another is sufficiently high to make an average profit on the whole. Should we happen to have the only stock that is low in price, we sustain an irreparable loss, and nothing to fall back on. We cannot see into the future sufficiently to guard against such contingencies, and the best plan is to have something to sell that will give us a profit; and in order to be sure we have that article, we must have all. We will then be sure to have something that will reward our labors. It is bad economy to be always shifting from one kind of stock to another, for in so doing we are never ready for anything. Some farmers have tried cattle; they have failed to make any money, and conclude that raising beef does not pay, but Mr. A. cleared five hundred dollars in wool, and so it's the sheep business, after all, that pays; so I will sell off all my scrawny cattle for just what they will bring and I will hire five hundred dollars and purchase me a flock of nice, profitable, merinoes,—make some money as well as Mr. A. By the time I get my flock of sheep fairly started and a crop of wool ready for market, the price is down below anything that would pay for their keep, and from my ignorance of their management, the flock has become diseased and will not sell for more than one-half the price paid for them, so I conclude that the sheep business is a humbug. Those beautiful Spanish merinoes are nothing but little, black, greasy animals, and no profit in them, so I will clean them out as soon as possible. I am not sure, after all, that what has been said of stock-raising being so profitable, is true. There is something more profitable, but I have hardly decided whether I will next try the wine-plant business, or hops. It is hardly necessary for me to say that the class of farmers I have pictured do not belong to the Volinia Farmers' Club. Another reason why we should keep all kinds of stock, is that we can use up all the surplus products to better advantage by having animals of different requirements. The number of each kind of animals that I have found to be the most profitable, or that I have ascertained can be kept on my farm, is to each improved acre as follows: One sheep, one pig, to two and a half acres; one cow to each five and a half acres, and a sufficient number of horses to carry on all the operations of the farm. But I do not wish to be understood as advocating just that number of animals under all circumstances, because, if wool should be so low as to be produced at a loss, and pork should be more profitable, I would reduce the number of sheep and increase the pork crop. I would give the preference to that branch that paid the best. But I should not close out the sheep, for, in all probability, that branch of husbandry would soon begin to pay. Just to have a certain number of animals is not sufficient. They should be of the best kinds. Good animals are as easily raised and cared for as inferior ones. It is also necessary that farmers understand the requirements of their animals, and see that their wants are supplied. Unless animals are properly cared for, no matter how good they are, they cannot be profitably raised. Raising animals requires more judgment than any other branch of farming, and to the want of proper management nearly all the losses may be attributed. But to give a full and explicit statement of the care and management of each kind of stock, in this essay, would be out of the question, as it would require an essay of considerable length for each. I would say, however, that it is necessary that they have warm, comfortable quarters in winter. Cattle should have warm stables, well ventilated, kept clean, and well bedded. They should have plenty of good, wholesome food to eat, and pure water to drink, at regular intervals. Sheep and swine the same. Pigs require a warm, dry place to sleep; and for the want of proper care in that respect, many pigs are permanently injured.

IMPLEMENTS.

On every well-regulated farm there is necessarily a large amount of capital invested in implements. This investment also requires judgment and experience. The best should always be procured. The implements used on the farm have much to do in economizing labor. The land cannot be properly fitted unless we have an implement suitable to the work. The plow is one of the most useful implements used upon the farm. There are several kinds of plows, of different construction, and adapted to different soils and different conditions of the same soil. I think we should have as many plows as we have different conditions for their use. For instance, a plow that is the best for turning a sod is not the best for plowing mellow or stubble land. I think it poor economy to have but one plow for both purposes. The investment would be more to commence with, but would not be in the end, and the work would be more perfectly done. It is bad economy to use a poor tool because we happen to have one. If the plow we have does not do quite so good work, or draws a few pounds heavier than is necessary, we had better dispense with it and procure another and better one. If the old plow was only to be drawn a few rods, the extra draft would not amount to much; but when we have to draw it for several weeks in each year, and for several years, it amounts to an enormous expense of muscle that has to be supplied at a cost. In purchasing implements, we should be guided by the principle of the greatest efficiency with the least expense of muscle. Drawing useless weight is an expense of muscle that is continually draining so much of the profits as it costs to supply the power. To further illustrate: Suppose we set a man at hoeing corn, and he has muscular power sufficient to make thirty strokes per minute with a hoe of two pounds weight, and with such an implement he would hoe one acre per day at the rate of $1.50 per acre. We will then suppose, instead of such a one, we furnish him with one of four pounds weight. With his muscular power he would only be able to wield it fifteen times per minute, and accomplish but one-half acre per day, at the same price per day. You will readily see that we lose on each day's work, 75 cents, or nearly the price of a good tool. The same principle applies with equal force to all the implements used upon the farm. We often see farmers driving two horses to an old-fashioned "A" drag, cutting at each passing about five or five and a half feet, and the land having to be twice passed over to get the preparation that once over, with an efficient tool, would accomplish. Now the man and team, with one extra horse and a suitable harrow, would do more than twice the amount of work, at the same price per day and expense of muscle, and do it in a better manner. Ask the farmer why he does not throw away the old drag and procure the best, and his reply generally is, "I am not able," when the fact is right the reverse. Some care should be exercised in purchasing new and untried implements. If venders of new implements refuse to let you put them to a practical test, do not purchase. Every farmer should bear in mind that not taking proper care of tools causes a leak, and much of the profits of the farm is lost. Many farm tools are injured more by exposure to the weather than actual use. It is bad economy to buy costly tools and winter them out-doors.

Selling the products of the farm is another part of farm management, and sometimes the judgment exercised in selling one year's products determines the loss or profit in farming. It is impossible to give any definite rules. I have generally found that when I have an article ready and in market condition, and make the best sale possible, that I have not often had cause to regret the sale, especially stock. A great deal of caution is necessary when dealing with sharp middle-men. They are buying for the purpose of making money, and all they make we lose. The most of our products we can just as well ship ourselves as to sell to the middle-men, and thus save the profits they make.

Properly saving seeds is another important part of farm management. And so long as farmers pay so little attention to saving and cleaning their seed-wheat, the opinion will prevail that wheat turns to chess. It is a very common thing to see fields of wheat badly mixed with rye, cockle, and chess. There is a great loss to the country by that cause alone, as well as to the individual farmers who pay so little attention to that part of farm management. It is very often the case that by neglecting to save seed-corn properly, a light crop is the result, by its failing to grow. It requires but little poor seed to cause two or three bushels to the acre less; and twenty acres, at three bushels to the acre, makes a loss of sixty bushels. I might point out many other leaks in farm management, but a word to the wise is sufficient.

BREEDING AND GENERAL MANAGEMENT OF HOGS.

ESSAY, BY MILTON J. GARD, TO THE VOLINIA FARMERS' CLUB.

Mr. President, Ladies and Gentlemen of the Club:—The subject for our consideration this evening is one of much importance, and worthy of the consideration of every member of this association, and all others who make any profession to pork-raising, especially if profit is the object. I think I may safely say that profit is the only object, especially with those who are raising pork for market; and no less so for those who only raise for their own consumption. The question that naturally presents itself to our minds is: "Can pork be raised and give us a good profit? And, if so, what is necessary to bring about so desirable a result?" I should answer the first question in the affirmative. Pork can be raised at a good profit, as I think, before I have done, I shall be able to prove. I consider pork-raising the most profitable of any one branch of agriculture. And not only the most profitable, but it requires the least care, labor, and capital. The pig grows right along up into money, and while thus growing, he has been busy doing us good by converting our grass and clover into suitable food for our wheat crop, and into meat and lard. He always leaves our field richer than he finds it. He only requires about eighteen months to mature. It is more profitable than wheat-raising, for several reasons: First and greatest: Instead of exhausting the fertility of the soil, it enriches it. Secondly: It is more certain in its results, having fewer enemies to contend with. In the pork crop there is no risk of smut, midge, rust or mildew, blight, winter-killing by freezing out, lodging, or by being beaten down by hailstorms, or sprouted by wet weather at harvest time.

Now, as to the question, How is it to be raised so as to make us a good profit? I would say, in answer, that to raise pork profitably, a certain amount of common sense, and a knowledge of the habits and requirements of the pigs are the most essential requisites, and to see that their wants are all supplied, a necessity. Then, if you have the right breed you certainly will succeed. And I now propose to say something in regard to breeds, of which there are several possessing more or less merit. All have their admirers, and it would be presumptuous in me to say which of the popular breeds are the best. Any of the following, and probably many others are good, and probably all are the best with their several advocates, to wit: The Polands (a black and white spotted hog), Chester White, Berkshire, Suffolk, Essex, and China. Some pork raisers contend that the large breeds are the most profitable; others, that the small breeds have the advantage in point of profitableness, but I prefer a medium breed. The question to be determined is not how large a hog may grow in a given time, or how small, but how much pork has he made from the food consumed? for that is what determines the profit or loss. Now, my opinion is that the hog that is hardy and not prone to disease, that grows rapidly, fattens kindly at any age, that thrives and grows best on grass or clover, and makes the most pounds of meat in proportion to food consumed and care bestowed, is the best and most profitable breed. All of the above qualities are possessed in an eminent degree by the Polands. Another recommendation in favor of the Polands is their freedom from disease. I have been raising them for the last six years, and during that time I have not had one become scurvy or diseased, and have lost none only by accident. It is a common belief that black, or black and white spotted hogs are more exempt from diseases of the skin than white ones, and so far as my observations go, they tend to establish that fact. But, notwithstanding, I am not sure that color has anything to do with disease. I am more inclined to believe that the diseases above named are constitutional defects inherent in breeds, and when the proper requirements of health are neglected, and favorable conditions for the disease are allowed by the breeders, the disease will show itself. The Chester Whites and Suffolks are more subject to the disease than any of the improved breeds with which I am acquainted. The Poland breed of hogs possesses another quality which I consider a valuable acquisition to any breed; that is, a gentle and quiet disposition, but little inclined to fight, or roam in quest of mischief; always quiet and happy, provided they are properly fed and cared for. They are great eaters, and have the requisite qualities of turning food consumed into pork in a greater degree than any breed with which I am acquainted. In discussing the good qualities of the Polands, I do not wish to be understood as arguing against the good qualities of other breeds of which I have had but little experience. Each pork-raiser will choose for himself as regards breeds.

After settling the question as to breeds, the next thing in order will be breeding, and general management. Breeding animals is a trade, or profession, and but few ever become successful. Breeding requires a great deal of judgment, skill, and practice. Hogs are no exception to other animals, and require constant care to keep them improving, and any carelessness or neglect by the breeder soon degenerates the herd. On the other hand, by careful and judicious selection they are quicker raised to a high standard of perfection than most other stock. Whoever undertakes the breeding of hogs, in order to be successful, must have in his mind the form of animal he wishes to breed. He must work to a pattern. He should have in his mind a perfect animal, possessing all the good qualities necessary to form a complete whole. I will here venture a description of such a hog as in my mind comes nearest the form of a perfect animal. The circumference of his body, back of his fore legs, should be the length of the hog, measuring from the butts of the ears to the tail. He should be broad across the breast, and a little less behind; should be deep in the chest, giving plenty of room for the vital organs. His legs should be short, but heavy bone. His neck should be short and thick, and of good width. His head should be wide between the eyes, and set on a level with the back, which should be straight. His snout should be large but short, with large nostrils; middling sized ears, with points a little lopped, and nicely curved. His ribs should be well arched, but not so much so as to give too much roundness to the body. The above description is as near a perfect hog as I am able to picture. In selecting breeders, care should be taken that they possess as many of the required points as possible to make up the perfect hog. Almost any form or color may be acquired by judicious selections. Size may be increased or diminished, always bearing in mind that like is more apt to produce like than otherwise. A great deal of care should be exercised in selecting the males for breeding purposes, and none used that possess any defect in regard to form or constitution, as those are increased in the offspring at a rapid rate, especially if the sows possess the same defects. In selecting sows for breeding, I choose those that are inclined to size and heavy bone; and to avoid too much size, choose a boar with finer bone and more compact build. I think this plan has advantages over the opposite course. The sows, being coarser, are apt to have larger and stronger pigs, and are more ape to prove good nurses, and have less trouble and risk in dropping their pigs. Breeding in-and-in should never be permitted if possible, as an indiscriminate breeding with near blood kin will soon destroy the health and vitality of the herd. But if it should be necessary to adopt that course to preserve the purity of a breed, the greatest of care and judgment should be used in selecting the animals, and none used that show the least defect in either form or constitution.

Cross breeding, or the mixing of breeds, seems to be a pet theory with some intelligent pork-raisers, but I must confess that I cannot see any good logic in the practice or theory. I fail to see the point, that there is anything gained by mixing one breed, not so good, with one that is better, to improve the better. If we become satisfied that there is a better breed than the one we are using, why not discard the one, and procure that which is better, unadulterated? When we cross or mix two breeds, we have neither one nor the other, and they are not reliable for breeding purposes. We create a race of mongrels, that are just as likely to produce the bad as good qualities. The only justification I can see for crossing, is to produce a new breed, and in that case should be conducted by one well skilled in the art of breeding, and would have to be continued for several generations before sufficient character would be established to produce reliable results. I have experimented some by crossing the Chester and Polands, with unsatisfactory results, the offspring not being so hardy and uniform as the pure breed of either kind, and I have discarded the practice, with the belief that the less Chester and more Poland the better. I have a theory in breeding to advance that is not generally conceded by swine-breeders, but which I have found to be true, and profitable in practice,—that is, breeding from young sows. The objections urged against the practice are that young sows do not produce as large and thrifty pigs as old sows, and do not make as large and profitable hogs as pigs from old sows; and, if continued, will have the effect of degenerating, and the extinction of the breed would only be a question of time. Now, if that theory is correct, each succeeding generation would become weaker, and soon become worthless. Men, and especially farmers, are too apt to jump at conclusions from a single experiment, without giving the subject sufficient thought or investigation. Too many conclusions are made from single experiments, like the following, which I copy from the Country Gentleman:

The correspondent to that journal, I believe, wrote from Massachusetts (as the paper has been mislaid, and I quote from memory). His experiment seemed to be satisfactory to himself, and probably to many of the readers of that journal. He says he had two sows,—one an old one, the other a young one of the same breed, and got with pig by the same boar, and had pigs about the same time. The old sow having overlaid and destroyed a portion of her pigs, he took a portion of the young sow's pigs and gave them to the old sow, which she readily adopted. The sows and their pigs were treated in every respect alike. He says there was a marked difference at first in the two litters,—the old sow's pigs being larger and stronger than the young sow's, which they maintained throughout the whole period of their growth,—the old sow's pigs making the largest and best hogs.

Now, if the above experiment proved anything, it was simply the fact that the old sow was the best breeder. It is a well-known fact that all sows of the same breed are not equally good breeders. I have been experimenting for a number of years, in order to determine the truth or falsity of the theory, and have come to the conclusion that young sows raise just as good pigs as old ones, and make just as large and profitable hogs. I have for the past six years wintered one or more old sows which raised pigs with the young sows, and were treated in every respect alike, and I have never been able to discover any difference in size or thrift in the young or old sows' pigs after a few weeks. It is true that old sows generally have more pigs at a litter than young ones, and will generally crowd them faster on the start than young sows, but as soon as the pigs commence to eat, the young sow's pigs will overtake them, and in the fall no person would be able to discover any difference as to size or appearance. Sows may be bred too young, and in that case I think an injury would be done to the sow, as well as the pigs. A sow should not be permitted to get with pig until she is eight months old. She would then have pigs at one year old. I do not approve of wintering old sows for the purpose of raising pigs, for the reason that I get nothing for the food consumed. It takes as much food to winter one old sow in good condition as two young ones that are continually gaining in size, and in the spring I have got something to show for the corn they have consumed; whereas, if I should winter old ones, after having fed them four or five months I have for credit the same old sows that I had in the fall, minus a large pile of corn. I never raise but one litter from a sow unless she should prove very extra, or for the purpose of experiment. I have another objection to wintering old sows. They are very disagreeable associates for the younger stock. They are sure to take the best and warmest part of the bed, and the choicest part of the food, and are continually chastising the younger portion of the herd, to their disadvantage.

Sows, for breeding purposes, should be kept in good order. I prefer keeping them in better condition than many breeders would recommend. But they should not be so fat and heavy as to injure the proper growth of bone. High feeding and want of exercise cause too much weight and heat in the system for the proper development of the bones, and the pigs seem to be small, weak, and more or less diseased. Pigs from sows kept lean are apt to be large and more active at first than pigs from sows in better condition; but the pigs from sows in good condition soon overtake and pass them in growth. The sows are easier kept in condition while raising their pigs, and fatten with less expense. I would recommend raising pigs from as many sows as possible; and if I find myself overstocked, it gives me a chance of culling out and making the pigs more uniform, and, by weeding out all poor and weak ones, make them more vigorous, and less liable to degenerate. I prefer to have the pigs come from the first of May until the last of June. I would have them all come in May, if I could, and not fat and sell them until one year from the following October or November. They would then be about eighteen months old. My object in having the pigs come at that time is that the clover is at that time large enough to be nourishing, and I am able to keep the sows and pigs with less expense than to have them come earlier in the season. They are also wintered with less feed and expense, and require less room for shelter and bedding, and are of good size and age to take on flesh the next season on the clover. My experience is, that a shoat well wintered, and turned on the clover about the tenth of May, will double his weight by the first of October, with no other feed. And here is where the profit comes in with the least expense. I am aware that some pork-raisers advocate and practice a different theory, and have their pigs come in March or April, and fatten and market at about nine or ten months old; but I question the propriety or profitableness of that practice, because a number of old sows would have to be kept to supply the pigs, at too great a cost. They would require constant care and the best of feed, from birth to the time of marketing, and would necessarily make less from clover, for they would not be old enough, during the pasture season, to convert the clover into pork to the same extent that older ones would.

About all the care necessary to give breeding sows is to see that they are removed from the herd, and from other stock, a few days before having their pigs, and to see that they have a dry place, protected from the cold rains that are common about the first of May, and fed plentifully. There should be but one sow in a place. They should be kept from other stock until the pigs are old enough to take care of themselves, which will be in a few days. For a week or two after pigging, the sows should be fed some bran and shorts in connection with corn, and as soon as the pigs are old enough to take care of themselves,—that is, not to be stamped upon by cattle or horses, —they should be allowed to run on the clover. Then would commence the summer management. All except such as are not sufficiently strong, should, about the tenth of May, be turned upon the clover; and if that crop is sufficiently large as to be nourishing, they will live and thrive with no extra feed, but I would advise feeding some corn until after the wheat is harvested, when they should have the privilege of the stubble-field. They should have access to water at all times, and should be salted regularly at least once a week, and twice-weekly would be better. They should have sulphur mixed with their salt in the proportion of one part sulphur to four parts salt. In case the pasture should fail from drought or over-stocking, they would require feeding more liberally, as it would be bad economy to permit them to become poor or stunted. No after-care would compensate the owner for such neglect. After they have cleaned up the stubble-fields, and before they commence falling off in flesh, they should be fed by cutting the new corn up at the root and feeding stalk and all. Until the stalks become dry and unpalatable, they will eat them, and thrive faster than at any other time. My experience is that it is more profitably fed at that time than after it has become dry and hard. The change is more easy and natural, and it is in a more favorable condition to make bone and muscle. Another mode which is adopted by some intelligent pork-raisers is to fence off a portion of a field and turn the hogs in and let them help themselves; but as I have never had any experience in that mode I could not recommend it. Feeding should then be continued in earnest, or, as General Grant would say, "it should be pushed," while the weather is warm and dry. The latter mode has the saving of labor to recommend it. Feeding in the field has another advantage: all the manure is made on the field where wanted, and ready spread. I am decidedly opposed to fattening in pens, and if I could use language strong enough to split all hog-pens into kindling-wood, and send the hog-pen theories to the other side of Jordan, I think I would confer a great deal of comfort to the pigs, and a greater blessing to humanity, in health. Why so many farmers will persist in this unnatural, cruel, and unprofitable course, I am unable to determine. Hogs shut up in a close pen suffer for want of exercise. They also suffer from the filth they are obliged to eat and breathe. It is impossible to keep a hog-pen clean enough to avoid a bad smell, unless in a few instances where they could be flooded, and the filth carried off in some stream of water, where other animals might get a portion of it in a diluted form, and the land be robbed of its just dues. But such pens are the exception. It is not an uncommon thing to see a lot of hogs shut up in a small pen and fed in their own filth, often ankle-deep; and the only cleaning the pen gets is an occasional shoveling out. The hogs are kept in that condition for several weeks, and it is no uncommon thing for them to lose the use of their limbs. Hogs thus treated become diseased, and the suffering they have to endure is not the worst feature in the case. They are slaughtered and converted into human food. It is just as necessary to healthy hogs and wholesome meat, that they have pure air to breathe, pure water to drink, and exercise, as it is that they have good food and a clean place to eat it. All the elements so necessary to the health of the hog cannot be obtained so easily and perfectly as to give him a grass lot to run in while going through the process of fattening. There is no place so natural to a pig as our mother earth, carpeted with a good sod of clover.

When and how to sell, I deem of sufficient importance to be considered in this essay. My opinion is that we should sell early, or as soon as the hog can be got into good condition, and before they have consumed a great amount of corn. The early market is generally the best, and, as a general thing, those farmers who have their hogs in condition to go into an early market make the most money. As a general rule it pays the best to sell on foot, but not always. If the early market should be low, with a fair prospect of a rise, and corn cheap, it would pay to keep and feed until the hogs are fully matured. Then the prices paid for live and dressed would determine which would be the most profitable. A well-fatted hog will not shrink over one-sixth in dressing.

Shall we sell to the middle-men, or ship our own hogs? My opinion is that farmers should ship and sell their own hogs. They will then know that they got all the hogs bring. The middle-men live, and some of them get rich; and their gain is just what they sell for more than they pay, and is just so much off of the profits of the producer. If they buy A.'s hogs and lose money, they endeavor to make it up on B.'s, by buying them below the market sufficient to make up the loss, and a profit beside. They are apt to look for money where they lose it. If they buy our hogs and lose money on them, the next time they come around they are pretty sure to be looking for the lost money. If one farmer has not a car-load, let two or more join and make up a load and ship them. Do not be scared if the middle-men should laugh at you for shipping your own produce, for they will try to discourage you.

Having disposed of the hogs, the next thing to consider will be the winter management. The requirements of the pig are few and easily supplied. 1st. Plenty of room for exercise. 2d. Plenty to eat, fed at regular intervals, and water that they can have access to, and a warm, dry place to sleep. The sleeping department for pigs is the most essential part of their management, and for the want of proper care in that respect more pigs are injured than from any other cause. They should have a warm house, well ventilated. The cracks of the siding should be battened, and the floor should be made double, to insure warmth. When pigs are allowed to sleep in a damp or wet bed, they will pile up and sweat, and when they leave the bed for their feed, they come out often into zero cold, and they take cold and cough, and very often contract consumption and rheumatism.

I will now conclude this essay by giving an experiment with 42 pigs. I said in the beginning of this essay that pork-raising was the most profitable branch of agriculture. Now for the figures:

My pork account commences Nov. 1, 1866, with 42 pigs about five months old. The 42 pigs consumed from that time until the 15th of May, 1867, 8 52-70 bushels of corn each, valued at 55 cts. per bushel, making $4.80 per pig.

Forty-two pigs, Nov. 1, 1866, worth $3 each	$126 00
Corn consumed from Nov. 1 to May 15, worth	201 60
From May 15 to Sept. 1, pasture estimated worth	54 09
Corn fed from Sept. 10 to Oct. 15, 180 bushels at 65 cents	117 00
Corn fed from Oct. 15 to Dec. 8, 50 bushels at 65 cents	32 50
Total cost of 42 hogs	$531 19

Oct. 15, sold 30 hogs for	$339 06
Nov. 3, sold 10 hogs for	143 51
Nov. 3, sold 2 hogs for	16 00
Raised from the 42 hogs, 35 pigs worth $3.50	122 50
Total sales	$621 07
	531 19
Profit	$89 88

You see that one hundred and twenty-six dollars invested in pigs, pays over 80 per cent. Practically, I made $143.97, as the land was renovated to the amount of the $54.09 charged for pasturage. I believe from practical experience that the land is benefited more by the clover being fed to the hogs than if it had been plowed under. You also see that corn was above the average price, and pork below, which, under more favorable circumstances, would have paid much better.

The thirty hogs sold in October average 234 pounds, and sold for $4.84; and the ten sold in November were sold for $5.25. The forty-two hogs were turned upon the clover the

15th of May, and were not fed anything until the first of September, except what they got on forty acres of wheat-stubble. The thirty-five pigs were also fed with the hogs until the first lot was sold, in October, and all the corn charged to the pork account.

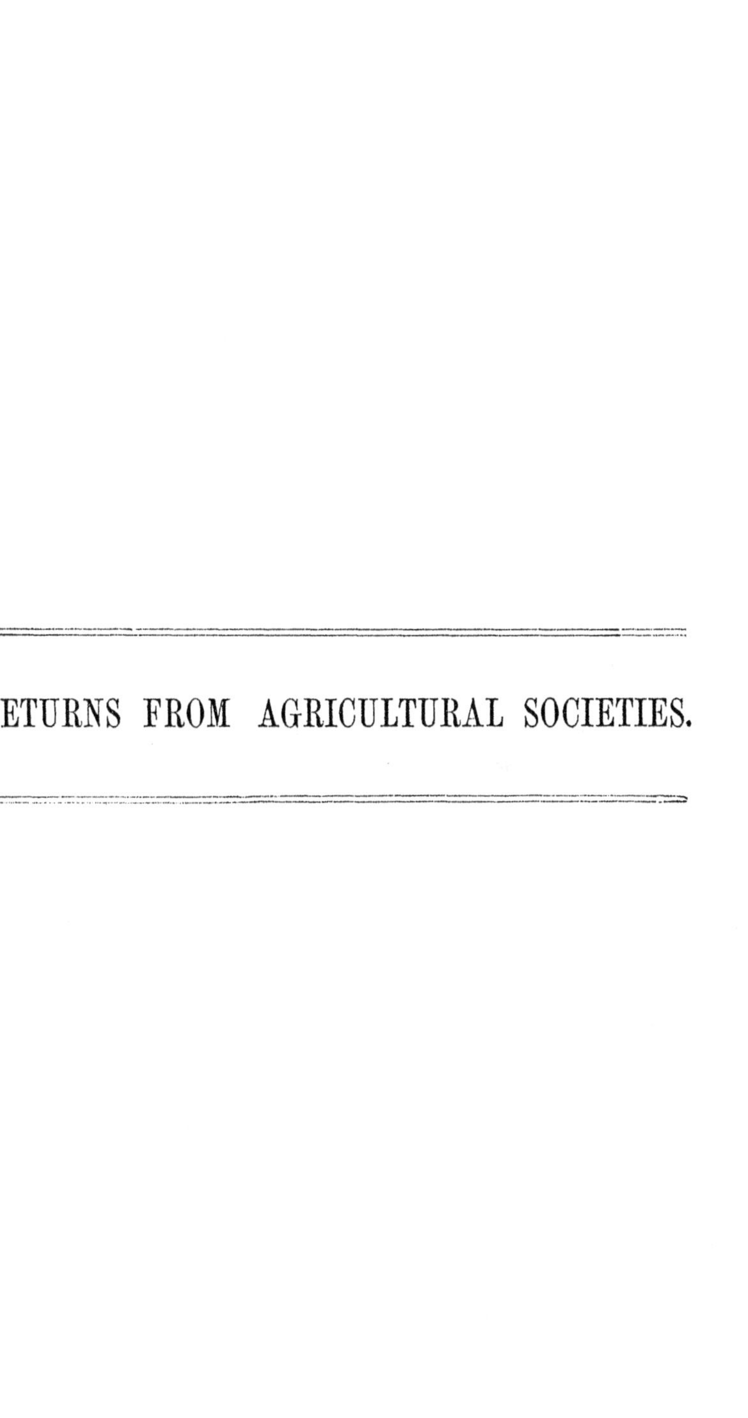

RETURNS FROM AGRICULTURAL SOCIETIES.

REPORT

OF THE

MICHIGAN STATE AGRICULTURAL SOCIETY,

FOR THE YEAR 1870.

In pursuance of notice given, the annual meeting of the members of the State Agricultural Society was held at the President's office, on the Fair Grounds, at 2 o'clock on Friday, the 23d of September, 1870. The President called the meeting to order, and stated that he had endeavored to gather together as many of the members of the Society as possible, and he was pleased to see so many present in answer to the call. The object of the annual meeting was to elect the officers for the coming year, and to transact such other business as the Society might deem proper and necessary.

Mr. John Starkweather of Ypsilanti moved that a committee of one from each Congressional District, and one at large, be appointed to submit the names of persons suitable for officers, to the meeting for its approval.

The President stated that it would be very difficult to name such a committee without the assistance of the members present, as he could not be sure of the residence of those who might be selected.

Mr. J. M. Sterling moved that the President be empowered to select a committee of seven members as a nominating committee, and Mr. Starkweather accepted the substitute.

The President then nominated as such committee:

J. M. Sterling of Monroe County; J. D. Adams of Kalamazoo; Lysander Woodward of Oakland; L. S. Scranton of Kent; R. Strickland of Clinton; C. C. Beahan of Genesee; D. M. Uhl of Washtenaw.

The committee having considered the business referred to it, reported as follows: That it recommends the election of the following officers for the ensuing year:

For President—W. G. Beckwith of Cass County;

For Secretary—R. F. Johnstone, Detroit;

For Treasurer—E. O. Humphrey, Kalamazoo;

For Members of the Executive Committee—G. W. Philips of Macomb County; A. J. Dean of Lenaweee County; E. W. Rising of Genesee County; Andrew Stout of Clinton County; C. W. Greene of Oakland County; M. Shoemaker of Jackson County; J. M. Sterling of Monroe County.

Mr. Seymour Brownell moved that the report be accepted, which was adopted.

Mr. Starkweather moved that the meeting now take up the names of the candidates separately, and vote on the same, which was adopted.

On taking the vote on President, the motion to adopt was unanimous.

On presenting the names of Secretary and Treasurer, it was voted that these officers be voted in by acclamation, which was adopted.

The names of the several members of the Executive Committee were then each called off and voted upon separately and adopted.

Mr. Brownell offered the following resolutions:

Resolved, That the thanks of the Society be tendered to the officers for the very able and efficient manner in which they have conducted the business of the Society for the past year.

Resolved, That the Executive Committee be directed to present to Mr. John Gilbert of Ypsilanti, a silver medal with a suitable inscription, as a testimonial of the appreciation in which his services are held, in making preparations for the annual exhibition.

Resolved, That the thanks of the Society be presented to the Dowagiac Band for its services during the Fair.

Mr. Baxter of Jonesville presented the following resolution:

Resolved, That the Constitution of this Society be so amended as to make the Executive Committee consist of twenty members, and that half be elected annually, to hold their office for two years; and that one-half of the additional members, elected under this change in the Constitution, shall be elected for one year, and the other half for two years.

Mr. Baxter stated, in explanation, that the State had grown so much in population, and that this population had extended into districts which were entirely unimproved when the Constitution of the Society was adopted, that it was necessary now to give it larger and fuller representation than fifteen members of the Executive Committee would allow; and that, after consultation with a large number of the members and friends of the Society, it was thought that if the Executive Committee was increased to twenty in number, it would be sufficient. The Constitution required that a year's notice be given if any change be made in that instrument, and this resolution was the notice of this change, which could be adopted at the next annual meeting, if it were then thought advisable.

The resolution was adopted, and the meeting then adjourned *sine die.*

W. G. BECKWITH, *President.*

R. F. JOHNSTONE, *Secretary.*

REPORTS

OF THE COMMITTEES APPOINTED TO AWARD PREMIUMS AT THE ANNUAL EXHIBITION OF THE MICHIGAN STATE AGRICULTURAL SOCIETY FOR 1870, HELD AT JACKSON, SEPT. 20, 21, 22, AND 23.

DIVISION A.

CLASS I—SHORTHORNS.

Wm. Curtis & Son, Wheatland, Hillsdale Co., bull, 4 years, J. E. B. Stewart, 1st, $50; S. Brownell, Utica, Lenawee Co., bull, 4 years, Sheldon Duke, 2d, $25; Wm. Smith, Detroit, bull, 3 years, Duke of Somerset, 1st, $40; Chas. Whittaker, Chelsea, bull, 2 years, Garnet, 1st, $30; J. Lyon, Pewamo, Ionia Co.. bull, 2 years, Duke of Elkford, 2d, $15; D. M. Uhl, Ypsilanti, bull, 1 year, Colonel Welch, 1st, $25; A. F. Wood, Mason, bull, 1 year, Belmont, 2d, $12; Wm. Curtis & Son, Addison, Lenawee Co., bull calf, Duke Hillsdale, 1st, $10; Wm. Smith, Detroit, bull calf, Duke of Somerset, 2d, $5; Wm. Curtis & Son, Addison, Lenawee Co., cow, 6 years, Lucy, 1st, $50; D. M. Uhl, Ypsilanti, cow, 10 years, Florence, 2d, $25; Wm. Curtis & Son, Addison, Lenawee Co., cow, 4 years, Duchess of Hillsdale, 1st, $40; D. M. Uhl, Ypsilanti, cow, 4 years, Florence 3d, 2d, $20; Wm. Curtis & Son, Addison, Lenawee Co., heifer, 3 years, Zenobia, 1st, $40; Chas. Whittaker, Chelsea, heifer, 3 years, Lizzie, 2d, $20; Wm. Curtis & Son, Addison, Lenawee Co., heifer, 2 years, Zenobia 11th, 1st, $30; William Curtis & Son, Addison, Lenawee Co., heifer, 2 years, Ella 10th, 2d, $15; D. M. Uhl, Ypsilanti, heifer, 1 year, Lallah 4th, 1st, $20; Wm. Curtis & Son, Wheatland, Hillsdale Co., heifer, 1 year, Miss Argyle, 2d, $10; Wm. Curtis & Son, Addison, Lenawee Co., heifer calf, Ella, 1st, $10; Charles Whittaker, Chelsea, heifer calf, 2d, $5.

IRA R. GROSVENOR,
FRED FOWLER,
ROBERT ROME,
DAVID W. FINDLEY,
Committee.

CLASS II—DEVONS.

Ira H. Butterfield, Lapeer, bull, 3 years, Batavia 159, 1st, $40; Orton Drake, Utica, bull, 3 years, Genesee, 2d, $20; John Allen, Coldwater, bull, 2 years, Prince of Wales, 1st, $30; W. B. Perkins, Durbur's Corners, Williams Co., Ohio, bull, 2 years, Prince George, 2d, $15; John Allen, Coldwater, bull, 6 years, Blucher, 1st, $50; Edwin T. Doney, Jackson, bull, 1 year, Vick, 1st, $25; John Allen, Coldwater, bull, 1 year, Wallace, 2d, $12; Ira H. Butterfield, Lapeer, bull calf, Model, 1st, $10; W. B. Perkins,

Durbur's Corners, Ohio, bull calf, Romeo, 2d, $5; Ira H. Butterfield, Lapeer, cow, 12 years, Helena 11th, 1st, $50; John Allen, Coldwater, cow, 7 years, Sophia 2d, 2d, $25; W. B. Perkins, Durbur's Corners, Ohio, cow, 4 years, Ruth, 1st, $40; W. B. Perkins, Durbur's Corners, Ohio, cow, 4 years, Adelaide 2d, 2d, $20; W. B. Perkins, Durbur's Corners, Ohio, cow, 3 years, Victoria, 1st, $40; Orton Drake, Utica, cow, 3 years, Pina, 2d, $20; John Allen, Coldwater, heifer, 2 years, Nettie, 1st, $30; W. B. Perkins, Durbur's Corners, Ohio, heifer, 2 years, Princess Helena, 2d, $15; John Allen, Coldwater, heifer, 1 year, Lucy, 1st, $20; John Allen, Coldwater, heifer, 1 year, Red Lady, 2d, $10; W. B. Perkins, Durbur's Corners, Ohio, heifer calf, Ruth 3d, 1st, $10; Orton Drake, Utica, heifer calf, Sophia 3d, 2d, $5.

WM. WHITFIELD,
WM. H. S. LOTHAM,
JAMES TAYLOR,
Committee.

CLASS III—HEREFORDS.

Wm. W. Crapo, Flint, bull, 3 years, Velvet Jacket, 1st, $40; Wm. W. Crapo, Flint, bull, 1 year, Willie, 1st, $25; Wm. W. Crapo, Flint, bull calf, Connie, 1st, $10; Wm. W. Crapo, Flint, cow, 4 years, Gentle 7th, 1st, $40; Wm. W. Crapo, Flint, heifer, 1 year, Rosie, 2d, $20.

MORGAN CASE,
ROBERT ROME,
HENRY WARNER,
Committee.

CLASS IV.

F. I. Tomkins, Girard, Branch Co., bull, 6 years, General, 1st, $40.

MORGAN CASE,
ROBERT ROME,
HENRY WARNER,
Committee.

CLASS V—ALDERNEYS, GALLOWAYS, AND HOLSTEIN CATTLE.

H. D. Court, Battle Creek, Alderney bull, 3 years, Jack, 1st, $40; H. D. Court, Battle Creek, Alderney cow, 3 years, Nettie, 1st, $30; H. D. Court, Battle Creek, Alderney heifer, 1 year, 1st, $20; H. D. Court, Battle Creek, Alderney heifer calf, 1st, $10; J. A. Smith, Bath, Galloway bull, 2 years, Napier, 1st, $30; J. A. Smith, Bath, Galloway cow, 5 years, Belle, 1st, $40; J. A. Smith, Bath, Galloway heifer, 2 years, Daisy Maid, 1st, $30; J. A. Smith, Bath, Galloway heifer calf, 1st, $10.

CLASS VI—HERDS.

John Allen, Coldwater, bull and 4 cows, Devon, 1st, $50; Wm. Curtis & Son, Addison, Lenawee Co., bull and 4 cows, Short-horns, 1st, $50; Wm. Curtis & Son, Addison, Lenawee Co., bull and 4 of progeny, 1st, $50; D. M. Uhl, Ypsilanti, cow and 4 of her progeny, Lellah, 1st, $50; S. Brownell, Utica, bull and 4 of his progeny, diploma; J. N. Smith, Bath, 3 cows and a heifer, Galloways, diploma.

H. ARNOLD,
H. SCOTT,
GEORGE FOX,
Committee.

CLASS VII.

J. S. Tibbits, Nankin, Wayne Co., grade cow, Spot, 10 years, 1st, $25; D. L. Cady, Mason, grade cow, Timid, 5 years, 2d, $12; J. L. Harris, Marshall, grade heifer, 2 years, 1st, $10; J. N. Smith, Bath, grade heifer, 2 years, 2d, $5; L. D. Wheeler, Jackson, grade heifer, Rose, 1 year, 1st, $8; L. D. Harris, Marshall, grade heifer, 1 year, 2d, $4; Wm. Smith, Detroit, grade heifer calf, 1st, $5; P. Morrison, Jackson, grade heifer calf, Bell, 2d, $3; J. S. Tibbits, Nankin, grade cow, Roan, 3 years, 1st, $15; C. J. Sprague, Farmington, 1 pair of grade steers, 2 years, 1st, $12; J. L. Tibbits, Nankin, Wayne Co., 3 milch cows, 3 years, 1st, $15; James Stone, Hillsdale, grade bull, 2 years, Billie (discretionary), $10.

T. J. TOBEY,
Chairman of Committee.

CLASS VIII—CATTLE.

F. M. & L. A. Cooley, Jackson, oxen, 7 years, 1st, $20; P. A. Corey, Hubbardstown, working oxen, 5 years, 2d, $15; Robert Gould, Ceresco, pair steers, 4 years, 1st, $15; T. Dinsmore, Dansville, pair steers, 2 years, 1st, $8; T. A. King, Parma, pair steers, 2 years, 2d, $5; T. A. King, Parma, pair steers, 1 year, 1st, $5; C. J. Sprague, Farmington, trainer of trained stock, 1st, $10; Albert G. True, Rives, working oxen, 5 years, discretionary premium; T. A. Corey, Hubbardstown, trainer of trained stock, discretionary premium.

PETER HARWICK,
DAVID WOODMAN,
DAVID WOODMARD,
LEWIS DRAKE,
Committee.

CLASS IX—CATTLE.

H. B. Jones, Dexter, 1 fat cow, 1st, $10; C. J. Sprague, Farmington, Oakland Co., pair fat oxen, 3 years, 1st, $20; T. Densmore, Dansville, Ingham Co., 1 fat steer, 2 years, 1st, $6.

J. R. MONROE,
NATHAN ALVORD,
A. BEEBE,
Committee.

DIVISION B.

CLASS X—HORSES, THOROUGHBRED.

Elliott Gray, Tecumseh, best stallion, with 5 colts, Morris, 1st, $100; George Quick, Flint, best stallion, 4 years old and over, 1st, $50; J. W. Bishop, Lima, 2d best stallion, 4 years old and over, Tycoon, 2d, $25; H. Chappell, Detroit, best mare, 4 years, no colt, Minnie, 1st, $12; H. Chappell, Detroit, 2d best mare, 4 years, no colt, Alice Pool, 2d, $8.

E. VAN VALKENBURGH, Hillsdale,
E. D. BEACH, Battle Creek,
A. STOUT, St. Johns,
Committee.

CLASS XI—HORSES HALF THOROUGHBRED.

J. Berry, Quincy, best stallion, 4 years and over, 1st, $50; Richard Lewis, Brady, 2d best stallion, 4 years and over, Touchstone, 2d, $25; H. A. Flint, Dexter, best stallion, 3 years, 1st, $15; E. Gray, Tecumseh, best stallion, 1 year, 1st, $10; E. Gray, Tecumseh, 2d best stallion, 1 year. 2d, $5; Richard Lewis, Brady, best gelding, 4 years, Dolby, 1st, $12; L. Fowler, Adrian, best stallion foal, Magna Morris, 1st. $6; J. C. Deyo, Jackson, best brood mare, 5 years and over, colt by side, Mary, 1st, $20; Richard Lewis, Brady, best mare, 4 years and over, no colt, Bessie, 1st, $12; H. A. Flint, Dexter, 2d best mare, 4 years and over, no colt, 2d, $8; Richard Lewis, Brady, best mare, 2 years, Beauty, 1st, $10; E. Gray, Tecumseh, best mare, one year, Lady Morris, 1st, $7; Richard Lewis, Brady, best filly foal, 1st, $6.

J. T. DOWNS, Marshall,
B. J. BIDWELL, Tecumseh,
JOHN CAMPBELL, Ypsilanti,
Committee.

CLASS XII—HORSES OF ALL WORK.

A. Clark, Coldwater, best stallion, 5 years and over, Barney Searcher, 1st, $50; J. D. Crouch, Spring Arbor, 2d best stallion 5 years and over, 2d, $25; W. Farnsworth, Battle Creek, best stallion 4 years, 1st, $30; B. F. Wade, Manchester, 2d best stallion, 4 years, Young Plover, 2d, $15; A. C. Fiske, Colwater, best stallion, 3 years, Lexington Chief, 1st, $15; William Bradford, Sandstone, 2d best stallion, 3 years, Prince, 2d $10; G. H. Clickner, Leslie, best stallion, 2 years, 1st, $12; M. J. Ellis, Springfield, 2d best stallion, 2 years, 2d, $8; J. C. Deyo, Jackson, 2d best stallion, 1 year, Bill Flushing, 2d, $6; J. C. Deyo, Jackson, best gelding, 3 years, Charley, 1st, $10; A. Finlay, Kalamazoo, 2d best gelding, 3 years. 2d, $6; J. C. Deyo, Jackson, best brood mare, 5 years and over, colt by side, 1st, $20; J. H. Graham, Columbia, 2d best brood mare, 5 years and over, colt by side, 2d, $12; J. S. Tibbets, Nankin, best brood mare, 4 years, no colt, 1st, $12; H. A. Flint, Dexter, 2d best brood mare, 4 years, no colt, 2d, $8; F. J. Randall, Columbia, best mare, 3 years, 1st, $12; Hiram Draper, Rives, 2d best mare, 3 years, 2d, $8; W. B. Reynolds, Jackson, best filly, 2 years, 1st, $10; E. D. Beach, Battle Creek, 2d best filly, 2 years, 2d, $6; Robert Gould, Ceresco, best filly foal, 1st, $6; J. C. Deyo, Jackson, 2d best filly foal, 2d, $4.

S. F. SPAFFORD, Des Moines, Ia.
S. CROUCH, Tecumseh,
B. H. CURTISS, Dundee,
Committee.

CLASS XIII—ROADSTERS AND TROTTERS.

R. Armstrong, Coldwater, trotting stallion, 5 years and over, 5 of his colts, Marshall Chief, 1st, $100; D. B. Hibbard, Jackson. trotting stallion, 5 years and over, Marshall Chief, 2d, $50; J. C. Deyo, Jackson, stallion 4 years and over, Andy Johnson, 1st, $50; A. Brownell, Quincy, stallion, 4 years and over, 2d, $25; R. Armstrong, Coldwater, stallion, 3 years, Marshall Chief, Jr., 1st, $20; N. G. Davis, Jackson, stallion, 3 years, Fred, 2d, $10; J. C. Deyo, Jackson, stallion, 2 years, Billy Bostwick, 1st, $12; A. C. Fiske, Coldwater, 1 year, Mamb. Patchin, 1st, $10; R. W. Chamberlain, Jackson, stallion foal, 1st, $6; P. Palmer, Liberty, stallion foal, 2d, $4; J. C. Deyo, Jackson, brood mare, 5 years and over, colt by side, Champion, 1st, $25; N. Crawford, Sandstone, brood mare, 5 years and over, colt by side, 2d, $15; H. A. Flint, Dexter, mare, 5 years and over, no colt, 1st, $20; J. C. Deyo, Jackson, mare, 5 years and over, no colt, Lucy, 2d, $20; L. Sterling, Canton, mare, 4 years and over, no colt, 1st, $12; R. Armstrong, Coldwater, mare, 2 years and over, 1st, $8; P. L. Schuyler, Marion, mare, 1 year and over, 1st, $8; N. Crawford, Sandstone, mare, 1 year and over, 2d, $5; N. Crawford, Sandstone, filly foal, 1st, $6; P. L. Schuyler, Marion, filly foal, 2d, $4; Jesse Hurd, Jackson, pair driving horses, 2d, $30; D. W. Arnold, Moscow, pair driving horses, 3d, $20; J. C. Deyo, Jackson, single trotter, 5 years and over, Pat Malloy, 1st, $30; S. S. Vaughn, Jackson, single trotter, 5 years and over, $20; H. Conant, Monroe, single trotter, 5 years and over, 3d, $10; A. C. Fisk, Coldwater, single trotter, 4 years old, Dexter, 1st, $20; M. E. McKercher, Moscow, single trotter, 4 years old, 2d, $15; Jacob West, Jackson, single trotter, 5 years old, Limber Jim, 3d, $10.

W. G. PATTISON, Kalamazoo,
G. H. GALE, Kalamazoo,
B. VOSBURGH, Galesburgh,
HENRY COLE,
A. J. DEAN, Adrian,
Committee.

CLASS XIV—DRAFT HORSES.

M. McGarvin, Chatham, Ont., Canada, stallion, 7 years, 1st, $50; Thomas Wilson, Greenfield, stallion, 9 years, Young Britain, 2d, $25; Thomas J. Sourman, Tompkins, stallion, 2 years, 2d, $8; S. A. Barnes, Jackson, stallion, 1 year, 1st, $10; S. A. Barnes, Jackson, mare, 7 years, no colt, dis., 2d, $8.

P. A. SUTTON, Sandstone,
S. CROUCH, Tecumseh,
E. D. BEACH, Battle Creek,
Committee.

CLASS XV—CARRIAGE AND BUGGY HORSES.

Daniel B. Hibbbard, Jackson, first pair matched carriage horses, 5 years and over, 16 hands and over, 1st, $50; J. C. Deyo, Jackson, 2d pair matched carriage horses, 5 y'rs and over, 16 hands and over, 2d, $25; M. Knapp, Jackson, best p'r matched carriage horses, 4 years and over, and 16 hands and over, 1st, $30; S. S. Vaughn, Jackson, best pair matched carriage horses, 4 years and over, under 16 hands, 1st, $40; C. E. Bennett, Jackson, 2d best matched carriage horses, 4 years and over, under 16 hands, 2d,

$20; S. H. Smithers, Detroit, best single carriage horse, 4 years and over, 1st, $8; Daniel B. Hibbard, Jackson, 2d best single carriage horse, 4 years and over, 2d, $6.

W. G. PATTISON, Kalamazoo,
A. H. HANFORD, Tecumseh,
E. VAN VALKENBURGH, Hillsdale,
Committee.

CLASS XVI—SADDLE HORSES.

W. W. Parshall, Kalamazoo, gelding, 9 years, Black Jack, 1st, $15; George Deyo, Jackson, gelding, 5 years, Master George, 2d, $10; Bowen W. Shoemaker, Jackson, gelding 6 years, Mack, dis. prem., $5; Dwight Merriman, Jackson, Shetland pony, dis. prem. $5; Charles Bostwick, Jackson, pony, 8 years, dis. prem., $5.

HENRY A. CONANT,
E. WAY,
OSCAR DOOLITTLE,
Committee.

CLASS XVII—MATCHED HORSES.

William Campbell, Parma, pair of matched horses, 3 years old, 1st, $15; A. S. Wing, Jackson, pair of matched horses, 3 years old, 2d, $10.

ANDREW CUTTER,
S. S. VAUGHN,
E. O. GROSVENOR,
Committee.

The Committee awarded the premiums to No. 399 and 134 in the section calling for horses 1,200 lbs. or over. But a protest having been entered on account of the horses answering to these numbers not being up to weight, an examination was ordered, and under the charge of the Executive Superintendent, also directed that no premiums should be paid previous to the meeting of the Executive Committee at its next session.

R. F. JOHNSTONE,
Secretary.

CLASS XVIII—JACKS AND MULES.

George Gibson, Wayne, Spanish Jack, 8 years, 1st, $30; Clark Durkee, Seneca, Lenawee County, Spanish jack, 8 years, 2d, $15; Henry Lyon, Adams, pair mules, aged 5 years, 1st, $10; C. E. Webb, Jackson, pair mules, aged 5 years, 2d, $5; Henry Lyon, Adams, show of young mules, 1st, $20; S. A. Strong, Liberty, mules, 2 years old, very fine, dis. prem., $3.

P. A. SUTTON, Sandstone,
S. CROUCH, Tecumseh,
B. H. CURTIS, Dundee,
Committee.

DIVISION C.

CLASS XIX—MERINOS AND FINE WOOLED SHEEP.

Geo. S. Wood, Saline, buck, 3 years, 1st, $30; N. & C. Chilson, Battle Creek, buck, 3 years, 2d, $15; Geo. S. Wood, Saline, buck, 2 years, 1st, $20; J. S. Wood, Saline, buck, 2 years, 2d, $10; Geo. S. Wood, Saline, buck, 1 year, 1st, $16; I. M. Whitaker, Lima Center, buck, 1 year, 2d, $8; J. S. Wood, Saline, 5 buck lambs, 1st, $16; I. M. Whitaker, Lima Center, 5 buck lambs, 2d, $8; Geo. S. Wood, Saline, 5 ewes, 3 years, 1st, $20; I. M. Whitaker, Lima Center, 5 ewes, 3 years, 2d, $10; J. S. Wood, Saline, 5 ewes, 2 years, 1st, $16; Geo. S. Wood, Saline, 5 ewes, 2 years, 2d, $8; Geo. S. Wood, Saline, 5 ewes, 1 year, 1st, $12; J. S. Wood, Saline, 5 ewes, 1 year, 2d, $6; J. S. Wood, Saline, 5 ewe lambs, 1st, $10.

J. C. DAYTON,
H. HUFF,
HARVEY HAYNES,
Committee.

CLASS XX—SOUTHDOWNS.

P. Waken, Madison, buck, 3 years, 1st, $30; R. Kimmins, Pontiac, buck, 3 years, 2d, $10; R. Kimmins, Pontiac, buck, 2 years, 1st, $20; R. Kimmins, Pontiac, buck, 2 years, 2d, $10; R. Kimmins, Pontiac, buck, 1 year, 2d, $8; R. Kimmins, Pontiac, pen of 5 buck lambs, 1st, $16; R. Kimmins, Pontiac, pen of 5 buck lambs, 2d, $8; R. Kimmins, Pontiac, 5 ewes, 3 years, 1st, $20; R. Kimmins, Pontiac, 5 ewes, 1 year, 1st, $12; R. Kimmins, Pontiac, 5 ewe lambs, 1st, $10; R. Kimmins, Pontiac, 5 ewe lambs, 2d, $5.

WM. WITFIELD,
J. SHEARER,
WM. H. S. LOTHAM,
Committee.

CLASS XXI—COTSWOLD, LEICESTER, AND OTHER LONG-WOOLED SHEEP.

Wm. Smith, Detroit, Leicester buck, 3 years, 1st, $30; E. S. Bryan, Marshall, Leicester buck, 2 years, 1st, $20; A. McDonald, Jackson, Leicester buck, 1 year, 1st, $16; P. Waken, Madison, Wis., pen of 5 buck lambs, 1st, $16; A. McDonald, Jackson, Leicester pen 5 ewes, 3 years, 2d, $10; P. Waken, Madison, Wis., Leicester pen of 5 ewes, 1 year, 1st, $12; A. McDonald, Jackson, Leicester pen of 5 ewes, 1 year 2d, $6; A. McDonald, Jackson, Leicester pen of 5 ewe lambs, 1st, $10; P. Waken, Madison, Wis., Leicester pen of 5 ewe lambs, 2d, $5; Wm. Smith, Detroit, Cotswold buck, 3 years, 2d, $15; A. M. Edwards, Detroit, Cotswold buck, 2 years, 2d, $10; Wm. Smith, Detroit, Cotswold buck, 1 year, 1st, $16; P. Waken, Madison, Wis., Cotswold buck, 1 year, 2d, $8; P. Waken, Madison, Wis., Cotswold 5 buck lambs, 2d, $8; Wm. Smith, Detroit, 5 ewes, 3 years, 1st, $20.

H. G. HOLT,
RICHARD FOGG,
WM. NEWTON,
Committee.

CLASS XXII—FAT SHEEP.

R. N. Chamberlain, Jackson, pen of fat sheep, 2d, $5; E. Driggs, Adrian, pen of Cashmere goats (dis. prem.), $10.

JOHN ALLEN,
NORMAN ALLEN,
ROBERT MORRISON,
Committee.

CLASS XXIII—SWINE.

H. D. Court, Battle Creek, Chester-White boar, 6 mos., 1st, $5; H. D. Court, Battle Creek, Chester-White boar, 3½ mos., 2d, $3; H. D. Court, Battle Creek, Chester-white pen of pigs, 1st, $5; A. F. Wood, Mason, Essex boar, 3 years, 1st, $5; E. F. Downy, Jackson, Essex boar, 2 years, 2d, $3; E. F. Downy, Jackson, Essex sow, 1 year, 1st, $5; A. F. Wood, Mason, Essex sow and 6 pigs, 2d, $3; A. F. Wood, Mason, Essex 4 pigs, 6 mos., 1st, $5; A. E. Brackett, Marshall, Poland boar, 14 mos., 1st, $5; A. D. Brackett, Marshall, Poland sow, 14 mos., 1st., $5; A. E. Brackett, Marshall, Poland lot of pigs, 2½ mos., 1st, $5; P. Waken, Madison, Wis., Berkshire sow, 1 year, 1st, $5; A. E. Brackett, Marshall, Big Spotted China sow, 14 mos., 1st, $5.

WM. STODDARD,
M. L. RAY,
J. McKAY,
Committee.

CLASS XXIV—POULTRY.

A. F. Wood, Mason, coop of Asiatic fowls, 1st, $3; W. K. Gibson, Jackson, coop Asiatic dark Brahmas, dis., $3; Mrs. E. W. Heaton, Jackson, coop of game fowls, 2d, $2; J. J. Walker, Ann Arbor, coop of White Dorkings, 1st, $2; Dr. Gibson, Jackson, coop of Black Polands, 1st, $2; J. J. Walker, Ann Arbor, coop of Silver-spangled Polands, 1st, $2; J. J. Walker, Ann Arbor, coop Black Spanish, 1st, $2; J. J. Walker, Ann Arbor, coop White Bantams, 2d, $1; J. J. Walker, Ann Arbor, coop Houdans, 1st, $5; A. F. Wood, Mason, coop domestic turkeys, 1st, $2; W. A. Holcomb, Jackson, coop common ducks, 1st, $2; E. Armstrong, Jackson, coop domestic pigeons, 1st, $3; J. J. Walker, Ann Arbor, coop of long-combed White Leghorns, dis.; W. A. Burnham, Ann Arbor, coop Light Brahmas, 2d, $2; W. A. Burnham, Ann Arbor, coop Black Red-wings, 1st, $2; L. D. Watkins, Manchester, Mad. Lop-eared rabbits, 1st, $2; W. A. Holcomb, Jackson, Jefferson County blue duck (dis.), $2.

JOSEPH WOOLHOUSE,
B. SULLIVAN,
J. W. HELME,
Committee.

DIVISION D.

CLASS XXV—PLOWS AND THEIR TRIALS.

O. C. Gale & Co., Albion, plow for sod or greensward, diploma and $10; O. C. Gale & Co., Albion, plow for general use, made in Michigan, 1st, diploma and $5; O. C. Gale & Co., Albion, attachment, with plow, for grass or manure, 1st, $5; O. C. Gale &

Co., Albion, gang plow, 1st, diploma; J. F. Bryan & Bro., Detroit, plow for turning stubble, 1st, $5; J. F. Bryan & Bro., Detroit, steel plow, 1st, diploma; F. C. Goff & Bro., Cleveland, Ohio, plow for general use, made in any State, 1st, diploma and $5; Allen Chaney, Detroit, self-cleaning plow coulter, 1st, $2; L. Miller & Co., Jonesville, lot of 13 plows, entered for exhibition without competition.

WM. TAFT, Plymouth,
FRANK R. BECKWITH, Cassopolis,
JAMES DEPUY, Jackson,
Committee.

CLASS XXVI—HARROWS, CULTIVATORS, ETC.

O. C. Gale & Co., Albion, harrow for general use, 1st, diploma and $5; O. C. Gale & Co., Albion, one-horse hoe, 1st, $3; O. C. Gale & Co., Albion, machine for hoeing gardens, 1st, $3; Eli Shupe, Middleville, two-horse cultivator for fallow, 1st, and silver medal; A. Dickerson, Manchester, roller, or clod-crusher, 1st, and silver medal; H. H. Pratt, Hart, Oceana Co., implement to do work of a harrow, 1st, diploma and $5; Ira Ladd, Adrian, two-horse cultivator for corn or drill, 1st, and silver medal; Ann Arbor Agricultural Co., Ann Arbor, combined wheel cultivator for corn or drilled crops, 1st, and silver medal.

WILLIAM TAFT, Plymouth,
FRANK R. BECKWITH, Cassopolis,
JAMES DEPUY, Jackson,
Committee.

CLASS XXVII—SEED-DRILLS, SOWERS, PLANTERS, ETC.

Hoosier Drill Co., Milton, Indiana, one-horse drill, diploma; R. A. Tyrrell, Detroit, seed-sower and cultivator combined, 1st, silver medal; S. A. Flower, Pontiac, ashes and plaster sower, for horse-power, 1st, diploma and $3; F. Van Doren, Adrian, hand corn-planter, 1st, diploma and $2.

WILLIAM TAFT, Plymouth.
FRANK R. BECKWITH, Cassopolis,
JAMES DEPUY, Jackson,
Committee.

CLASS XXVIII.

W. S. Penfield, Detroit, Excelsior hand lawn-mower, 2d, $2; Withington, Cooley & Co., Jackson, 6 hand-cutters, 1st, $1; Withington, Cooley & Co., Jackson, 2 grain cradles, diploma; Withington, Cooley & Co., Jackson, 6 scythe-snaths, diploma; W. F. Cawham, Jackson, horse-rake, Tompkins' two-horse power, 1st, $3; A. B. Clark, Grass Lake, horse pitchfork, and rigging for loading and unloading hay, diploma; J. F. Bryan & Bro.. Detroit, hand-power mowing-machine for lawn, diploma and $5; F. Nagley, Fillmore, hay rigging for double wagon, 1st, $5; J. H. L. Luck, Joliet, Ill., corn-harvester, 1st, diploma and $5; Joshua Palmer, Jackson, potato-digger, 1st, $3; S. P. Hitchcock, Lake Mills, Wis., one-horse hay-rake, 2d, $2; J. F. Bryan, Detroit, hay-loading machine, 1st, diploma; W. W. Platt, Jonesville, double harpoon fork, recommended.

The Committee further report: The exhibition of reapers and mowers exceeded anything of the kind ever witnessed at our State Fair, and did great credit to the inventors and manufacturers of those useful machines. Thirty-eight entries were made, and the machines were all upon the ground, many of them having improvements, in the opinion of the Committee, of great value. The workmanship, finish, and mechanical skill exhibited in their construction were all that could be desired. The trial of horse hay-forks was unsatisfactory to the Committee, as no conveniences were provided for testing the merits of the several forks. One fork, the double harpoon, was misplaced upon the ground, and escaped the notice of the Committee until after the awards had been made; but the Committee are of the opinion that it contains merits which should entitle it to a premium.

J. J. WOODMAN, Paw Paw.
Z. COOK,
LEWIS BROWN,
Committee.

CLASS XXIX.

J. S. & M. Peckham, Utica, N. Y., barrel, furnace and boiler, 1st, dip.; Jason Surk, Marshall, machine for preparing fodder for stock, 1st, dip.; Nichols, Shepard & Co., Battle Creek, thresher and separator, 1st, dip.; Nichols, Shepard & Co., Battle Creek, horse-power for general use, diploma and $5; Ira C. Hibbard, Rochester, N. Y., adjustable harvester knife, dip.; Green, Townson & Co., Brooklyn, hay, straw, and stalk cutter, horse power, 1st, med.; Green, Townson & Co., Brooklyn, corn-sheller for horse power, 1st, dip. and $3; J. Sedgebeer, Painesville, O., corn and cob crusher, 1st. dip.; Younglove, Massey & Co., straw and stalk cutter, hand power, 1st, dip. and $3; Younglove, Massey & Co., Cleveland, O., portable cider-mill, 1st, dip.; George Connor, Mishawaukie, Ind., portable corn-mill for grinding food for stock, 1st, dip.; J. F, Bryan & Bro., Detroit, corn-sheller for hand power, 1st, dip. and $2; W. H. Wiley, Fredonia, N. Y., horse-power for general use, special, dip.; J. T. Weeks, Napoleon, mill for separating sorghum sugar, 1st, med; Giphart & Rigging, Dowagiac, elevating bag-holder for fanning-mill, 1st, dip.; Giphart & Rigging, Dowagiac, fanning-mill, 1st, dip.; F. J. Buck, Adrian, corn-sheller, hand power, dip.; Birdsell Mfg. Co., South Bend, Ind., clover-thresher, huller, and sheller, 1st, med; F. D. Prouty, Jackson, threshing machine, combined with horse-power and separator, 1st, diploma and med.;

JOHN ALLEN,
E. T. WALKER,
P. DENSMORE,
Committee.

CLASS XXX.

Arthur Wood, Grand Rapids, spring market wagon, $10; C. J. Whitney, & Co., Detroit, one dray, $6; A. Bedford, Coldwater, trotting wagon $10; A. Bedford, Coldwater, trotting sulky, $6; P. Kidney & Co., Olmstead Falls, O., set of carriage felloes, dip.; Raymond & Scranton, Grand Rapids, one-horse carriage, $10; Raymond & Scranton, Grand Rapids, top buggy, $10; Raymond & Scranton, Grand Rapids, buggy without top, $10; George C. Fitch, Grand Rapids, single sleigh or cutter, $5; C. C. Lawrence, Marengo, adjustable carriage-top rest, dip.; C. C. Pond, Jackson, one phæton, $10; James Martin, Morenci, pair of bob sleds, dip; A. O. Colburn, Wayne, sand cap for wagon, dip.; H. S. Harris, Paw Paw, farm wagon, $10; J. J. Deal, Jonesville, two-horse family carriage, silver medal and $10; J. J. Deal, Jonesville, two-seated buggy, $10; J. N. Toplif, Elyria, Ohio, case of steel tubular band sockets, dip. J. J. Deal, Jonesville, best and greatest variety of carriages and buggies, dip. and $25.

JOHN T. CLAPP,
G. W. CLARK,
H. C. PHILLIPS,
R. EMERSON,
A. RANDALL,
Committee.

CLASS XXXI.

Catlin & Wright, Coldwater, Page's Grub-Puller, 1st dip.; O. C. Gale & Co., Albion, three-horse whiffletrees, 1st, dip.; O. C. Gale & Co., Albion, three-horse clevis, 1st, dip.; O. C. Gale & Co., Albion, road scraper, 1st, $1; Spicer & Crossman, Marshall, three-horse equalizer, 1st, dip.; Withington, Cooley & Co., Jackson, 6 hoes, 1st, dip.; Withington, Cooley & Co., Jackson, assortment garden tools, 1st, $3; Withington, Cooley & Co., Jackson, 6 manure forks, 1st, $3; Withington, Cooley & Co., Jackson, 6 potato-hooks, $1; W. H. Pendleton, Kalamazoo, wind-mill and pump, 1st, medal; Merrit & Burwell, Chesaning, hand sawing-machine, dip. and $3; Catlin, Wright & Co., Coldwater, stump and grub extractor, dip. and $5; B. F. Dickey, Marshall, automatic carriage-gate, dip. and $1; P. Kidney & Co., Olmstead Falls, O., portable farm fence. dip.; J. H. Verity, Portland, fruit ladder, dip.; Henry C. King, Forrester, force-pump for house purpose, dip. and $2; S. H. Dickerson, Hudson, improved well-pump and auger, dip.; O. H. Belden, Deerfield, grub-puller, dip.; John Ainsley, Marshall, pruning tool, dip.; Mills & Ballard, Litchfield, pruning tool, dip.; C. S. Cannon, Battle Creek, fruit gatherer, dip.; M. D. Beardsley, Detroit, horse-power for wood-saw, etc., dip.; P. B. Perkins, Detroit, jack and self-feed wood-saw, dip.; Grand Rapids Wood Co., Grand Rapids, ox yoke, $1; A. H. Springer, Eaton Rapids, one doz. axes, $1; Buffalo Scale Co., Buffalo, N. Y., hay or cattle scale, dip.; Buffalo Scale Co., Buffalo, N. Y., counter and platform scale, medal; Buffalo Scale Co., Buffalo, N. Y,, improved grain scale; W. J. Campbell, Jackson, farm gate, dip.; J. Longshore, Mansfield, Ohio, collection of wooden ware, dip.; N. Parish, Kalamazoo, water lifter and hoisting machine, dip.; N. Parish, Kalamazoo, automatic gate, dip.

A. M. FITCH,
J. E. LELAND,
JOHN ASHLEY,
Committee.

CLASS XXXII.

Dwelly, Robinson & Co., Jackson, assortment stoneware, 1st, $5; Goffe, Eddy & Co., Jackson, assortment stoneware, 2d, $3; Allen Chaney, Detroit, dog-power churn, dip.; R. S. Janney, Toledo, O., one Red Jacket churn, 1st, $5; Saffel & Baldwin, Tiffin, O., union churn, 2d. dip.; R. S. Janney, Toledo, O., churn, 3d, $1; Geo. E. Beebe, No. 120 Broadway, N. Y. City, foot and bed warmer, dip.; O. S. Burgess, Battle Creek, steam washing-machine, dip.; J. Sedgebeer, Painesville, O., hand coffee-mill, dip.; J. S. Peabody, Verdon, Ill., mel-extractor for honey, dip.; R. S. Janney, Toledo, O., clothes-horse, 1st, $1; R. S. Janney, Toledo, O., clothes-rack, 2d, dip.; T. D. Ingersoll, Monroe, culinary boiler, dip.; Iron Clad Can Co., 42 and 44 Murray st., N. Y., iron-clad milk cans, 1st, dip.; T. D. Ingersoll, Monroe, dust-pan, dip.; T. H. Crossman Ypsilanti, potato-washing machine, 2d, dip.: Hanson Fitch, Hanover, clothes-washer, 2d, dip.; M. H. V. Young, Cleveland, O., meat and vegetable chopper, 3d, $1; Buffalo Scale Co., Buffalo, N. Y., weight and scales for dairy purposes, $12; H. Turner, Detroit, paper pails for house and dairy purposes, $12; Wm. Harriman, Elkhart, Ind., milk and provision safe, dip.

W. DEVINE,
MRS. GEO. WEBB,
MRS. JOHN ASHLEY,
MRS. W. DEVINE,
Committee.

CLASS XXXIII—STOVES AND HOLLOW WARE.

Perry & Co., 15 Hudson st., Albany, N. Y., cooking stove for wood, 1st, medal: Perry & Co., Albany, N. Y., cooking stove for coal, 1st, or medal; Perry & Co., Albany, N. Y., parlor or hall stove, 1st, or medal; E. H. Camp, Jackson, furnace for warming houses, silver medal.

ORSON INGALLS,
E. H. WHITING,
TITUS DOAN,
Committee.

DIVISION E.

CLASS XXXIV.

John Gilbert, Ovid, winter red wheat, 1st, $5; M. Sharp, Jackson, winter red wheat, 2d, $3; M. M. Miller, Coldwater, winter white wheat, 1st, $5; John Gilbert, Ovid, winter white wheat, 2d, $3; S. Fenno, Coldwater, best one bu. of rye, 1st, $2; John Gilbert, Ovid, best one bu. four-rowed barley, 1st, $3; Ira Ladd, Adrian, best one bu. winter barley, 1st, $3; H. J. Grego, Liberty, second one bu. of winter barley, 2d, $2; A. McDonald, Liberty, best bu. of oats, 1st, $2; John Gilbert, Ovid, second bu. of oats, 2d, $1: M. Sharp, Jackson, best bu. of shelled corn, $2; J. F. Townley, Parma, second best bu. shelled corn, $1; Thomas Mayett, Jackson, one bu. of white beans, 2d, $1; Oscar Burras, North Fairfield, O., best bu. of timothy seed, 1st, $2; John Gilbert, Ovid, best bu. of buckwheat, 1st, $2; Oscar Burras, North Fairfield, O., general assortment of seeds, 1st and diploma, $5; M. M. Miller, Coldwater, best sample of hops, 1st, $3; M. M. Miller, Coldwater, sample of hops, 2d, $2; M. M. Miller, Coldwater, best seeds for field crops, 1st and diploma, $10; M. M. Miller, Coldwater, best heads of wheat, 1st, $5; M. M. Miller, Coldwater, best heads of oats, 1st, $5; M. Sharp, Jackson, best assortment of heads of corn, 1st, $5.

R. G. WHITNEY,
A. E. PARDEE,
B. B. MOSHIER,
F. G. BISHOP,
A. HARLOW.
Committee.

CLASS XXXV—VEGETABLES.

Oscar Burras, North Fairfield, Ohio, culinary vegetables, $10; Oscar Burras, North Fairfield, Ohio, 12 turnips, beets, 2d, $1; Oscar Burras, North Fairfield, Ohio, ½ peck white onions, 1st, $2; Oscar Burras, North Fairfield, Ohio, Windsor beans, 1st, $2; Oscar Burras, North Fairfield, Ohio, sweet potatoes, 2d, $1; Oscar Burras, North Fairfield, Ohio, 12 summer radishes, 1st, $2; Oscar Burras, North Fairfield, Ohio, 12 winter radishes, 1st, $2; Oscar Burras, North Fairfield, Ohio, best ears of late sweet corn, 1st, $2; E. Armstrong, Jackson, best Scotch cabbage, 1st, $2; Oscar Burras, North Fairfield, Ohio, 2d best Hubbard squashes, 2d. $1; Oscar Burras, North Fairfield, Ohio, orange carrots, 2d, $1; Oscar Burras, North Fairfield, Ohio, best vegetable eggs, 1st, $2; Oscar Burras, North Fairfield, Ohio, best red peppers, 1st, $2; Louisa Dart, Jackson, 2d best red peppers, 2d, $1; J. S. Tibbets, Nankin, 2d best Lima beans, 2d, $1; Thomas Mayett, Jackson, best celery, 1st, $2; Thomas Mayett, Jackson, best parsnips, 1st, $2; Thomas Mayett, Jackson, best carrots, 1st, $2; Thomas Mayett, Jackson, imperial lettuce, 1st, $2; Thomas Mayett, Jackson, 2d best white turnips, 2d, $2; Thomas Mayett, Jackson, 2d best Windsor beans, 2d, $1; D. C. Griggs, Jackson, best crooked necked squash, 1st, $2; Oscar Burras, North Fairfield, Ohio, best yellow onions, 1st, $2; D. C. Griggs, Jackson, best Hubbard squash, 1st, $2; J. S. Tibbets, Nankin, 2d best yellow onions, 2d, $1; John Gilbert, Ovid, best varieties potatoes, 1st, dip. and $10; John Gilbert, Ovid, 2 best early potatoes, 2d, $1; John Gilbert, Ovid, best 5 varieties potatoes, 1st, $5; John Gilbert, Ovid, 2d 5 varieties late potatoes, 2d, $1; George F. Perry, Ypsilanti, best winter squash, 1st, $2; George F. Perry, Ypsilanti, sauce and pie melon, $1; A. B. Smith, Napoleon, best red onions, 1st, $2; D. H. Ranney, Rives, 2d cream squash, 2d, $1; D. H. Ranney, Rives, best pumpkins, 1st, $2; Henry Purdy, Jackson, best orange tomatoes, 1st, $2; Henry Purdy, Jackson, Prussian crumpled lettuce, 2d, $1; B. W. Rockwell, Jackson, best early rose potatoes, 1st, $2; B. W. Rockwell, Jackson, best rhubarb, 1st, $2; J. C. Smith, Charlotte, 2d best peck tomatoes, 2d, $1; J. C. Smith, Charlotte, 2d best rhubarb, 2d, $1; J. C. Smith, Charlotte, best sweet potatoes, 1st, $2; A. G. True, Rives, best peachblow potatoes, 1st, $2; A. G. True, best white turnips, 1st, $2; C. G. Seeley, California, Branch Co., mammoth squash, 1st, $2; C. G. Seeley, California, Branch Co., 2d, red onions, 2d, $1; A. M. Chapin, Eden, Ingham Co., 2d, white onions, 2d, $1; S. Woodward, Grass Lake, second best pumpkins, 2d, $1; G. B. Palmer, Jackson, best Lima beans, 1st, $2; C. D. Webb, Jackson, best early sweet corn, 1st, $2; Thomas Mayett, Jackson, best early sweet corn, 2d, $1; J. E. Smith, Charlotte, best late sweet corn, 2d, $1; H. H. Terpening, Jackson, best pop-corn, 1st, $2; Thomas Mayett, Jackson, best pop-corn, 2d, $1; Mrs. Amos Van Aiken, Hillsdale, 1 peck dried apples, 2d, $1.

GEORGE WEBB, Jackson,
G. K. HURD, Monroe,
J. H. ENGLISH, Boston,
S. E. BEIREN, South Boston,
Committee.

DIVISION F.

CLASS XXXVI.

F. Fowler, Reading, 36 cheeses, 1 lot, 1st, medal; Mrs. D. O. Barnard, Sandstone, 25 lbs. butter, 1st, $8; A. R. Smith, Napoleon, 25 lbs. butter, 2d, $5; A. F. Ward, Mason, 15 lbs. butter, June, 1st, $6; D. O. Barnard, Sandstone, 15 lbs. butter, June, 2d, $4; F. Fowler, Reading, 1 cheese, 50 lbs. or over, 1st, $10; L. W. Simmons, Farmington,

1 cheese, 50 lbs. or over, 2d, $5; L. W. Simmons, Farmington, 1 cheese under 50 lbs., 1st, $5; F. Fowler, Reading, 1 cheese under 50 lbs.. 2d, $3.

RUFUS BAKER,
MRS. MORGAN CASE,
NELSON PALMER,
M. D. BAKER,
Committee.

CLASS XXXVII.

J. F. Townley, Parma, Jackson Co., beehive and Italian bees, method swarm, dip. and $5; C. J. Sprague, Farmington, Wayne Co., 10 lbs. maple sugar, $1; C. J. Sprague, Farmington, Wayne Co., 1 gallon maple molasses, $1; David Cady, Indianapolis, Ind., Mitchell's buckeye beehive, dip.; O. Callier, Ann Arbor, 10 lbs. of honey, 1st, $2; F. Butler, Jackson, 2 boxes honey, $1; George Engle, Ypsilanti, Washtenaw Co., 1 beehive, recommended; W. C. Weeks, Napoleon, Jackson Co., 1 gallon sorghum syrup, 1st, $3; J. F. Weeks, Napoleon, Jackson Co., 5 lbs. sorghum sugar, 1st, $5; A. Youngs, Saranac, 10 lbs. maple sugar, 1st, $2; A. Youngs, Saranac, 1 gallon maple syrup, $1; A. Youngs, Saranac, specimen maple candy, dip.; H. Huff, Jonesville, Hillsdale Co., beehive, Thomas' patent, 1st, dip.

JAMES PORTMAN,
JOHN T. ROSE,
LORENZO NOWLIN,
Committee.

CLASS XXXVIII.

Miss Louisa Dart, Jackson, sample brown bread made by a girl 16 years old, 1st, $3; Mrs. J. V. Cookingham, 3 loaves, milk rising bread, 1st, $4; Mrs. J. V. Cookingham, 3 loaves brown bread, 1st, $3; Mrs. H. A. Stetson, preserved peaches and citron, 2d, $2; Mrs. R. Stevenson, preserved peaches, 1st, $3; Mrs. D. H. Ranney, Rives, Jackson Co., can blackberries and raspberries, 2d, $2; S. Fenno, Coldwater, Calhoun Co., rye and Indian bread, 2d, $2; S. Fenno, Coldwater, Calhoun Co., tomato catsup, 1st, $2; E. Armstrong, Jackson, Jackson Co., currant jelly, 1st, $2; Mrs. Mary Cook, Grass Lake, Jackson Co., 3 loaves hop rising bread, 2d, $2; Mrs. Morgan Case, Napoleon, Jackson Co., 1 loaf corn bread, 1st, $2; Mrs. H. G. Cole, Rives, Jackson Co., 3 loaves salt rising bread, 2d, $3; Mrs. S. Woodward, Grass Lake, Jackson Co., display of small fruits, 2d, $2; Miss Anna Shopman, 17 years of age, Jackson, Jackson Co., 3 loaves hop rising bread, 1st, $3. The best display of small fruits in cans, was by Mrs. H. A. Stetson, who is entitled to the first premium, $3; Mrs. H. A. Stetson, for the best display of pickles, 1st, $5; Mrs. E. W. Heaton, of Jackson, for the best display of canned large fruits, 1st, $3.

OREN STONE,
H. JOHNSON,
MRS. ANN M. JOHNSON,
Committee.

CLASS XXXIX.

C. F. Allen, Paw Paw, specimens Delaware grape wine, 2d, $5; C. F. Allen, Paw Paw, spec. blackberry wine, 1st, $2; C. F. Allen, Paw Paw, spec strawberry wine, 1st, $2; C. F. Allen, Paw Paw, specimen raspberry wine, 1st, $2; C. F. Allen, Paw Paw.

specimen elderberry wine, 1st, $2; C. F. Allen, Paw Paw, spec. red, white, and black currant wines, 1st, $2; C. F. Allen, Paw Paw, display bottle native wine, 1st, $25; Pleasant Valley Wine Co., Hammondsport, N. Y., American brandy, 1st, dip.; Pleasant Valley Wine Co., Hammondsport, N. Y., Delaware grape wine, 1st, $10; Pleasant Valley Wine Co., Hammondsport, N. Y., Catawba grape wine, 1st, $5; Mrs. J. B. Tuttle, Jackson, specimen cider vinegar, 1st, $2; Henry Dessenberg, Jackson, specimen cider vinegar, 2d, $2; Point de Peau Co., Monroe, specimen Concord wine, 2d, $5; Point de Peau Co., Monroe, specimen Catawba wine, 2d, $2; J. M. Sterling, Monroe, specimen Concord wine, 1st, $10; J. M. Sterling, Monroe, display of native wines, 2d, $15.

The committee on wines have performed their arduous duties, and after a very critical test as to fineness, boquet-aroma, light, heavy, and quality, have awarded to J. M. Sterling the first premium on Concord; and the Point de Peau Co. the second on a very fine sample Red Concord, racked but once. On four very fine samples of Delaware still-wine, the first premium to the Pleasant Valley Wine Co., Hammondsport, New York; the second to C. F. Allen. On Catawba, the premium to the Pleasant Valley Wine Co. for a very excellent sample of three year old wine containing all the elements of a first class sherry; second premium to Point de Peau Wine Company. Your committee would recommend a special premium to the Pleasant Valley Wine Co., for a very fine sample of sweet Isabella, and also to the same Company for a very excellent sample of native grape brandy. To C. F. Allen the first premium on blackberry, strawberry, raspberry, elderberry, red, white and black currant wine. The first premium on display to C. F. Allen, and second to J. M. Sterling. On cider vinegar, first premium to Mrs. J. B. Tuttle, second to Henry Dessenberg. All of which is respectfully submitted.

E. S. BARTHOLOMEW,
D. PAGE,
H. S. PARMELEE,
Committee.

CLASS XL.

James Pyle, 350 Washington st., N. Y., specimen O. K. soap, dip.; James Pyle, 350 Washington st., N. Y., specimen O. K. saleratus, dip.; M. A. Watson, Flint, specimen prepared pop-corn, dip.; W. B. Davis, Adrian, Rose patent burning fluid, dip.

FRANK BROWNELL,
ALANSON YOUNG,
JOB HOBART,
Committee.

DIVISION G.

CLASS XLI.

Mrs. Dwight Merriman, Jackson, 1 woolen scarf, 2d, $1; Mrs. J. B. Tuttle, Jackson, pair woolen stockings, 1st, $2; Mrs. J. B. Tuttle, Jackson, silk patch-work quilt, 1st, $3; G. A. Baldwin, pair seamed woolen stockings, 2d, $1; Mrs. W. K. Gibson, Jackson, 1 patch work quilt, 2d, $2; Miss Mary L. Lowell, Jackson, 1 pair linen stockings, 1st, $2; Mrs. B. Carson, Jackson, 1 log-cabin quilt, 2d, $2; Mrs. Amos Vanaiken, Hillsdale, 2 lbs. stocking yarn, 1st, $2; Mrs. E. A. Mantle, Jackson, 2 cradle quilts, 2d,

$1; J. W. Welch, Jackson, 30 yards rag carpet, woven by a lady 74 years old, 2d, $2; J. W. Welch, Jackson, 1 pair socks, 1st, $2; J. W. Welch, Jackson, 1 pair socks, 2d, $1; Mrs. B. M. Davis, Jackson, 1 silk quilt, 2d, $2; Mrs. Ann M. Johnson, Clinton, 2 p'rs woolen blankets, 1st, $4; Mrs. Ann M. Johnson, Clinton, 2 woolen coverlets, 1st, $2; Mrs. Ann M. Johnson, Clinton, 1 pair linen sheets, dis. prem., $1; Mrs. Ann M. Johnson, Clinton, 1 pair linen tablecloths, 1st, $2; Mrs. P. A. Sutton, Sandstone, 1 pair woolen blankets, 2d, $2; Mrs. P. Sutton, Sandstone, 2 white bed spreads, 1st, $3; Mrs. W. D. Thompson, Jackson, 1 woolen scarf, 1st, $2; Mrs. Talman Case, Manchester, 1 knitted cradle spread, 1st, $2; Mrs. Talman Case, Manchester, 1 knitted bed spread, 2d, $1; Mrs. S. S. Smith, Jackson, 1 yarn carpet, 1st, $3; Miss Jennie D. Wood, LaSalle, Monroe Co., 1 patch-work quilt cover, 2d, $2; Mrs. J. P. Hawley, Napoleon, 10 yards rag carpet, 1st, $3; Mrs. D. B. Blanchard, Northville, 1 patch-work quilt, 1st, $3; Mrs. D. B. Blanchard, Northville, 1 bed spread, 2d, $2; Mrs. M. R. Reed, Liberty, 1 bed quilt, 1st, $3; Mrs. A. Brink, Perrinsville, Wayne Co., 1 double coverlet, 2d, $1; Mrs. A. Brink, Perrinsville, Wayne Co., 1 woolen blanket, 2d, $2; Mrs. A. Brink, Perrinsville, Wayne Co., display of home-made cloths, 2d, $5; Mrs. A. Brink, Perrinsville, Wayne Co., pair woolen sheets, 2d, $1; Mrs. A. Brink, Perrinsville, Wayne Co., 2 lbs. woolen yarn, 2d, $1; Mrs. A. Brink, Perrinsville, Wayne Co., 10 yards fulled cloth, 1st, $4; Mrs. A. Brink, Perrinsville, Wayne Co., 1 pair woolen mitts, 1st, $2; Mrs. P. Palmer, Liberty, 1 log quilt, 1st, $3; Mrs. H. A. Stetson, Jackson, 1 pair cotton stockings, 1st, $1.

B. D. PADDOCK,
MRS. H. A. HAYDEN,
MRS. J. P. SHOEMAKER,
MRS. B. D. PADDOCK,
MARY J. CAMP,
Committee.

CLASS XLII—DOMESTIC MANUFACTURES.

Wm. M. Bennett, Jackson, display of carpets, 1st, dip.; Wm. M. Bennett, Jackson, display of 5 hearth rugs, 1st, dip.; Wm. M. Bennett, Jackson, display of 5 door mats, 1st, dip.; Wm. M. Bennett, Jackson, display of oil cloths, 1st, dip.; Wm. M. Bennett, Jackson, display of wool blankets, 1st, dip.; Wm. M. Bennett, Jackson, display of mars. quilt, 1st, dip.; Clinton Woolen Manufacturing Co., Clinton, display of goods from any woolen factory in Michigan, 1st, $25 and medal; Clinton Woolen Manufacturing Co., Clinton, 1 piece fancy cassimere, 1st, $5 and dip.; Clinton Woolen Manufacturing Co., Clinton, 1 piece plain cassimere, 1st, $5 and dip.; Clinton Woolen Manufacturing Co., Clinton, 1 piece overcoat cloth, 1st, $5 and dip.; Clinton Woolen Manufacturing Co., Clinton, display of flannels, 1st, $5 and dip.

J. C. E. HOSFORD,
G. W. REMINGTON,
C. S. WEBSTER,
Committee.

CLASS XLIII—ARTICLES OF DRESS.

Stiles Brothers, Jackson, 1 pair beaver gloves, diploma and $1; Stiles Brothers, Jackson, 1 pair lambskin gloves, dip. and $1; Stiles Brothers, Jackson, 1 pair buck gloves, dip. and $1; Stiles Brothers, Jackson, 1 pair buck mittens, dip. and $1; Stiles

Brothers, Jackson, 1 pair calf mittens, dip. and $1; Stiles Brothers, Jackson, 1 case of gloves and mittens, dip. and $1; Mrs. H. S. Curtis, Jackson, 1 case of millinery goods, 1st, dip. and $10.

J. A. BROOKIN,
G. H. BOTSFORD,
N. T. WATERMAN,
Committee.

CLASS XLIV.

E. A. Webster, Jackson, half dozen horse collars, 1st, $1; E. A. Webster, Jackson, one roll finished upper leather, 1st, $2; E. A. Webster, Jackson, one roll collar leather, 1st, $2; S. Cox & Son, Jackson, one roll of finished calf, $1; S. Cox & Son, Jackson, one roll of finished kip, 1st, $2; Stiles Brothers, Jackson, display of dressed buckskin, 1st, $2; M. M. Miller, Coldwater, one leather halter, 1st, dip.; J. W. Welch, Jackson, six buffalo robes, 1st, $2; D. W. Bean, Tecumseh, 12 sides harness leather, 1st, $2; D. W. Bean, Tecumseh, six calfskins, 1st, $1; Geo. J. Bailey, Jackson, one pair stoga boots, 1st, $1; Geo. J. Bailey, Jackson, one pair ladies' summer walking boots, 1st, $1; Geo. J. Bailey, Jackson, one pair ladies' winter walking boots, 1st, $1; Geo. J. Bailey, Jackson, one pair gents' slippers, 1st, $1; Geo. J. Bailey, Jackson, one pair ladies' slippers, 1st, $1; Geo. J. Bailey, Jackson, display (best), dis. prem., medal.

ALFRED BAILEY,
J. B. ABBOTT,
S. V. HAWKES,
SABENS GARDNER,
Committee.

CLASS XLV.

Gilbert, Ransom & Knapp, Jackson, set chamber furniture, 1st, diploma and $10; Gilbert, Ransom, & Knapp, Jackson, 1 bureau, 1st, diploma and $3; Gilbert, Ransom & Knapp, Jackson, 1 sofo, 1st, diploma and $3; Gilbert, Ransom & Knapp, Jackson, set parlor chairs, 1st, diploma and $3; Gilbert, Ransom & Knapp, Jackson, office desk, 1st, diploma and $3; Gilbert, Ransom & Knapp, Jackson, book-case, 1st, diploma and $5; Gilbert, Ransom & Knapp, Jackson, child's carriage, 1st, diploma; George Smith, Detroit, set of fine parlor furniture, 1st, $10; C. G. Herrington, Northville, double and single, single and improved school seat and desk, $5; R. Beal, Commerce, window-curtain fixtures, $1; B. Partello, Detroit, 2 spiral-spring beds, 1st, diploma; George Smith, Detroit, 1 French-carom billiard table and fancy cues and balls, diploma and $10.

W. W. WITHINGTON,
C. S. TERREL,
J. C. E. HANFORD,
G. W. REMINGTON,
Committee.

CLASS XLVII.

Davis Sewing Machine Co., Watertown, N. Y., double-thread sewing machine for family use, 1st, medal; Davis Sewing Machine Co., Watertown, N. Y., sewing machine for manufacturing purposes, 1st, medal; Davis Sewing Machine Co., Watertown, N. Y., display of sewing machines, 1st, medal; A. M. Leslie, Chicago, Ill., ruffler for sewing machines, diploma; B. Eldridge, Detroit, double-thread sewing machines for family use, diploma; B. Eldridge, Detroit, hemmer-attachment sewing machines, diploma; B. Eldridge, Detroit, display of family sewing work, 1st, medal; Lamb Knitting Machine Co., Chickopee Falls, Mass., 2 Lamb knitting machines, 1st, diploma; C. Hill & Co., Laingsburg, work done by knitting machine, medal.

ALLEN WARREN, *Chairman.*
HARRIET E. RUSSELL,
MISS E. M. STERLING,
DELIA HAZELTON.

DIVISION H.

CLASS XLVIII.

Western Art Association, Detroit, collection of oil paintings (41 paintings), 1st, silver medal and $5; Western Art Association, Detroit, collection of chromos, 1st, $5; Western Art Association, Detroit, painting in oil, Below Silver Cascade (professional), 1st, $8; Western Art Association, Detroit, painting in oil, Wreck Cliff (Mich. artist), 2d, $3; Western Art Association, Detroit, landscape, Summer (Michigan artist), 1st, $5; Western Art Association, Detroit, landscape, Hill Glenn, Scotland (Mich. artist), 2d, $3; Western Art Association, Detroit, animal portrait, Caught at Last (Michigan artist), 1st, $5; Western Art Association, Detroit, animal portrait, At Home (Michigan artist), 2d, $3; Western Art Association, Detroit, water-color, View in Antwerp (Mich. artist), 2d, $2; Western Art Association, Detroit, oil, Morning Prayer (amateur), 2d, $3; Western Art Association, Detroit, oil, Winter (amateur), 1st, $8; Western Art Association, Detroit, animal piece, oil, Rich and Poor (amateur), 1st, $5; Western Art Association, Detroit, landscape, in oil, Morning in the Alps (amateur), 1st, $5; Western Art Association, Detroit, fruit and flower pieces, in oil (amateur), 1st, $3; Western Art Association, Detroit, fruit and peaches, in oil (amateur), 2d, $2; Western Art Association, Detroit, water-colors, original, Gardner's Shed (amateur), 2d, $2; Western Art Association, Detroit, water-colors, bird painting, Golden-Crested Wren (amateur), 1st, $3; Western Art Association, Detroit, pencil drawing, Fawn (amateur), 1st, $1; Western Art Association, Detroit, water-color, Bluebird (amateur), 2d, $2; Goldsmith's Bryant & Stratton School, Detroit, 3 specimens business penmanship, 1st, diploma; Mrs. D. Merriman, Jackson, specimen wood carving, 1st, $3; Mrs. D. Merriman, Jackson, specimen mosaic wood-work, table and chairs from Switzerland, 1st, $3; Mrs. W. R. Gibson, Jackson, original water-color painting, 1st, $3; Mrs. Dr. Nims, Jackson, oil landscape, Swiss Scenery (amateur), 2d, $3; Mrs. Dr. Nims, Jackson, oil, The Squirrels (amateur), 2d, $3; Mrs. R. Geer, Detroit, portrait from life (professional) dis., $3; M. H. Kerr, Jackson, best colored portraits, oil (professional), 1st, $3; J. S. Bird, Ann Arbor, specimen of architectural drawing, diploma; Western Art Association, Detroit, water-colors, Admiration (amateur), 1st, $3; Mrs. Dr. Hunt, Jackson, landscape, colored crayon (amateur), 1st, $5; Mrs. Dr. Hunt, Jackson, black crayon portrait (amateur), 2d, $3; A. C. Parsons, Ann Arbor, ornamental penmanship, diploma; T. Y. Homes, Camden, collection of stuffed birds, diploma; Mrs. J. W. Hough, Jackson, collection of photographs, $5; J. K. Trego, Ypsilanti, group of children medal; E. B. Smith & Co., Jackson, display of elegantly-bound books and stationery, diploma.

A. D. PERKINS,
W. W. WHITTINGTON,
M. E. WALDEN,
A. B. BAXTER,
MRS. W. R. GIBSON,
MRS. W. D. THOMPSON,
Committee.

CLASS XLIX.

A. H. Brown, Jackson, reed organ, church and school, 2d, diploma; A. H. Brown, Jackson, two-reed organ, general use, 1st, medal, $5; H. Badger & Bro., Jackson, reed organ, church or school use, 1st, medal, $5; H. Badger & Bro., Jackson, reed organ, parlor use, 2d, diploma; H. Badger & Bro., Jackson, melodeon of any make, reported for 1st premium; C. J. Whitney & Co., Detroit, orchestra concert grand piano, 1st, medal, $10; C. J. Whitney & Co., Detroit, best square grand piano for parlor use, 1st, medal, $10; C. J. Whitney & Co., Detroit, best square grand piano for home use, 1st, medal, $10; C. J. Whitney & Co., Detroit, combination organ, church and school, 1st, medal, $5; C. J. Whitney & Co., Detroit, organ for parlor use, 2d, diploma; C. J. Whitney & Co., Detroit, organ for parlor use, 1st, medal, $5; C. J. Whitney & Co., Detroit, pipe organ, church use, 1st, medal, $10; C. J. Whitney & Co., Detroit, portable pipe organ, church use, 1st, medal, $10; C. J. Whitney & Co., Detroit, portable pipe organ, parlor use, 1st, medal, $10; C. J. Whitney & Co., Detroit, melodeon any make, reported for 1st premium; C. J. Whitney & Co., Detroit, assortment of instruments for orchestra, 1st, medal, $5; C. J. Whitney & Co., Detroit, assortment of instruments for band, 1st, medal, $5; E. A. Jones, Sturgis, one square piano, reported for 1st premium.

G. T. Ganthur & Co. entered their church organ, manufactured at Ann Arbor, after the Examining Committee had closed the examination of Class 49. We would earnestly call the attention of the Executive Committee to their organ as one worthy of merit and their consideration, and would recommend a discretionary premium to them.

DR. TOMPKINS, Cassopolis,
MRS. WORTH, Jackson,
MRS. LOVELL,
Committee.

CLASS L.

Sermon & Beebe, Jackson, case hardware, bronze and plain, dip.; Sermon & Beebe, Jackson, one exhibit of plain and plated ware, dip.; S. F. Bancroft, Almont, artificial teeth, 1st, and dip., $2.

The committee find no competition in the list of articles in this class. Would name a first prize for artificial teeth,—and that a diploma be awarded for a case of bronze hardware and silver-plated ware.

T. H. GLENN,
L. I. LAING.
MRS. A. W. V. PANTLIND,
Committee.

CLASS LI—MISCELLANEOUS.

Mrs. S. Brownell, Utica, crochet chair cover, 1st, $3; Mrs. Chas. Ford, Jackson, specimen wax-work, 1st, $3; Mrs. Chas. Ford, Jackson, specimen wax-work, 2d, $2; Mrs. Chas. Ford, Jackson, crochet shawl, 1st, $3; Mrs. H. A. Curtis, Jackson, velvet sofa pillow, dip.; Cornelia Jameson, Jackson, group wax flowers, 2d, $2; Miss Josie M. Trefrey, Jackson, grotto, or Swiss cottage, 2d, $3 and dip.; Miss Josie M. Trefrey, Jackson, basket tissue flowers, $3; Mrs. D. Merriman, for Maria Miller, Jackson, specimen plain sewing by girl 11 years old, 1st, $3 and dip., Mrs. D. Merriman, Jackson, American flag, knitted, dip.; Mrs. D. Merriman, Jackson, ornamental wreath, leaves grapes, 1st, $2; Mrs. D. Merriman, Jackson, Lord's Prayer, German text, needle-work, $1; Louisa Dart, Jackson, specimen darned stockings, 1st, $3 and dip., Mrs. J. B. Tuttle, Jackson, velvet patch-work chair bottom, $1; Mrs. J. B. Tuttle, Jackson, display feather fly-brushes, dip.; Mrs. M. B. Blintenbach, Jackson, pair embroidered pillow-cases, dip.; Mrs. M. B. Blintenbach, Jackson, infant's embroidered silk skirt, 2d, $2; Mrs. D. W. Maltby, Jackson, specimen of worsted and bead-work pincushion, $2; Mrs. D. W. Maltby, Jackson, specimen worsted-work, album mat, dip.; Miss Hattie Stern, Jackson, Mexican needle-work, dip.; Mrs. H. A. Curtis, Jackson, pair embroidered slippers, $1; Mrs. W. K. Gibson, Jackson, embroidered cushion cover, $1; Mrs. W. K. Gibson, Jackson, chair cover, worsted-work, $1; Mrs. W. K. Gibson, Jackson, silk corded tea mat, made by lady 71 years old, $1; Mrs. W. K. Gibson, Jackson, cross of autumn leaves, dip.; Mrs. W. K. Gibson, Jackson, phantom bouquet, $1; Henrietta Prell, Detroit, embroidered night-dress, 2d, $2; Miss Lizzie T. Beebe Jackson, infant's linen dress and infant's Nansook dress, dip.: Miss Lizzie T. Beebe, Jackson, embridered skirt, 1st, $3; Miss Sylvia Palmer, Jackson, Cornell's self-fitting dress cloak, dip.; Mrs. Dr. Nims, Jackson, childs' Affghan robe, $1; Mrs. Dr. Nims, Jackson, embroidered crochet shawl, 2d, $2; Mrs. Dr. Nims, Jackson, specimen lace embroidery, 1st, $3; Mrs. Dr. Nims, Jackson, specimen silk embroidery, dip.; Mrs. Dr. Nims, Jackson, specimen hair jewelry, 1st, $3; Mrs. Dr. Nims, Jackson, two specimens embroidered undergarments, 1st, $3; Mrs. S. J. Springer, Farmington, crochet chair cover (wool), 2d, $2; Miss Hattie Sterne, Jackson, specimen braided work, 1st, $3; Mrs. John McConnell, Jackson, specimen worsted-work, 2d, $2; Mrs. Charles Ford, Jackson, group wax flowers, 1st, $3; Miss Julia Fish, Jackson, child's crochet Affghan, dip.; Miss Ruth Root, Jackson, case butterflies, dip.; Mrs. E. F. Whitmore, Jackson, specimen shell and hair-work, 1st, $3; Mrs. L. Eggleston, Jackson, embroidered tidy, $1; Mrs. L. Eggleston, Jackson, worsted horse embroidery, dip.; Mrs. L. Eggleston, Jackson, chenille embroidered chair seat, $1; E. Kellerman, Grand Rapids, collection moss goods, dip. and $3; Mrs. W. K. Gibson, Jackson, gent's crochet scarf, 1st, $3; Mrs. James Ferguson, Parma, worsted wreath, 1st, $3; Miss P. M. Sterns, Detroit, collection silk embroidery, 1st, dip. and $3; Miss Ritta Orr, Jackson, worsted tidy, dip.; Mrs. E. P. Ranney, Lansing, silk embroidered long shawl, 2d, $2; J. W. Collins, Jackson, collection fancy articles, dip.; Eugene Frink, Wenona, Japanese wool-work, dip.; Mrs. M. Shoemaker, Jackson, finest exhibition of lace, 1st, dip. and $5; Mrs. D. Peabody, Albion, Affghan robe, 2d, $2; Clara Hering, Jackson, tuffted worsted-work, dip.; Mrs. W. D. Thompson, Jackson, specimen Berlin-wool work, 1st, $3; Mrs. Irena A. Gross, Jackson, crochet embroidered woolen shawl, 2d, $2; Mrs. W. Hough, Jackson, embroidered table cover, 2d, $2; J. Titus, Dansville, agritultural wreath, dip.; S. Fenno, Coldwater, fancy wool wreath, dip.; E. Blanchard, Jackson, insect tree, dip.; A. M. Lersly, Chicago, Ill., fancy ruffling, dip.; Clara Grinnell, Jackson, pair bead brackets, dip. and $1; Mrs. O. G. Hunter, Jackson, piece needle-work, dip.; Mrs. Dr. Kent, Jackson, crochet table cover, 1st, $3; Mrs. Dr. Kent, Jackson, embroidered collar and cuffs, dip.; Mrs. W. H. Kimball, Adrian, carriage Affghan, 1st, $3; Mrs. W. H. Kimball, Adrian, gent's wool scarf, 2d, $2; Mrs. O. H. Heath, Adrian, embroidered ottoman cover, 1st, $3; Mrs. Carrie Thompson, Ann Arbor, feather wreath, dip.; Mrs. Whallen, Fitchburg, case wax fruit and flowers, 2d, $2; Mrs. R. D. Hendee, Jackson, fancy wool lamp mat, 2d, $2; Mrs. D. J. Workum, Detroit, ornamental needle-work, 1st, dip. and $2; E. B. Raney, Lansing, linen handkerchief embroidery, 2d, and dip.; Miss Mess, Detroit, infant's dress, 1st, $3; Miss Mess, Detroit, collection of embroidery, $6; Miss Mess, Detroit, ladies' sack embroidered, 2d, $2; Miss Mess, Detroit, hardkerchief, dip.; T. J. Holmes, Camden, 4 boxes stuffed birds and animals, 1st, dip. and $3; Mrs. M. H. Kimball, Adrian, otoman cover, 2d, $2.

Your committee would respectfully report that a very fine display was made in Division H, Class 57, of needle, shell, wax, and fancy work. A great variety of meritorious work is entirely unprovided for in the published premium list,—which your committee suggests needs a careful revision,—making it necessary to recommend many discretionary premiums and diplomas. Very many articles of real merit were wrongfully entered, and where competition is close, and the entries numerous, as was the case with many articles, the committee could only follow the rule, and exclude the articles improperly entered. Your committee would be glad, if time would permit, to notice specially many articles entered, which did not receive a premium. Among such, were some beautiful specimens of Mexican needle-work, shown by Miss Hattie Sterne of Jackson; linen under-garments, by Mrs. Curtiss; a collection of Rocky Mountain curiosities, by J. W. Welch, not properly entered, and many articles of embroidery and crochet-work, shown by different persons. The labor and responsibility of awarding premiums in this class are by no means light, and while your committee are not sure that their judgment has been entirely correct in all cases, they can confidently say it has been impartial, and without fear or favor. All of which is respectfully submitted.

W. K. GIBSON,
MRS. H. KIMBALL,
MRS DR. NIMS,
MRS. A. V. PANTLIND,
Committee.

DIVISION I.

CLASS LII.

Simpson, Loud & Trask, Adrian, 20 greenhouse plants, 1st, $10; Simpson, Loud & Trask, Adrian, 10 greenhouse plants, 2d, $3; Simpson, Loud & Trask, Adrian, variety roses, 1st, $5; Simpson, Loud & Trask, Adrian, dahlias, 3d, $1; Simpson, Loud & Trask, Adrian, German asters, 1st, $3; Simpson, Loud & Trask, Adrian, perrennial phloxes, 2d, $2; Simpson, Loud & Trask, Adrian, seedling verbenas, 2d, $2; Simpson, Loud & Trask, Adrian, collection of gladioli, 2d, $2; Simpson, Loud & Trask, Adrian, collection of pansies, 2d, $2; Simpson, Loud & Trask, Adrian, hardy annual flowers, 1st, $5; Simpson, Loud & Trask, Adrian, 6 ornamental-leaved plants, 1st, $5; Mrs. M. J. Treffey, Jackson, 20 greenhouse plants, 2d, $5; Mrs. M. J. Treffey, Jackson, 10 greenhouse plants, 1st, $5; Mrs. M. J. Treffey, Jackson, seedling phloxes, 1st, $2; Mrs. M. J. Treffey, Jackson, collection antirrhinums, 1st, $3; Mrs. M. J. Treffey, Jackson, collection of pansies, 1st, $3; Mrs. M. J. Treffey, Jackson, round bouquet, dis., $1; Mrs. M. J. Treffey, Jackson, 6 varieties of caladium, dis. premium recommended, $5; E. Armstrong, Jackson, variety of flowers, skillfully grouped, 1st, $10; E. Atmstrong, Jackson, variety of dahlias, 2d, $2; E. Armstrong, Jackson, verbenas, 1st, $3; E. Armstrong, Jackson, floral designs, 2d, $3; E. Armstrong, Jackson, seedling verbenas, 1st, $2; John Ford & Son, Detroit, variety roses, 2d, $3; John Ford & Son, Detroit, variety dahlias, 1st, $4; John Ford & Son, Detroit, 12 varieties dahlias, 1st, $2; John Ford & Son, Detroit, col. German asters, 2d, $2; John Ford & Son, Detroit, variety phloxes, perennial, 1st, $3; John Ford & Son, Detroit, variety verbenas, 2d, $2; John Ford & Son, Detroit, col. of antirrhiums, 2d, $2; John Ford & Son, Detroit, round bouquet, 1st, $1; John Ford & Son, Detroit, basket flowers, 1st, $2; John Ford & Son, Detroit, flat bouquet, 1st, $1.

S. O. KNAPP,
MRS. E. H. REYNOLDS,
MRS. W. H. WITHINGTON,
MRS. D. M. UHL,
MRS. C. FORD,
Committee.

CLASS LIII.

Mrs. E. Cooley, Jackson, 20 green hot-house plants, 1st, $10; Mrs. E. Cooley, Jackson, 20 green hot-house plants, 2d, $5; Mrs. E. Cooley, Jackson, 10 green hot-house plants. 1st, $5; Mrs. E. Cooley, Jackson, 6 ornamental-leaved plants, suitable bedding, 1st, $5; Mrs. Dwight Merriman, Jackson, flowers skillfully grouped, 2d, $6; Mrs. Dwight Merriman, Jackson, German asters, 2d, $2; Miss Louisa Dart, Jackson, hardy annual flowering plants, 1st, $5; Mrs. J. B. Tuttle, Jackson, basket flowers, 1st, $2; J. B. Tuttle, Jackson, 2 Acacia trees, (discretionary), $1; Mrs. Amos Van Aiken, Hillsdale, collection China asters. 1st, $3; Mrs. Amos Van Aiken, Hillsdale, collection zinneas, 1st, $2; Mrs. Amos Van Aiken, Hillsdale, everlasting flowers, (discretionary), $1; Peter Coller, Adrian, dahlias, 1st, $2; Peter Coller, Adrian, 6 pinks, 1st, $2; Mrs. James Vale, Jackson, flowers grouped, 3d, $3; Mrs. Nathan Smith, Adrian, floral design, 2d, $3; Mrs. Nathan Smith, Adrian, variety of flowers grouped, 1st, $10; Mrs. Nathan Smith, Adrian, 6 stocks, 1st, $2; Mrs. Nathan Smith, Adrian, 6 petunias, 1st, $2; Mrs. Nathan Smith, Adrian, tropcolumne, 1st, $2; Mrs. Nathan Smith, Adrian, seedling verbenas, 1st, $2.

S. O. KNAPP.
MRS. D. M. UHL,
MRS. E. H. REYNOLDS,
MRS. C. FORD.
Committee.

CLASS LIV.

Joseph Gridley, Kalamazoo, Eaton County collection of fruits, 1st, $100; E. H. Reynolds, Lewis & Co., Monroe, Monroe County collection of fruits, 2d, $50; N. & C. Chilson, Battle Creek, Calhoun County collection of fruits, 3d, $25.

J. P. WILLIAMS,
J. SHORTWELL,
GEO. WRIGHT,
B. W. VERREY,
Committee.

CLASS LV.

N. & C. Chilson, Battle Creek, best summer seedling apples, 1st, Downing's Fruit Book; Geo. Webb, Jackson, everbearing raspberry, of the Antwerp variety, discretionary premium; Seedling peach of Peter Coller, Adrian, grown from seed, seems to be hardy and productive, ripening late. Committee report, large size, fair quality, possessing no special merit to recommend its culture; No. 235, exhibited by N. & C. Chilson, Battle Creek, history of the tree—set for early Crawford, broken down by accident, tree sprouted below the place of budding, which this year, the first fruiting, produced the specimens exhibited. Report by Committee: description—medium size, yellow flesh color, red-mottled and yellow, clingstone, considered best seedling peach and worthy of cultivation. Committee named it Chilson's Cling; No. 207 is worthy of further trial and discretionary premium.

The Committee on Seedlings beg leave to report that they have found quite a numerous representation of seedlings on exhibition, of apples and peaches, but on examination found no apples worthy the notice, and much less the cultivation, and but one, exhibited by N. & C. Chilson, of Battle Creek, which for color and flavor merits the attention of fruit culturists, but its merits are modified by its being a clingstone.

D. W. ABRAMS, Paw Paw,
ISRAEL PENNINGTON,
I. M. HARWOOD, Jackson,
Committee.

CLASS LVI.

J. Brown, Battle Creek, exhibit winter apples, 1st, $10; J. Brown, Battle Creek, 12 specimens single variety winter apples, 2d, $5; O. C. Clements, Richfield, 20 varieties apples, 1st, $20; O. C. Clements, Richfield, 12 specimens single variety winter apples, medal; W. A. Holcomb, Jackson, 12 varieties apples, 1st, $10; W. A. Holcomb, Jackson, 6 varieties apples, recommend 2d premium; W. A. Holcomb, Jackson, 12 specimen summer apples, 1st, $3; W. H. Eddy, Bay City, 20 varieties apples, successive, 1st, $20; Peter Coller, Adrian, 26 varieties apples, successive,—recommended 2d premium; Peter Coller, Adrian, 6 varieties apples, successive, 1st, $7; Peter Coller, 12 specimens single variety autumn apples, 2d, $3; Peter Coller, Adrian, exhibit winter apples, 1st, $10; Peter Coller, Adrian, 12 specimens apples, single variety, medal; H. J. Crego, Liberty, exhibit autumn apples, 1st, $4; H. J. Crego, Liberty, 22 specimens single variety summer apples, 2d, $2; H. J. Crego, Liberty, 12 specimens single variety autumn apples, 2d, $2; O. H. Hicks, Sandstone, exhibit autumn apples, 2d, $2; Benj. Cook, Grass Lake, exhibit summer apples 1st, $4; Mrs. S. Dilly, Napoleon, 12 single specimens autumn apples, 3d, $1.

Your committee report that they found a very large and fine assortment of apples. In many cases it was difficult to find some of the entries on account of their being so mixed up and not properly labeled.

C. C. BEAHAN,
H. B. CHAPMAN,
H. S. PARMELEE,
Committee.

CLASS LVII.

S. O. Knapp, Jackson, 12 varieties pears, 1st, $10; S. O. Knapp, Jackson, collection autumn pears, 2d, $5; Peter Coller, Adrian, collection autumn pears, 1st, $7; Peter

Coller, Adrian, 6 specimens single variety autumn pears, 1st, $2; Mrs. Sarah Woodard, Grass Lake, 6 specimens summer pears, $2.

DARIUS BOYNTON, Benton Harbor,
HENRY ALCOTT, Sandstone,
F. F. RICHARDSON, Parma,
Committee.

CLASS LVIII.

J. T. Weeks, Napoleon, best six varieties of peaches, successive, 1st, $3; Mrs. T. Woodard, Grass Lake, best 4 varieties of clingstones, 1st, $2; J. G. Hartman, Benton Harbor, best 6 specimens freestones, 1st, $1; W. H. Eddy, Bay City, single variety, 6 specimens, clingstones, 1st, $1; Mrs. Dwight Merriman, Jackson, 12 quinces, 1st, $3; W. A. Holcomb, Jackson, 12 quinces, 2d, $2; Joseph Gridley, Kalamazoo, 12 quinces, 3d, $1; A. Crittenden, Leoni, 2 specimens white melon, 1st, $2; J. C. Smith, Charlotte, 2 specimens white melons, 2d, $1; J. Brown, Battle Creek, two musk-melons, 1st, $2; Henry Purdy, Jackson, two musk-melons, 2d, $2; J. C. Smith, Charlotte, 4 nutmegs, 1st, $2; J. C. Smith, Charlotte, best collection melons, 1st, $5.

J. L. MITCHELL,
JOSEPH C. WOOD,
Committee.

CLASS LIX.

N. & C. Chilson, Battle Creek, 20 varieties of apples (succession), 2d, $10; N. & C. Chilson, Battle Creek, exhibit of summer apples, 2d, $3; N. & C. Chilson, Battle Creek, exhibit of autumn apples, 2d, $3; N. & C. Chilson, Battle Creek, exhibit of winter apples, 2d, $3; N. & C. Chilson, Battle Creek, 12 specimens of single vartiey summer apples, 2d, $3; N. & C. Chilson, Battle Creek, 12 specimens of single variety autumn apples, 1st, $3; N. & C. Chilson, Battle Creek, 12 specimens of single variety winter apples, 1st, $10; Israel Pennington, Macon, 20 varieties of apples (succession), 1st, $20; Israel Pennington, Macon, exhibit of summer apples, 1st, $6; Israel Pennington, Macon, exhibit of autumn apples, 1st, $6; Israel Pennington, Macon, exhibit of winter apples, 1st, $10; Israel Pennington, Macon, 12 specimens of single variety summer apples, 2d, $2; Israel Pennington, Macon, 12 specimens of single variety of autumn apples, 2d, $2; Israel Pennington, Macon, 12 specimens of single variety of winter apples, 2d, $2.

Your Committee would say that the fruit in this department was very fine, and, in the opinion of this Committee, hardly surpassed by any exhibition of former years, and especially the specimens of apples exhibited by Mr. Pennington of Macon, and Messrs. Chilson of Battle Creek.

All of which is respectfully submitted.

H. DALE ADAMS, Clinton,
A. F. MOON,
J. L. MITCHELL,
Committee.

CLASS LX.

Israel Pennington, Macon, 6 specimens of single variety of summer pears, $2.

DARIUS BOYNTON, Benton Harbor,
HENRY ALCOTT, Sandstone,
F. F. RICHARDSON,
Committee.

CLASS LXII.

Charles Bowman, North Lansing, 1 peck cranberries, 1st, $2; J. M. Sterling, Monroe, 5 varieties of native grapes, 1st, $2; Point De Peau, Monroe, 8 varieties of native grapes, 1st, $3; Point De Peau, Monroe, collection native grapes, 1st, $6; D. O. Burras, North Fairfield, Ohio, collection of native grapes, 2d, $3; John Crane, Lockport, N. Y., 12 varieties native grapes, 1st, $5; W. D. Thompson, Jackson, 4 varieties native grapes, adapted to vineyard or general culture, 1st, $10; L. P. Reynolds, Hillsdale, best single variety, 1st, $2.

We, as your Committee on Grapes, think Mr. Crane should have the first premium for twelve varieties. He started with twelve varieties, but two varieties became damaged so badly that they were not fit for exhibition. We think he should have something for bringing such a good collection so far.

D. C. DRIGGS, Jackson,
JOSEPH WEIR, Monroe,
NATHANIEL CHILSON, Battle Creek,
E. H. REYNOLDS, Monroe,
Committee.

DIVISION K.

CLASS LXIII.

Jackson Fire Clay Co., Jackson, display of sewer pipe, dip.; Jackson Fire Clay Co., Jackson, display of Fire brick, dip.; W. H. Pendleton, Kalamazoo, iron grist-mill for horse-power, or steam, dip.; Francis M. Reasner, Jackson, specimen white lead in oil, dip.; Stillwell, Bierce & Co., Dayton, Ohio, lime extractor, heater and filter, silver medal; Frank & Co., Buffalo, N. Y., pony planer, silver medal; Frank & Co., Buffalo, N. Y., panel planer, silver medal; F. G. Chapman, room 9, M. E. C. block, Chicago, machine for making sash, doors, and blinds, silver medal; F. G. Chapman, room 9, M. E. C. block, Chicago, sand-papering machine, silver medal; E. Dennis, Jackson, engine lathe, silver medal; James Jenks, Detroit, double-cylinder planer and matcher, with beading attachment, silver medal; James Jenks, Detroit, four-sided moulding machine, silver medal; James Jenks, Detroit, self-feeding blind-slat tender, silver medal; James Jenks, Detroit, saw-gumming machine, dip.; James Jenks, Detroit, lot of emery wheels, dip.; E. Whiteside, Buffalo, N. Y., one ditching machine, dip.; E. M. Benster, Detroit, barrel-heading machine, silver medal; G. S. Wormer & Son, Detroit, one improved shingle and heading machine, silver medal; G. S. Wormer & Son, Detroit, one Blake & Co. patent steam pump, No. 5, silver medal; G. S. Wormer & Son, Detroit, one Sturtevant patent blower, dip; G. S. Wormer & Son, Detroit, one steam, gauge, dip.; G. S. Wormer & Son, Detroit, one Dobson's patent reflecting strainer, silver medal; G. S. Wormer & Son, Detroit, scroll saw, silver medal; G. S. Wormer & Son, Detroit, one compression coupling, dip.; G. S. Wormer & Son, Detroit, one patent self-oiling arbor, dip.; G. S. Wormer & Son, Detroit, one improved relishing machine, silver medal; G. S. Wormer & Son, Detroit, one side-sash stitcher, dip.; G. S. Wormer & Son, Detroit, Dreyfo's patent self-oilers, dip.; G. S. Wormer & Son, Detroit, one improved fanning machine, silver medal; G. S. Wormer & Son, Detroit, one improved foot-joiner, dip.; G. S. Wormer & Son, Detroit, one large patent Champion, dip.; G. S. Wormer & Son, Detroit, moulding machine, silver medal; L. Wilcox, Jackson, one coil rope, dip.; J. M. Groat, Almont, Lapeer Co., combined stove-hole, carpet-tool, etc., dip.; J. Adams, Jackson, 2 bbls. salt, dip. and $5; Thos. Baylis, Tecumseh, Am. Tur. water-wheel, Stout, Mills & Temples, dip.; H. Collis, Coldwater, one patent carriage-step, dip.; Mark Flanigan, Detroit, circular saws, smooth, dip.; National Fence Co., Detroit, 7 specimens combined fence, dip.; Willis F. Geer, Marshall, patent buggy-thill coupling, dip.; H. Brewer & Son, Tecumseh, machine for manufacturing drain tile, 1st, dip. and $10; S. Heiser & J. Harding, Jackson, machine for raising door-panels, silver medal; R. H. Kimball, Garrison House, Detroit, improved low-water indicator for boilers, dip.; Franklin McCounel, Dowagiac, model of piston-packing, dip.; H. M. Dresser, Jonesville, sample comp. packing, dip.; H. M. Dresser, Jonesville, preservative paint for shingles, dip.; J. C. McKenzie, Adrian, brick machine, dip.; Patterson & Comfort, Tecumseh, specimen drain tile, dip.; John H. Latimer, Jackson, mineral paint, dip.; H. G. Porter, Grand Rapids, sectional bent barrel, dip.; D. E. Rice, Detroit, 1 steam-governor, silver medal; G. S. Wormer & Son, Detroit, 1 flue-scraper, silver medal; D. L. Calkins, Waterloo, Indiana, artificial stove, dip.; D. E. Rice, Detroit, sample barrel-heading, dip.; S. H. Dickerson, Hudson, 1 imp. well-point and augur, dip.

GEO. C. MUNROE,
E. B. ROOT,
Committee.

CLASS LXIV.

P. Daniels, Jackson, railroad car dumper, dip.; E. H. Camp, Jackson, brass model of furnace for warming houses, dip.; A. F. R. Arndt, Cleveland, Ohio, passenger railroad check-box, dip.; S. T. Hayward, Chicago, Ill., Babcock fire extinguisher, medal; H. Earl Cole, Sully, Onondaga Co., N. Y., model of fruit-gatherer, dip.; D. G. Miller, Bucyrus, Ohio, one sash-lock, dip.; Sylvester Taylor, Coldwater, model of lumber wagon, dip.; O. W. Bean, Tecumseh, model of machine for handling leather in liquor, dip.; H. Collins, Coldwater, one Daggett's patent clothes-frame, dip.; J. H. Verity, Portland, one pair window-blinds, dip.; James Newman, Portland, model for water-wheel, 2d, dip.; Edward Racine, Chicago, Ill., model of wagon-wheel, 2d, dip.; John Skinner, Hadley, model of machine to connect gearing without stopping motive power, dip.; James York, Monroe, model of York's re-sawing attachment, dip.; G. C. Winters, Onondaga, model of self-supporting extension quilting-frame, dip.; T. H. Breed, Dundee, model of flood-gate, dip.; Benjamin Baker, Addison, model of grain measure and tally, dip.; Stillwell, Bierce & Co., Dayton, Ohio, model of eclipse turbine water-wheel, dip.; F. Van Doren, Adrian, one shingling-stool, dip.; D. P. Griffes, Flint, model of excelsior log-carting machine, dip.; H. M. Dresser, Jonesville, model of shifting-rail for top buggy, dip.; H. Phinney, Lansing, model of horse hay-fork, 1st, dip.; W. Adams, Charlotte, bench shears for cutting iron, dip.; Theodore Munger, Detroit, model of farm gate, dip.; E. Warren, Ceresco, pick for dressing mill-stones, dip.; J. Vincent, Rome City, Ind., model of self-acting brake, dip.; A. H. Bryant, 103 West Lake street, Chicago, Ill., one champion egg-carrier, dip.; A. H. Bryant, 103 West Lake street, Chicago, one corn-husking glove, dip.

F. M. FOSTER,
Chairman of Committee.

SPECIAL REPORT ON FIRE EXTINGUISHER.

Your committee on miscellaneous articles having viewed the fire extinguisher, after having seen the trial of two respective machines on exhibition, have fully come to the conclusion that 59, Babcock machine, is entitled to the premium, whatever it may be.

JOHN STARKWEATHER,
HENRY UNDERWOOD,
ALLEN WARREN,
Committee.

CITIZENS' PREMIUMS.

1. Green horses that never started for money—mile heats, best 3 in 5: No report of this race has been returned, owing to protests against the winners, which are yet unexamined.

2. Horses four years old—mile heats, best 3 in 5 in harness: J. C. Deyo's Andy Johnson, b. s., 1st, $100; A. C. Fisk's Dexter, Jr., br. g., 2d, $50; A. J. Derby's Limber Jim, 3d, $25. Time—2:58, 2:58, 3:00.

3. Three-year-old colts—mile heats, in harness: R. Armstrong's b. s., Membrino Chief, Jr., 1st, $100; A. C. Fisk's b. g., Coldwater Billy, 2d, $25. Time—3:10¼, 3:07.

4. Horses of any age; running race. Two mile heats; weight for age: Henry Chappel's Alice Pool, 1st prize, $200; W. H. Chapel's Minnie, 2d prize, $50. Time—4:06¾, 4:06.

5. Horses that have never beaten 2:25—mile heats, best three in five: Charles Skinner's Dictator, 1st prize, $200; W. S. McLaughlin's Byron Chief, 2d prize, $70. D. B. Hibbard, Fanny, 3d prize, $30.

A. J. DEAN,
Executive Superintendent.

ANNUAL MEETING OF THE EXECUTIVE COMMITTEE AT JACKSON, FEBRUARY 14TH, 1871.

On Tuesday, February 14th, in pursuance of the call of the President, the members of the Executive Committee of the Michigan State Agricultural Society met at the Hibbard House, Jackson.

The President, W. G. Beckwith, of Dowagiac, called the Committee to order, and the following gentlemen answered the call of the roll: George W. Phillips, Romeo; C. W. Greene, Farmington; H. O. Hanford, Plymouth; John Gilbert, Ypsilanti; J. M. Sterling, Monroe; A. J. Dean, Adrian; W. J. Baxter, Jonesville; A. C. Fisk, Coldwater; A. V. Pantlind, East Saginaw; M. Shoemaker, Jackson; F. M. Manning,

Paw Paw; L. S. Scranton, Grand Rapids; A. Stout, St. Johns; E. W. Rising, Richfield; E. O. Humphrey, Treasurer; R. F. Johnson, Secretary.

The President then made his address to the Committee.

ADDRESS OF THE PRESIDENT, W. G. BECKWITH, ESQ., OF DOWAGIAC.

The Executive Committee of the State Agricultural Society met in Jackson on Tuesday, the 14th, and herewith we present the annual address of the President. It contains many suggestions both of interest and importance to the farmers of Michigan, and will well repay a perusal:

GENTLEMEN OF THE EXECUTIVE COMMITTEE—Pursuant to notice, you have assembled to close up the business of the State Agricultural Society for the year just passed, and to arrange its affairs for the new one upon which we have just entered.

How much of the success of this organization depends upon your deliberation, it would seem unnecessary for me to say. I take it for granted you fully comprehend the responsibility resting upon you, and will acquit yourselves to the satisfaction of those who represent the agricultural and mechanical interests of the State. Custom requires the presiding officer, at each of your annual meetings, to make such suggestions in relation to those interests as he may deem necessary or proper. Having discharged that duty annually during the last seven years, I do not feel that I have much to add that is new, or would prove really valuable. I must rely, therefore, mainly upon your practical good sense to carry the association successfully through the year 1871.

The Secretary not having furnished me with his annual report, showing the condition of our finances at this time, I am unable to speak very definitely in regard to them. For all details connected with the working of the society during the past year, I would refer you to that document when it shall be laid before you, entertaining no doubt but that you will find it in all respects satisfactory.

So far as the permanent location of the place or places for holding State Fairs are concerned, I entertain the same views expressed by me at our last annual meeting, and to which I would again call your attention. You will find them in the Eighth Annual Report of the Secretary of the State Board of Agriculture, page 341.

In regard to the premiums to be offered at the next State Fair, I would recommend a thorough revision of the list, with a view to its more complete equalization. The value and importance of the article of product should, as a general rule, guide you in fixing the amount; yet at the same time, those articles of product which enter into general use, and have always been regarded as indispensable to the happiness and comfort of the human family, though of small market value when compared with some kinds of stock and grain, should receive at your hands more encouragement than formerly. Such, in my opinion, are the garden products usually brought to perfection in this latitude. In examining the awards of premiums for last year, I find that $72 only was paid for grain, flour, and seeds; and for roots and vegetables, covering the entire products of the field and garden, only $106. This amount, when compared with the aggregate premiums in other classes on the list, is much too small. If Spain, Holland, France, and the British Isles can send to our shores their vegetable products, and undersell us at our seaboard towns, it would be well that we give market gardening more attention and encouragement than formerly.

In this State the demand for garden products will increase in a ratio hitherto unknown to our people. The reasons are obvious. Within the last few years a number of long lines of railroads have been projected and are in process of completion, reaching out in every direction from the great lakes on either side of us, and every few miles along each line of road will spring up a thriving village, to absorb all the garden products that the adjacent neighborhoods can furnish. As the country grows older, the tendency of our population is toward cities and villages. This disposition of the people to centralize will vastly increase the number of consumers, and consequently increase largely the demand for all garden products. I would recommend more liberal premiums in this department, and at the same time require larger quantities on exhibition of each article offered for premiums.

In regard to wheat and other grains, I would require not less than five bushels on exhibition of each kind offered for premium by one individual, the wheat to be grown on land producing not less than 20 bushels to the acre. I am of the opinion that upon all grains and valuable seeds, the amount of premium should be considerably higher than formerly.

On occasions similar to this, I have, at some length, called attention to the propriety of encouraging the farmers of Michigan to breed that class of cattle which experience has fully proved to be best adapted to our climate and general wants, rather than a class of fancy cattle, such as Alderney and Ayrshire, which, while they possess some excellent points, are upon the whole less profitable to the general farmer than larger and more vigorous breeds. What an intelligent farming community have settled down upon as the best and most profitable cattle to raise in their midst, may safely be taken as a guide by you in making up the premium on that kind of stock.

I hold to views much the same in regard to horses. For years, the capital, time, and skill of breeders have been directed to the so-called improvements of our horses, and yet if we look back for the past twenty years, or since the organization of this Society, we have not produced a race, trotting, or turf horse that has repaid the breeders, in either pocket or reputation, for the time expended. Meanwhile the horse of all work has been materially improved, and I think it would have been much more had the same amount of time and money been given to him. I am of the opinion that this Society should use its intelligence and its premiums to promote the establishment of a vigorous race of horses; one full of stamina to withstand the vicissitudes of climate, of a constitution that would protect it against disease, with bone, substance, power, adapted to all the various services that may be required of it; with size and proportion as well as form sufficient to satisfy the demand for first-class horses in the various markets, and of a style and action that will be satisfactory alike to those who are engaged in business in our towns and cities, or upon the farm.

I would recommend a considerable reduction from the amount usually expended upon race-tracks, by the Society; indeed, I can see no good results from the use of so large an amount of funds of the Society for such purposes. Every valuable point in the horse can be made apparent to good judges without such outlay.

I would also recommend the dispensing with the so-called citizen's premiums.

The proceeds of the sale of swine, both live and dressed, have become a very important item on the credit side of the farmer's account, in many parts of the State; especially is this the case in southern and western counties, where much care and attention is given to their breeding. I am of the opinion that no kind of stock kept by the farmers of Michigan show so marked an improvement within the last twenty years as swine. And, so far as I am able to judge, the pork of Southern Michigan is not second in point of value to any single commodity put into the market by its farmers. You, doubtless, will do all that seems necessary to stimulate the introduction of the best breeds throughout the State.

My attention has been called to the fact that there have existed during the past year combinations of individuals, having for their object the control and purchase of certain leading farm products, not only in Michigan, but in other States of the North and West, at prices below their actual value; that these combinations, in most cases, reach from the great emporiums of trade East to the Mississippi river, and not unfrequently do they count as their most faithful allies, long lines of railroad.

I am not aware that your committee can do more to repair the mischief than to express an opinion in regard to it; at the farthest, if the charges are found to be true, as stated, recommend to our farmers a course of action which may eventually cure the evil complained of.

I do not know of any better means for reaching it than for farmers to form counter combinations, with a view to control, when practicable, not only the price, but the time of sale and manner of transportation. It would be well for farmers of each neighborhood to meet and discuss the situation, in order to settle down upon some plan of action to circumvent the arrangements of those who seek to buy without rendering a fair equivalent. Of course, no general rule can be laid down applicable to all parts of the State.

REPORT OF THE SECRETARY FOR 1870.

The Secretary submitted his annual report, as follows:

To the President and Executive Committee of the State Agricultural Society:

The Secretary submits the following report of the business of the Michigan State Agricultural Society for the year ending December 31, 1870:

At the close of the year 1869, the Treasurer had in his hands, after all accounts had been settled and checks paid that had been presented, the following sums:

Cash on hand		$557 63	
Note due on lumber	$679 84		
Less discount	13 03		
		666 81	
			$1,224 44

During the year there has been received from the following named sources, as follows:

From proceeds of the exhibition of 1870	$10,039 44	
From Jackson subscriptions	419 70	
From rents of grounds	1,120 00	
From entries of horses for citizen's premiums	452 00	
From membership tickets	355 00	
From railroad subscriptions	903 53	
		13,289 67
		$14,514 11
Less amount in hands of Secretary uncollected		139 53
Total receipts		$14,374 58
Balance due Treasurer		1,539 37
		$15,913 95

The expenditures during the year have been as follows:

Paid business checks of 1869		$348 02	
Paid outstanding premiums of 1869		278 00	
Business checks of 1870 issued	$9,146 35		
Less amount not presented	193 42		
		8,952 93	
Premium checks of 1870 issued	$6,916 00		
Less amount not presented	576 00		
		6,340 00	
			$15,913 95

The details of the expenditures of the Society for the year are classified under the following heads, and the items will be found in the schedule of the accounts submitted with this report:

1. Expense of winter meeting of 1870	$200 00	
2. Two meetings of Executive Committee in summer	274 05	
		$474 05

CONSTRUCTION AND PREPARATION OF FAIR GROUNDS.

3. Lumber for buildings	$611 48
4. Glazing and windows	98 48
5. Hardware, nails, iron-work, etc.	159 10
6. Labor in clearing up fallen buildings	185 97
7. Labor on Floral Hall building	476 42
8. Labor on Manufacturer's Hall	223 55
9. Labor on halls, stables, and grounds	184 67
10. Whitewashing expenses	149 62
11. Clearing up and securing buildings	187 25
12. Trimmings and decoration of Floral Hall	341 45

13. Evergreens and work-women		$98.00	
14. Team work, hauling lumber and material		102 50	
15. Engine and freight		322 97	
16. Engineer in Mechanics Hall		40 87	
17. Fixtures for water supply		203 92	
			$3,386 25
18. Dowagiac band, music		$110 00	
19. Hay		256 74	
20. Straw		50 00	
21. Police, watch, and gate-keepers		313 35	
22. Marshals and Assistant Superintendents		89 50	
23. Feed expenses		23 50	
24. Advertising and publishing lists of awards		97 67	
25. Free Press, for printing premium lists		199 75	
26. Advertiser and Tribune, for large posters, horse-bills, and circulars		396 95	
27. Daily Post, posters and tickets		248 00	
28. Daily Union, entry cards, etc.		56 30	
29. Jackson Patriot, fair programmes		58 10	
30. Bill posting		22 50	
31. Telegraph, express, cartage		29 95	
32. Fair expenses, President's room, premium ribbons, drawbacks, crockery for Fruit Hall, watering track, etc.		386 10	
33. Expenses for Viewing Committees		431 80	
34. Assistants in Secretary's office		232 00	
35. Clerk in Secretary's office		70 30	
36. Treasurer's expenses at Fair	$224 50		
37. Interest, postage, and exchange account	77 58		
		302 08	
38. Insurance on buildings		60 00	
39. Stationery and blank books		79 10	
40. Postage and stamps		178 80	
41. Rent of office for 1869 and 1870		120 00	
42. Salary of Secretary		1,000 00	
			4,812 44
Total expenditures for the year			$8,622 74

The total amount of premium checks drawn and issued for 1870 are reported as follows:

DIVISION A—CATTLE.

Class 1. Shorthorns	$497 00	
Class 2. Devons	492 00	
Class 3. Herefords	135 00	
Class 4. Ayrshires	40 00	
Class 5. Alderneys, Galloways, etc.	210 00	
Class 6. Herds of thoroughbreds	200 00	
Class 7. Grades and natives	124 00	
Class 8. Working oxen and steers	78 00	
Class 9. Fat cattle	136 00	
		$1,912 00

DIVISION B—HORSES.

Class 10. Thoroughbreds	$195 00	
Class 11. Half thoroughbreds	186 00	
Class 12. Horses of all work	285 00	
Class 13. Roadsters and trotting stock	554 00	
Class 14. Draught horses	101 00	
Class 15. Carriage and buggy horses	179 00	
Class 16. Saddle horses	40 00	
Class 17. Matched horses	25 00	
Class 18. Jacks and mules	83 00	
		$1,648 00

DIVISION C—SHEEP, SWINE, AND POULTRY.

Class 19. Merinos and fine wools	$205 00	
Class 20. Southdowns and Middlewools	154 00	
Class 21. Cotswolds, Leicesters, and Longwools	202 00	
Class 22. Fat sheep	16 00	
Class 23. Swine	59 00	
Class 24. Poultry	37 00	
		$673 00

DIVISION D—FARM IMPLEMENTS.

Class 25. Plows and their trials	$32 00	
Class 26. Harrows, cultivators, etc	16 00	
Class 27. Seed drills, sowers, planters, etc	2 00	
Class 28. Haying and harvesting implements	26 00	
Class 29. Apparatus and machines for cleaning crops	13 00	
Class 30. Wagons, carriages, and sleighs	122 00	
Class 31. Barnyard articles	21 00	
Class 32. Dairy and household articles	18 00	
Class 33. Stoves and hollow ware	00 00	
		$250 00

DIVISION E—FIELD AND GARDEN PRODUCTS.

Class 34. Grain, flour, meal, seeds	$72 00	
Class 35. Roots and vegetables	104 00	
		$176 00

DIVISION F—DAIRY AND HOUSEHOLD MANUFACTURES.

Class 36. Butter and cheese	$46 00	
Class 37. Sugar, honey, beehives	21 00	
Class 38. Bread, pickles, preserves	46 00	
Class 39. Wines, vinegars, and cordials	96 90	
Class 40. Bitters, tobacco, soaps, groceries	00 00	
		$209 00

DIVISION G—DOMESTIC MANUFACTURES.

Class 41. Home-made goods	$78 00	
Class 42. Factory goods	45 00	
Class 43. Articles of dress	16 00	
Class 44. Articles of leather and India rubber	20 00	
Class 45. Articles of furniture	53 00	
Class 46. Ornamental iron work	00 00	
Class 47. Sewing machines	00 00	
		$212 00

DIVISION H—FINE ARTS AND NEEDLE-WORK.

Class 48. Paintings and statuary	$115 00	
Class 49. Musical instruments	90 00	
Class 50. Clocks, jewelry, plated ware	2 00	
Class 51. Needle, shell, wax, and other fancy work	130 00	
		$337 00

DIVISION I—FLOWERS AND FRUITS.

Class 52. Flowers, professional	$106 00	
Class 53. Flowers, amateur	75 00	
Class 54. County collections of fruits	175 00	
Class 55. Seedling fruits and new varieties	00 00	
Class 56. Apples, amateur	103 00	
Class 57. Pears, amateur	26 00	
Class 58. Peaches, plums, and other fruits, amateur	26 00	
Class 59. Apples, professional	88 00	
Class 60. Pears, professional	2 00	
Class 61. Peaches, plums and other fruits, professinal	00 00	
Class 62. Grapes	33 00	
		$634 00

DIVISION I—MISCELLANEOUS.

Class 63. Miscellaneous articles	$15 00	
Class 64. Models and small articles	00 00	
		$15 00
Citizens Premiums		850 00
Medal and diplomas		473 70
		$7,339 70

The total premiums, classified by divisions, show the following amounts, and compare with those paid in 1869 as follows:

	1869.	1870.
Division A. Cattle	$1,772 00	$1,912 00
" B. Horses	1,367 00	1,648 00
" C. Sheep, swine, and poultry	670 00	673 00
" D. Implements and their trials	263 00	250 00
" E. Field and garden products	81 00	176 00
" F. Dairy and household productions	194 00	209 00
" G. Manufactures	277 00	212 00
" H. Fine arts, needle-work, etc.	321 00	337 00
" I. Flowers and fruit	476 00	634 00
" K. Miscellaneous	14 00	15 00
" L. Winter premiums	....	....
Trials of speed for horses	955 00	850 00
Total premiums, cash	$6,390	$6,916 00
Medals and diplomas	499	473 70
Total premiums	$6,889 00	$7,389 70

The following table exhibits the number of entries in the several divisions for the past eight years:

	1863.	1864.	1865.	1866.	1867.	1868.	1869.	1870.
Cattle	155	66	86	140	132	118	190	174
Horses	252	227	269	263	269	230	308	401
Sheep, swine, and poultry	200	290	218	174	187	172	246	156
Implements	208	207	289	227	479	428	503	423
Seeds and vegetables	14	139	117	181	72	161	123	285
Dairy and household products	78	149	122	156	110	107	151	172
Manufactures	133	108	97	78	152	229	242	178
Fine arts	167	197	278	232	283	220	272	343
Flowers and fruits	256	75	231	303	169	246	258	290
Miscellaneous	21	59	72	87	277	240	228	133
Crops, farms, and orchards	1	2	18	14	16	17	9	....
	1,485	1,429	1,797	1,835	2,166	2,168	2,530	2,555

The annual exhibition for the year 1870 was held at Jackson, and though not so fully attended as the exhibition of 1869, which was held in the same city, it was generally regarded as very successful, and one of the best in some respects which had ever been held. The Society, however, was obliged to make larger expenditures than were expected to be necessary. During the preceding winter some of the buildings were injured, and the largest, Floral Hall, was crushed under the weight of snow. In the rebuilding and preparing the several structures for the exhibition, it will be seen that there was expended $3,386 25, while for the various expenses of the Fair and the Society for the year the amounts paid have been $5,236 49. The full amount of all premiums issued is $7,389 70, making the total expenses of the year as follows:

Cost of buildings and fixtures in 1870	$3,386 25
General expenses of exhibition and of the Society	5,236 49
Premiums awarded in 1870	7,389 70
	$15,012 44
The total receipts of the year	14,374 58
Excess of expenditures over receipts	$1,673 86

The Society has during the past two years invested its surplus largely on the grounds at Jackson for the purpose of making them of a more permanent character. During the year 1869, the Society invested in buildings and fixtures on the grounds at Jackson:

The lumber and buildings taken from Detroit, estimated at	$4,000 00
In new lumber and construction of buildings and track	8,411 98
	$12,411 98
During the year it has added to this	3,000 00
Making a total investment at Jackson of	$15,411 98
There is due the Society from the State for amount paid towards the construction of the Normal School museum at Ypsilanti	$3,250 00
There remains, therefore, as representing the property of the Society	$18,661 98

Should the annual exhibition be again held at Jackson, the buildings and fixtures are now in the very best order, and would relieve the Society from any very heavy outlay on that account, but should the system be adopted of holding exhibitions at several points in the State, these buildings at Jackson are in such a condition that they will remain in good order, and be ready for use whenever it is for the interest of the Society to use them. Hence they may be estimated as property that represents to a large extent the surplus funds of the Society. The city of Jackson from year to year is becoming one of the great central points to which railroads reach from all parts of the State, and also having very direct communication with other States, and being besides in the midst of a section of Michigan that is rich in agricultural resources, and which must be developed as progress is made in the improvement of farms and live stock. It is therefore a point to which the Society may look as one of those which a true regard for its future interests must select, as being adapted to hold the great annual exhibitions at such periods as may be deemed best for the interests of the agriculture of the State. The Jackson County Agricultural Society have placed their grounds at the command of the State Agricultural Society, free of expense, whenever they may be needed for the occupancy of the Society's buildings, and for the holding of the annual exhibitions. The investment made on these grounds, therefore, may be regarded as permanent.

During the past year the subject of establishing some definite system by which the benefits of the State Agricultural Society should be more generally diffused over the whole State, engaged the attention of the members of the Executive Committee, and of many friends of the Society, and a meeting of the committee was held during the summer to hear certain propositions which were sent in from several localities, the citizens of which were desirous that the State Agricultural Society should bring the annual exhibition within the reach of the people in the region which they represented. No definite result was reached, beyond the development of a friendly regard for the Society, and a desire to participate in its benefits, from various quarters.

It is easily to be seen that the eastern and southern sections of the State have been largely benefited by the State Society during the twenty years of its existence. The annual exhibition of 1870 indicates, by its register of the entries made in the divisions comprising the live stock, that though there were a few animals exhibited among the classes of the improved stock as coming from the west and northwest, yet that the great proportion of the entries of stock were from that section of the State where the State Society had for many years exerted the most commanding influence. The rapid development and progress in population and the improvement of land in the north and northwest, and the important railroad connections which have been made within the past two or three years, indicate that the State Agricultural Society should extend its benefits as much and as soon as possible in that direction. The benefits it has conferred on those sections of the State where its influence has been most felt, is to be seen in the fact that the best herds of improved cattle are mostly confined to such counties as lie east and south of Jackson, and that comparatively little attention has been given to the improvement of live stock wherever the Society has made its influence felt. Other divisions of the agriculture of the State might be referred to as exemplifying in some measure that the encouragement given to improvement by the State Society has been in some degree local, though it must be admitted that this has been practically owing to the position of the State, which for many years was without the advantage of railroad communication. Now, however, railroad communication, population, and the improvement of land, all combine to suggest that the time is coming when the State Agricultural Society will require locations that will permit its exhibitions to be changed every year, as it is in other States where State agricultural societies have been most successful, and that it will be sustained in its efforts by the State itself.

The agriculture of Michigan is undergoing a gradual change, and becoming of a much higher order in many sections. Still, a great proportion of labor and capital is expended in clearing the surface and in reducing unimproved lands to a condition that will fit them to become productive. But the interest that has sprung up during the past five years in regard to tile-draining, improved breeds of the domestic animals, and especially of cattle, implements and machinery that are better adapted for profitable farm work, and in regard to the production of fruit, is very marked, and ought to be encouraged by every means within the ability of the State Agricultural Society. In this respect it has kept pace with the organizations of other States much older, which have been aided by their own State Legislatures, as well as with the progress among its own members.

During the past year the Society has paid out larger premiums on its cattle classes, than any previous year, and the result has been that the quality of the cattle shown has been of a much higher and better grade. The amount paid for the exhibition of cattle has been greater than any other division of the premium list.

Last year the Society offered large premiums for the exhibition of fruits, and a new class was formed for collections of fruits from the several Congressional Districts of the State. The exhibition was very satisfactory, and brought out a very excellent show, which was highly reputable to the State, and in some degree presented a fair example of what the State is doing in this direction.

In swine, there has been much done during the past year to introduce the best and most profitable breeds; and what has been done has given a more thorough acquaintance with the quality of this stock. Still there is much remains but imperfectly understood, and it might be well to offer a class of premiums which would call out more information than is now prevalent. For instance, would it not be well to offer a series of premiums that would call for statements showing at what expense the several kinds of swine have been kept, or the amounts of food they have consumed up to a certain age, or to the time when they were exhibited? Size is not of so much importance in the hog as quality to produce pork.

Sheep have not become of less importance, but the sheep breeders are not so numerous as they were, and at the last exhibition there were not so many shown as usual. There is growing, however, a better knowledge of the necessity of growing the finer-wooled breeds pure and free from inferior blood, and also a more intimate knowledge of the properties and methods of treatment of the long-wooled breeds. The premiums now offered in this division are liberal, and require but little change.

Implements and machinery occupy such a prominent position in connection with the labor of the farm, that it is important that every facility should be given for their examination and trial at the State fairs. In the arrangements for the future, provision should be made for ample steam power and working gearing, so that all exhibitors may have such means placed at their disposal as will render the trials of the machines valuable tests. There have been many suggestions made on this subject, and hence the propriety of directing attention to it.

Respectfully submitted.

R. F. JOHNSTONE.

TREASURER'S REPORT.

The Treasurer made his annual report, as follows:

To the President and Executive Committee:

I herewith submit a statement of the business transactions of my office for the fiscal year ending December 31, 1870:

There was remaining on hand January 1, 1870			$557 63
Received from collection of note of H. N. Strong of Detroit			666 81
Total funds from 1869			$1,224 44
Receipts from annual exhibition of 1870		$10,039 44	
Receipts from collections by Secretary	$3,250 23		
Less amounts uncollected with Secretary	139 53		
		3,110 70	
			13,150 14
Total receipts			$13,374 58
Loaned by Treasurer to meet demands			1,539 37
			$15,913 95

PAID OUT BY TREASURER.

For business vouchers of 1869	$348 02	
Por premium checks of 1869	273 00	
For business checks of 1870	8,952 93	
For premium checks of 1870	6,340 00	
		$15,913 95

All of which is respectfully submitted.

E. O. HUMPHREY, *Treasurer*.

The reports of the Secretary and Treasurer were accepted.

Ordered, That the President's address and the Secretary's report be referred to a special committee of three members, and that so much of the Secretary's report as refers to the finances, and the Treasurer's report, be referred to the Committee on Finance.

The President named the following committees:

Special Committee on Address—Mr. Baxter, Mr. Shoemaker, Mr. Sterling.

Finance Committee—Mr. Dean, Mr. Rising, Mr. Gilbert.

Premium List—Mr. Phillips, Mr. Hanford, Mr. Manning, Mr. Stout, Mr. Greene.

The Committee then adjourned till Wednesday morning at 9 o'clock.

WEDNESDAY, February 15.

The Committee met at 9 o'clock.

Mr. Shoemaker, from the special committee on the address, made the following report:

The committee to which was referred the address of the President and the report of the Secretary, respectfully report:

First. That so much of the President's address and the report of the Secretary as refers to the finances of the Society, be referred to the Committee on Finances.

Second. That so much as refers to the premiums to be offered on animals and articles, and the relative amounts for each, with the remarks of the President relative to garden products, horses, swine, and other matters, be referred to the Committee on the Premium List.

Third. That so much as refers to the permanent location of the State Fair be referred to the committee now having that matter in charge, with instructions to report at as early a day as practicable.

Fourth. That so much as refers to the location of the State Fair for 1871 be referred to the committee of the whole.

Fifth. That so much as refers to the permanent property of the Society be referred to the Committee on Finances, and that all other subjects referred to in the address and report be referred to the committee of the whole for consideration.

The report was accepted, and its recommendations adopted.

Mr. Dean reported on the protests referred to him in connection with the awards of premiums made to the class of citizens' premiums for green horses. The report read that after having made as thorough an examination as possible, he found that only three horses were entered to compete in the race for green horses. This class was designed to afford an opportunity for those breeders who had young, untried horses, to compete in the matter of speed against each other, and to keep them outside and away from the sporting community, who had all the facilities for fitting and training horses for the track. The design of the Society should be carried out, and those who entered green horses in good faith should be protected and encouraged. While no unbred horse should be shut out because he belonged to the owner and breeder and trainer of trotting stock, still, all horses that were not up to the standard, or down to it, should be excluded. There were many breeders of good horses who would willingly bring their horses to the State Fair and exhibit them there, but they must be sure that they would have some chance for a real trial, and not be placed where they would have to yield to the superior dexterity and smartness of the skilled and experienced professional trotting man, who made it his business to carry his stable of horses from place to place, and did nothing else for a living. The breeders from the interior brought their horses in good faith to try them; and were they assured that they would have a fair competition among themselves, and not with the trained and hardened professional trotter, some of them would come out, and encouragement to the breeding of good horses be given by the State Society. It was from this class of horses that most of the stock was raised that was sold and went to the great markets East. The breeders on the farm raised the stock, and if we proposed to encourage improvement, and to increase the value of the roadsters of the State, the rules and requirements of the State Society at our annual exhibitions must be sustained. The Executive Committee had referred this whole subject to him to be investigated and reported upon, as he had given it considerable of his attention. From the correspondence he had had with a number of gentlemen throughout the State, he was satisfied there was a general feeling in favor of all efforts by the Society to promote improvement in the stock and in the breeding of the horses of the State; but the great question was how to do it without descending into the arena of the race-course; and it was here that the State Agricultural Society could use their commanding influence to advantage, and give that encouragement of premiums, popular applause, and general distinction which was so valuable and encouraging to the horse-breeder, and which animated him almost as much, if not more, than the profits of the business. At any rate, the distinction gained at State and county fairs helped to give eminence and profits that were anxiously looked for by the breeder of horses, and were an encouragement to introduce into the State the best stock. With these views, he would report that out of the eight entries made in the class of citizens' premiums for green horses at the State Fair of 1870, he had been able to find but three that were fairly entitled to be considered as within the meaning of the list, which provides that the premiums were offered "*for green horses that have never started for any premium whatever.*" To these three horses he had awarded the premiums, and therefore offered the following resolution:

Resolved, That the following awards be made in the class of citizens' premiums:

To J. Stephens, of Hillsdale, bay gelding, Bismarck, first premium, $100.

To A. J. Bennett, of Mason, cream gelding, second premium, $30.

To W. H. Sutton of Marshall, bay gelding, third premium, $20.

Respectfully submitted. A. J. DEAN.

Mr. H. B. Chapman was permitted to explain the circumstances of the protest in the class of matched horses, which was that the horses shown were not up to the standard weight.

Mr. Deyo explained that his horses had been weighed previous to the fair, and had then weighed more than 1,200 pounds each, but that one of them had been taken sick, and that it had lost weight without his knowledge.

Mr. Dean reported on this case that the awards had been suspended; and the horses of Mr. Deyo not coming up to the rules and directions of the Superintendent, there had been no award made, as it was impossible to make new awards after the fair passed.

On motion of Mr. Baxter the following resolution was adopted:

Resolved, That the premiums in Class 17, for horses weighing 1,200 pounds, be withheld until full evidence be produced before this committee that the entries are of the required weight, and that the gentleman to whom the second premium was awarded be requested to certify to Mr. Dean, Superintendent of Horses, relative to their weight.

Mr. Deyo presented a claim for a premium on four-year-old horses in Class 12, which had not been reported by the Viewing Committee. He accompanied his claim with a letter from the chairman of the committee, showing that the award was made but not recorded on the committee books.

Ordered, That the premium be paid to J. C. Deyo.

A letter was read from A. V. Pantlind, tendering his resignation on account of his removal from East Saginaw to Jackson.

On motion of Mr. Baxter the letter was read, and the resignation of Mr. Pantlind was not accepted.

A communication from the Hon. Jonathan Shearer, of Plymouth, on the subject of the preservation of the forest trees of the country, was received and read, and referred to the Committee on the Premium List.

A communication from J. J. Bagley, Esq., of Detroit, was received and read, relative to the propriety of offering a large number of premiums for agricultural produce. Referred to the Committee on the Premium List.

The committee then took a recess till seven o'clock, for the purpose of allowing the several committees to make up their reports.

At seven o'clock P. M. the Committee met.

The Committee on the Premium List reported and asked for more time, as they had not been able to get through the list, which was granted.

The subject of the locality of the next State Fair was then introduced, and the President announced that the Committee was ready to hear any propositions and suggestions from the members of the Society then present.

Mr. Scranton, of Grand Rapids, said the county society there have a fair ground of 24 acres, every foot of which is available, in addition to which they propose to lease a sufficient quantity adjoining for standing room for any number of horses and carriages of visitors; also have a good half-mile track, a floral hall in good condition, a large carriage-shed, large sheep-pens, and a good row of stabling, a grand stand

100 feet long and twenty-four rows of seats. The citizens think they can and will agree to raise $4,000 as an inducement to the committee to locate the fair in that place. The surrounding country will turn out well. Nearly a quarter of the area of the State which has never enjoyed any advantages of the State Fair, now desire that it shall be held in this, the second city of the State. The railroad facilities are as extensive as could be desired.

Mr. George W. Griggs, President of the Kent County Agricultural Society, supported, in a brief address, the claims of the northern and northwestern sections of the State to a participation in the benefits of the State Fairs, and called attention eo the rich agricultural counties lying north of the Milwaukee railroad, and the large number of inhabitants in those sections who would be attracted to such a meeting. An impetus was needed there to improve the stock and cattle, such as the southern counties had heretofore enjoyed, which had been worth thousands of dollars to them, and which, if similar facilities were extended to the northern counties, would equally benefit them. He referred to the location of the fair grounds, and the facilities that would be provided for reaching the same, and said that at least 15,000 people would attend a State Fair there who would not and could not come to any other locality proposed. He invited the Society to come there, and pledged every facility on behalf of the citizens and of the county society.

Dr. Foster Pratt, of Kalamazoo, submitted a statement on behalf of Mr. Reed, proprietor of the fair grounds in that city, which embrace sixty-five acres, of which twelve acres front upon the principal street, the rest being included within one of the best mile race tracks in the conntry. The hall bnilt by this society eight years ago is still in good condition; there are 150 or 200 stalls for horses; a grand stand 500 feet in length; and water-pipes leading to all portions of the ground. The railroad facilities are largely increased over those existing formerly, there being three different lines in operation; besides the Grand Rapids & Indiana Railroad, whose track runs within a few rods of the inclosure, will arrange to allow cars on all lines to land machinery and all articles as well as passengers immediately at the grounds. The grounds, with inclosures, buildings, etc., will be given by the proprietor of the same, who will make all necessary additions to buildings in consideration of receiving the lumber now in the buildings at Jackson at a fair valuation, or an equivalent in lumber at Kalamazoo. The citizens will subscribe at least $3,000 or $3,500.

Mr. Adams, of Kalamazoo, supported the claims and facilities of that place, but raised the question whether it would pay to put the lumber now in the buildings at Jackson into permanent structures.

Mr. Shoemaker, of Jackson, said that if the Committee should decide to remain in Jackson, the citizens would do whatever the Committee might deem right and proper as a money consideration, but, having had the fair two years, would not insist on holding it here against the desire of the Committee to remove it.

Mr. S. O. Knapp, of Jackson, called the attention of the committee to the interest of the State, the Society, and the Agricultural College at Lansing, where is located a State farm of 700 acres, all of which might in the not distant future be turned to good account in the establishment of permanent buildings for occasionally holding State fairs on those broad acres.

A member inquiring whether there was a race track there, Mr. Knapp began to say: "No, I am sorry there is n—" when, suddenly recalling himself, amid roars of laughter, he being somewhat opposed to racing as such, he concluded by saying that "in that respect the Society could *begin from the stump* and lay out to suit themselves."

Mr. Jonathan Shearer, of Plymouth, being present, responded in a hearty manner to a call for suggestions, paying a high compliment to the efficiency of this and similar societies in developing the resources and industrial pursuits of the country, stating that as an original member of the oldest Agricultural Society in the State, he had always had a deep interest in their success. He argued strongly for the protection of shade and forest trees.

Mr. Scranton said he was glad to have this subject brought before the Committee, and wished it might be presented in all it force and importance to the attention of the people. Timber is rapidly disappearing. At about the latitude of Grand Rapids is the limit of oak, hickory, and other woods useful for manufacturing purposes; black walnut, once so plenty, had nearly all disappeared, and will soon be worth its weight in gold unless means are taken to plant and protect a new growth So of other woods.

Mr. Adams wished that some means might be devised to prevent injury, by cattle running at large, of shade trees planted by the roadside according to the law passed by the State Legislature.

Mr. Wm. Taft, of Plymouth, in some brief remarks, thought the State Agricultural Society should be "passed around," that each section of the State might be benefited, as well as the Society itself. After several years at Detroit, it had some years ago gone to Kalamazoo, with good pecuniary results, and this might be done again with good effect. But he favored going to Grand Rapids this year, for the reasons so well stated by the gentleman from that city, and in view of the fact that the fair had never been held in that section of the State. Go to Kalamazoo next, perhaps. With respect to cattle running at large, he argued for a law totally prohibiting it, and proceeding to show that, instead of a benefit, it was a loss and damage to the owners of cattle themselves to allow the practice, and the sooner they ascertained this truth the better for them. As to shade trees by the roadside, he had, after much trouble and expense, succeeded in securing such in good condition across his premises.

Mr. Baxter stated that the law to restrain cattle from running in the highways was very defective, even where the supervisors attempted to carry it out. It was suggested that a resolution be passed by the Committee asking the Legislature to amend and make effective the law in question.

The President called up the matter of the claim of the Society for $3,250 invested some years ago in the Normal School building at Ypsilanti for the purposes of an agricultural museum, and stated that he had secured the introduction of a bill in the Legislature to refund that amount to the Society.

Mr. Sterling moved that the President and Mr. Baxter go to Lansing in the interest of the claim, and see that it is properly considered. The motion was carried, and Mr. Sterling added to the committee, on motion by C. W. Greene.

The committee then adjourned till Thursday morning at 8 o'clock.

Mr. Baxter called attention to the absence of a member who heretofore had never been absent from the counsels of the Committee, and took occasion to submit the following resolution, which, on motion, was unanimously adopted, with the request that it be published in the Detroit daily papers, and a copy be furnished to the family of the deceased:

Resolved, That we have heard with deep regret of the decease of Edwin Smith, late of Clinton, Lenawee County, in whom we have ever recognized an earnest friend of this Society, and one who has done very much to improve and advance the agricultural, mechanical, and manufacturing interests of the State. He was one of the pioneers of Michigan, and in the course of his long and useful life did much to develop the resources of our State, especially in the reclaiming of our low and wet lands by drainage, of the introduction of improved stock, seeds, and farm implements, and of late years in advancing the manufacturing interests of the State. In all the relations of life he was faithful to his ideal,—a man loved and respected by all who knew him.

THURSDAY, February 16.

Mr. Baxter submitted the draft of a memorial to the Legislature on the claim of the Society in the Normal School building matter, which was ordered engrossed, signed by the Executive Committee, and forwarded to Lansing.

The Committee on the Premium List, by Mr. Phillips, reported, and submitted their report to the Committee. The report was read, and then taken up for consideration and amended.

In Division A—Cattle—the premiums for Shorthorns (Class 1) and Devons (Class 2) are as follows:

Best bull, four years or over	$40 00
Second	20 00
Best bull, three years old	30 00
Second	15 00
Best bull, two years old	20 00
Second	10 00
Best bull, one year old	10 00
Second	5 00
Best bull calf	7 00
Second	5 00
Best cow, five years old and over	40 00
Second	20 00
Best cow, four years old	30 00
Second	15 00
Best heifer, three years old	20 00
Second	10 00
Best heifer, two years old	15 00
Second	8 00
Best yearling heifer	10 00
Second	5 00
Best heifer calf	7 00
Second	4 00

Herefords and Thoroughbreds (Class 3) have only first premiums, and are the same in amount, respectively, as those in Classes 1 and 2.

Ayrshires (Class 4) and Alderneys, Galloways, and Holstein Cattle (Class 5), commencing at "three years old," instead of four, are given the same relative premiums for the same ages, as above.

Class 6—Herds of Thorough-bred Cattle—is stricken out entirely.

Class 7—Grade and Native Cattle—the "ten best head of grade stock," "best pair grade steers," "best pair yearling steers," and "best three milch cows," etc., are stricken out, the class remaining as follows, with slight changes from last year:

Best grade or native cow, four years or over	$20 00
Second	10 00
Best heifer, three years old	15 00
Second	7 00
Best heifer, two years old	10 00
Second	5 00
Best yearling heifer	8 00
Second	4 00
Best heifer calf	5 00
Second	8 00

Class 8—Working Oxen and Steers—remains as last year, beginning with $20 for yoke of oxen five years or over, and reducing the premium to the "trainer of best exhibition of trained stock" from $10 to $5.

Class 9—Fat Cattle—remains as last year, the first premium being $50, except that herds are required to consist of four or more, instead of five, for the reason that feeders raise their animals generally in matched pairs.

At the suggestion of Mr. D. M. Uhl, of Ypsilanti, exhibitors showing cows as breeders will be required to exhibit proof that they *are* breeders, or have produced calves,—to prevent merely fat and barren cows, as has been the case, from competing with actual breeders and carrying off the prizes from the proper and worthy animals.

In Division B, Class 10, best and second best stallion of any age, shown with five of his colts, that has made a season in the State, with premiums, $100 and $50, are stricken out; the age of single stallions raised from four to five, at same premium as last year ($50 and $25), and the four-year-olds reduced to $40 and $20, respectively; the balance of this class to remain as last year.

The class for half thorough-breds was stricken out, as it was claimed that it had served its purpose of bringing to the observation of breeders the utility of the thorough-bred as a means of improvement in the horses of all work. Horses heretofore entered in this class will come under the class of all work.

To horses of all work, the class heretofore known as "matched horses," is added, so that the class of all work will include teams.

On motion of Mr. Dean, the following rule was adopted to govern the committee on this class:

"All entries in this class, where weight is one of the conditions, shall be accompanied with a reliable certificate, of date not exceeding one week previous to the opening of the exhibition, showing satisfactorily the weight of the animals that are entered for competition."

In Roadsters and Trotting Stock, the word "trotting" is stricken out wherever it occurs, and the two first entries,—best trotting stallion, five years old or over, with colts, and second best trotting Stallion, five years old or over,—with the premiums of $100 and $50, were stricken out, the balance remaining without change.

A new class was made up of a portion of this class, entitled "Driving Horses," including all horses and teams shown as roadsters, and the following rule was adopted:

"In this class no teams or single horses can be shown that are in use as sporting horses, or horses that have been used for public racing or trotting matches. Speed is not designed to be the sole test. Style, size, color, action, elegance, and general qualifications as first-class horses for driving, as well as speed, are to be considered. The owners are to drive; professional drivers will not be admitted to occupy that position. The teams and single horses are to be driven to road wagons only."

The Viewing Committee may require an affidavit as to ownership, if necessary to establish, to compete in this class.

Messrs. Fisk and Shoemaker urged the incorporation of premiums for stallions five years old or over in this class, but it was not agreed to.

All of Class 14—Draught Horses—was stricken out except the first two entries, to which a third was added, as follows:

Best stallion, four years or over	$50
Second best	30
Third best	20

Class 15—Carriage and Buggy Horses—remains as last year.

The following rule was adopted:

"All horses in this class shall be measured by a standard, on the ground, and by the committee, and 'speed' shall not be considered the sole test of merit. The several points of form, general beauty, style, action, matching, and evenness of gait, are to be considered"

Saddle horses remain unchanged, and the rule requires that they shall not be less than fifteen and a-half hands high under the standard, and trained to the three gaits appropriate to the saddle horse, viz: the walk, the trot, and the canter.

Jacks and mules remain unchanged.

In the classes of Merinos, Southdowns, and Long-wooled sheep, the highest premiums offered are to be $20, and pens of lambs shall consist of three, and not of five, as heretofore required.

Class 23—Swine—has had the premiums for "imported" boars stricken out, and one of $20 added for best boar of any breed; the other premiums being raised from $5 to $10, and $3 to $5, respectively.

Class 24—Poultry—remains as last year, as do also 25 to 33, —Farm Implements and their Trials,—except that Class 27 is combined with 28, after having its cash premiums stricken out, leaving medals and diplomas; also striking out the cash premium of $25 from the best and greatest variety of carriages, buggies, etc., in Class 30.

In Class 34,—Grain, Four, Meal, and Seeds,—two bushels instead of one are required to be exhibited, and in case of wheat, etc., the average yield must not be less than twenty hushels per acre. The assortments and specimens of heads of wheat, etc., are stricken out, and a diploma added to the ten dollar cash premium for the best display of the different kinds of grain in heads, arranged and named.

In Class 35,—Roots and Vegetables,—larger quantities are required to be exhibited of each, with the design of giving an opportunity to visitors at the fair to purchase seeds or specimens of premium varieties.

Additional premiums of one and two dollars are offered for samples of twenty-four ears each of eight-rowed red blaze, yellow, white, and smoked dent, and yellow flint corn.

In the class of Butter and Cheese, the premiums were made as follows:

Best and largest collection of cheese from any one county	$50
Best display of butter made by any one factory or person, of not less than fifty pounds, if deemed worthy	25
Best twenty-five pounds of butter, made at any time	10
Second best	5
Best display of cheese by any factory or private dairy, silver medal and	10
Second	5

The balance, including cheeses by weight, stricken out.

Classes 37 and 38—Sugar, Honey, and Bee-hives, Bread, Preserves, and Pickles—remain as last year. In Class 39,—Wines, Vinegars, Cordials, etc.,—the second premiums for six bottles of wine from Concord, Delaware, and Catawba grapes, are added to the value of the first premiums in each case, and divided into three varieties, so as to include in each sweet, dry or sour, and sparkling wines. The word "grape" is inserted between the words "native" and "wine," for the best display of bottled native *grape* wine.

On Musical Instruments—Class 49—the cash premiums

were stricken out; and for sewing machines, all the premiums were made to read "diploma," where the word "medal" stood last year.

In Domestic Manufactures, Class 41—Home-made; Class 43—Articles of Dress; Class 48—Paintings, Statuary, Photographs, etc., and Class 51—Needle, Wax, Shell, and other Fancy Work, were referred to the President, with an additional discretionary sum of $100, to be given in premiums if deemed advisable, after revision of these classes by a committee of five ladies, whom he is to associate with himself for that purpose.

In Class 46—Ornamental Iron-work—a premium of $3 is added for "galvanized iron cornice."

The remaining classes were all adopted without change from last year.

Mr. Baxter moved that a series of premiums of $25 each shall be awarded to the county agricultural society of the county whose citizens shall take the largest number of premiums in any one division of the premium list from A to K. For example, the county whose citizens shall take the largest number of premiums on cattle will be awarded $25; and so in horses; the same in sheep, swine, and poultry.

The Committee then took a recess till 2 o'clock.

At 2 o'clock the Committee met.

Mr. Phillips introduced the following propositions relative to an exhibition by breeders, of their stock of Short-horn cattle:

"Mr. D. M. Uhl hereby proposes to show, at the State Fair of 1871, his Short-horn cow Florence, and such a number of her progeny, owned by him at this date, as he pleases, against any Short-horn cow and her progeny, owned by any breeder in the State at this date, the money value of the whole of the stock to be taken into consideration, for a sweepstakes premium of $100, to be deposited with the Treasurer of the State Society thirty days before the State Fair, by each party making an entry, and to be paid to the party whose stock is decided to be the most valuable, by a committee appointed by the State Society.

"He also proposes to show the Short-horn two-year-old bull Colonel Welsh, against any Short-horn bull of a like age bred by Mr. Curtis of Addison, for the sum of $50 a side, the money to be deposited with the Treasurer of the State Agricultural Society thirty days previous to the next State Fair; the whole of the money deposited to go to the party whose animal is decided to be the most worthy by a committee appointed by the State Society."

After some discussion as to the usefulness and practicability of the proposals, they were withdrawn.

Mr. Gilbert moved that a special committee be appointed, to consist of the President, Mr. Baxter, and Mr. C. W. Greene, whose duty it shall be to confer with the parties making propositions for holding the next State Fair, and that said committee shall report their action within sixty days from the adjournment of this meeting. The said committee also to have authority to decide upon such location, or to submit the subject to the Executive Committee, as may be deemed most expedient.

The resolution was agreed to, and the committee appointed, after some discussion and inquiry, during which Mr. Stout thought the question should be more fully discussed in General Committee, and proceeded to advocate the claims of the northern portion of the State upon the Society.

Messrs. Shoemaker of Jackson, Pratt of Kalamazoo, Griggs and Scranton of Grand Rapids, and others submitted remarks and suggestions, from which it was decided that the committee would forward to each competing locality plans and descriptions of the buildings required for the fair, previous to calling upon the citizens for propositions.

Dr. Pratt said it would be desirable to know whether the Society contemplated the erection of permanent buildings, with the purpose of using them for future fairs, or only such temporary affairs as it had been the custom to provide.

Mr. Johnstone, the Secretary, said this phase of the question was a very important one. Permanent buildings were desirable—a necessity—for the future success and efficiency of the Society, and the best results would be reached by having them in all the principal sections of the State, so that fairs could be held in them at each, for a single year at a time, and returning regularly every four or six years, according to the number of places provided. But it was simply an impossibility for the Society now, or even in the next ten years, to erect such buildings. They might, however, with the assistance of each locality, erect one at a time,—a floral hall at first,—using temporary structures, as now, for the rest, and adding others in succeeding years as they might be able.

Mr. Gilbert moved that the Secretary prepare a schedule of the buildings and fixtures necessary for the next State Fair, and furnish it to the special committee on location.

Mr. Dean, from the Finance Committee, made the following report:

The Committee on Finance respectfully submit the following report, after having examined the Treasurer's and Secretary's reports, which have been referred to them:

We find that the Treasurer had on hand at the beginning of the year		$1,224 44
That he received during the year		13,150 14
		$14,374 98
He has paid out on account of 1869	$621 02	
On business checks of 1870	8,952 53	
On premium checks of 1870	6,340 00	
		$15,913 95
There was due him for advances on January 1		$1,539 37
Besides this balance we find that there have been issued business checks not yet presented	$193 42	
Premium checks not presented	576 00	
Premiums reported at this meeting	125 00	
		894 42
		$2,433 19

The resources which the Society have to meet this, consist of the balance of unpaid subscriptions.

Whole amount subscribed	$574 70	
Of which there has been paid	429 70	
Balance due and unpaid of Jackson subscription		$145 00
Uncollected accounts of 1869 at Jackson		70 29
Uncollected balances of 1870		189 53
Amount invested towards museum building at Normal School		3,250 00
Estimated value of buildings at Jackson		3,500 00
Total assets of Society		$7,104 82

Your committee believe this to be a fair statement of the financial condition of the Society's affairs. It is not so promising as we hoped it would be, and we have therefore given some attention to the details of its expenditures, and of its receipts. We found that its receipts during the year have been considerably less than those of the previous year, while its expenditures, in some respects, have been larger than ever expected. The amount of construction and outlay in preparing the grounds and buildings is stated by the Secretary's report at $3,386 25, and of this we find that there was fully $1,500 expended in the work of fitting up Floral Hall, but of this there are many articles which are on hand, and we may mention that it took $90 worth of glazing and 1,400 yards of muslin to do the work that was deemed necessary. So with regard to the water supply. It was found necessary to conduct the water to several points on the grounds, and the outlay for water was $153 60, a large portion of which was for the gas-pipe through which the water was conducted, and which has been taken up and is stored away for use. Last year the digging of the wells, the sinking of the fountain in Floral Hall, and the work necessary to give a supply of water, cost the Society $402. Another expense which was deemed necessary was that of making the halls look respectable, and the cost of lime and work amounts to $230. Another item is that for meals for officers and Viewing Committees. These expenditures include the cost of the gate-keepers, police, and watchmen, and other assistants needed on the grounds during the days of the Fair, as well as that of the members of the viewing committees. The printing is another large item of expense, reaching altogether nearly to the amount of $900; all of which seems to be necessary, but which should be done with the greatest economy.

In looking over the returns of income, and comparing them with those of the past year, your committee finds that there has been considerable falling off. Last year it was requested by the Executive Committee that the citizens' subscriptions here should be $1,500, but of this amount there was subscribed only $574 70, and only $419 70 of this was paid, while of this there was $144 70 that came from the 10 per cent allowed the Society by the State Prison authorities. There is always a falling off in subscriptions, and though promised $5,000 in 1869, yet the Society never realized over $4,000, a large amount being unpaid. In the sale of tickets, also, and in the returns of percentage by the only two railroads which have assisted the Society, there has been during the past year a falling off of $2,032 67, indicating that the attendance at the last Fair was not as large as that of the year previous.

In addition to this we may call attention also to the fact that the amount of premiums awarded in 1870 is $500 in excess of the amount of premiums paid in 1869.

We do not find that the ordinary expenses of the Society for business purposes have been very different from those of previous years. The general expenses have averaged, exclusive of mere construction and lumber accounts, from $4,000 to $5,500 per year. In 1869 the expenses were $5,357. During the past year they have been about $4,700.

The Society depends altogether on its receipts from the proceeds of its annual exhibitions, and hence it is very proper that it should have every facility possible to make its great annual meetings satisfactory. Your committee are satisfied that the receipts of the Society could have been largely increased, had the grounds at Jackson been large enough to have permitted the admission of teams and persons in carriages, and to have had more stall room for horses. That there was a very great loss in this respect we feel assured.

All of which is respectfully submitted.

A. J. DEAN,
E. W. RISING,
JOHN GILBERT.

On motion, the report of the Finance Committee was accepted.

Then a general discussion arose concerning the expenditures.

Mr. Dean called attention to what he considered an abuse of the tickets given to attendants and the persons who were required as help, especially by those who kept refreshment stands on the grounds. He was of the opinion that it would be better to let the grounds for less rents, and to make each person pay their admission.

Mr. Shoemaker explained that in renting the grounds, each party who had a lease was, by the possession of that lease, required to state the number of assistants he required. The number ranged according to the size of the establishment. Some required but two or three assistants, others required as many as ten or fifteen. One of the provisions of the lease was that these assistants should be admitted once each day before 8 o'clock in the morning. The tickets for their admission were delivered to the lessee in person every evening, between the hours of 5 and 6 o'clock, or about the time of the closing for the day. The names and persons of these lessees were taken down by the Secretary in a list, and the tickets were delivered to them in strict accordance with the lease. Many of these lessees, not being acquainted with the rules and regulations, were obliged to go out to procure supplies during the day, and would go out without tickets, and when they came to the gate were clamorous for free entrance; but they had no right to enter. It was explained to them that for all entrance after 8 o'clock they would be obliged to pay as any visitor would, and tickets were refused them by all the officers when they made application. He thought that if there was not provision made for such assistants, there could be no grounds rented, or the price would be so small that it would diminish the receipts very much from that source, without adding any return in its place. Last year the applications for grounds were not by any means as numerous as they were the year before, because nearly all those engaged in the business had laid in so much stock that they had lost money, and the chief part of the grounds rented was only leased a few days previous to holding the fair. He was in favor of renting the grounds on terms that would permit the lessees a fair opportunity to do business. We needed them for the accommodation of the people, who must have something to eat and drink. The lessees ran a large risk, for they had to put up buildings and to furnish stores of provisions for numbers who would be dissatisfied if not supplied. So far as he had had information of this matter, he thought there was but very little attempt at imposition beyond the difference of opinion that would arise from ignorance on the part of lessees as to what they were entitled to. He thought the interets of the Society were pretty well guarded in this respect.

Mr. Dean, also, had considered some other matters, which he had not introduced into his report, but simply because they had been called to his attention. Amongst them was the use of complimentary tickets. He had been furnished with a general list of the numbers used. There had been quite a number given out, but he found they had been given out under the general usage and custom of the Society. There was a certain number presented to the press, to the railroads, to the officers of the Society, and to such guests as the Executive Committee and the President deemed proper. There was no indiscriminate use of them, or any that could reasonably be cut off.

Mr. Gilbert, chairman of the Business Committee, said that

he had been very careful to place at the gates men who were adapted to that position. It was very difficult in the rush that was made, and the reception of from twenty to thirty thousand persons, always to discriminate, or to allow the passageways to be blocked up. During the past year he had alluded to the work done himself, and it had been as faithfully and economically done as if the whole had been done for his own house. The bill of lumber was not large or excessive, nor was the lumber high-priced, as he could show by reference to the items of the account which he held in his hand. So with the hardware, every item of which could be shown. The work had been done by a most efficient workman and a trustworthy man, whose bill was not excessive. But the work had to be done, and not any part of it was of such a nature that it could be neglected or left undone, without spoiling the effect of the whole.

Mr. Greene said he was gratified with the discussion, as it satisfied him that every effort had been made by those who were entrusted with the arduous duty of preparing the fair, to render it a success. The true position of affairs was, that the Society had been very generous in its premiums, and the several interests that clustered around it had grown so great that its receipts, when compared with the expenditures, was a satisfactory exhibit to him and to all his brother members of this committee. From his experience he was satisfied that the general management had been entirely such as was necessary, and the success of the exhibition had been owing in a great measure to it. It was impossible, in the endeavor to provide for the reception of such a crowd and the transaction of so much business within the limited period of two or three days, to provide for every contingency. All the straws could not be laid with regularity and be kept pointed in one direction.

Mr. Beckwith moved that the printing be taken from the Business Committee, and that the Treasurer be a committee to attend to it. No one objecting, the motion was passed.

Mr. Beckwith moved that the renting of the stands and grounds be placed in the hands of the Business Committee.

Mr. Shoemaker explained that the renting of the grounds had always been in the hands of the Business Committee, but he being the local member of that committee, much of this business had been left with him, and he had attended to it, though sometimes when he had been obliged to be absent, the Chairman or the Secretary had assisted him in the performance of the duties that had devolved upon him.

Mr. Gilbert moved the following resolution, which was adopted:

Resolved, That each Executive Superintendent is hereby requested to be on the grounds ready to take charge of his department the Monday preceding the opening of the exhibition.

Mr. Beckwith moved that any member issuing complimentary tickets write his name on the back of the same.

Mr. Baxter moved that any officer issuing complimentary tickets report the number issued.

The motion was discussed, but no action was taken.

The Committee then adjourned to meet at half-past 7.

At half-past 7 the Committee met.

On motion the Committee proceeded to nominate the Viewing Committees for the next annual exhibition.

The Committee adjourned to Friday morning.

FRIDAY, February 17.

The Committee met.

On motion of Mr. Greene, a Committee on Pedigrees of Cattle, and a Committee on Pedigrees of Horses, was ordered to be appointed.

The President appointed as the Committee on Pedigrees:

Cattle—Mr. Jasper Barber of Bellevue; Mr. I. H. Butterfield of Lapeer; and Mr. G. W. Phillips of Romeo.

Horses—W. G. Pattison, Kalamazoo; E. Van Valkenburgh, Hillsdale.

Mr. Greene moved that no member of the Committee on Cheese be appointed from any county in which there is a cheese factory. The motion was discussed and adopted.

On motion of Mr. Phillips, it was ordered that the Committee on Short-horns be made seven in number, so that all sections of the State may be represented on that important committee.

Mr. Gilbert moved that the Special Committee on Location of the State Fair be instructed to confer with the officers of the several railroads, relative to the amount of percentage allowed on their receipts, and their facilities at the several points for the accomodation of visitors and exhibitors.

Mr. Sterling moved that each member of the Executive Committee be requested to address the Members and Senators in aid of procuring the passage of the bill refunding to the Society the moneys expended upon the Normal School building.

Mr. Baxter offered the following resolution, which was adopted:

Resolved, That the next annual State Fair shall be held on Tuesday, Wednesday, Thursday, and Friday, September 19, 20, 21, and 22, 1871.

Mr. Baxter moved that the Committee on Rules and Regulations be instructed to draw up the proper rules for the formation of a cavalcade of the live stock on the last day of the Fair. Adopted.

Mr. Phillips moved that the Committee on Rules draw up and adopt a rule for the government of the Committees on Pedigrees of Cattle and Horses.

Mr. Griggs of Grand Rapids presented an invitation to the

members of the Executive Committee to visit that city, and tendered to them its hospitalities.

Mr. Greene moved that the Committee on Miscellaneous Articles be increased to seven in number, so that the duties of examining the various entries may be divided. Adopted.

The President announced the following appointments:

Business Committee.—Mr. Gilbert, Mr. Baxter, Mr. Sterling.

Rules and Regulations.—Mr. Shoemaker, Mr. Manning, Mr. Stout.

Executive Superintendents for the Annual Exhibition:

Cattle.—Mr. G. W. Phillips, Mr. E. W. Rising.

Horses.—Mr. A. J Dean, and Mr. C. W. Greene.

Floral Hall.—Mr. Baxter, Mr. Pantlind, and Mr. Shoemaker.

Implements.—Mr. H. O. Hanford.

Mechanics' Hall.—Mr. L. S. Scranton.

Manufacturers' Hall.—Mr. A. C. Fisk.

Fruit Hall.—Mr. J. M. Manning.

Agricultural Hall.—Mr. A. Stout.

Police and Gates.—Mr. J. M. Sterling.

Grand Marshal.—Mr. G. H. Gale, Kalamazoo.

On motion of Mr. Baxter, the thanks of the Committee were tendered to the reporters of the daily press of Detroit, and to the morning and evening papers of Jackson, for their careful reports.

On motion of Mr. Greene, the thanks of the Committee were tendered to the proprietors of the Hibbard House, Messrs. Pantlind and Robinson, for the accommodations and attentions which the members have received at their hands.

Mr. Sterling moved that a vote of thanks be tendered to Colonel M. Shoemaker for the hospitalities extended to the Committee.

Mr. Baxter moved that the whole subject of an annual address at the next State Fair be referred to the President, with power to extend invitations to such gentlemen as he may deem proper, and provide for such address if he judges it expedient.

On motion, the members of the Special Committee on Location, and the Business Committee were notified to meet after the close of the session of the Executive Committee.

On motion, the Executive Committee adjourned to meet on the 19th day of next September.

W. G. BECKWITH, *President.*

R. F. JOHNSTONE, *Secretary.*

FIRST ANNUAL REPORT

OF THE SECRETARY OF THE MICHIGAN STATE POMOLOGICAL SOCIETY, CONTAINING REPORTS, ADDRESSES, ARTICLES OF ASSOCIATION, COMMUNICATIONS, AN ACCOUNT OF ITS FIRST ANNUAL FAIR, ETC., FOR 1870.

THE ORGANIZATION OF THE SOCIETY.

To the Members and Officers of the Michigan State Pomological Society:

The soil, climate, and geographical position of the State of Michigan have shown that she is a favored region, and well adapted to the cultivation and growth of all fruits suitable to a Northern and temperate clime. For this, the State is now attracting the attention of thousands throughout the country, and her own horticulturists seem to be awaking to a realization of their location. Within the recollection of many in mature life, there were but few varieties of large or small fruits cultivated in the State; but to-day, as we look over a vast country, from lake to lake, we see large and flourishing orchards of apple, pear, peach, and cherry, besides thousands of acres devoted to the growing of strawberries, grapes, and other small fruits. This new industry is rapidly assuming important proportions, and it is desirable that those engaged in fruit culture should seek a closer connection with each other, and should establish an organization which should directly represent their common interests.

THE INFORMAL MEETING.

Such was the growing importance of this interest in the State, and particularly in that portion bordering upon Lake Michigan, that it was thought best by leading horticulturists to call a public meeting with a view to the organization of a permanent State Society. Consequently, notices were issued by A. T. Linderman and others, and a meeting was called at Sweet's Hotel, Grand Rapids, Feb. 11, 1870. This meeting was attended by gentlemen from different portions of the State, and S. L. Fuller of Grand Rapids was elected President, Sluman S. Bailey and L. S. Scranton were elected Vice Presidents, A. T. Linderman was chosen Secretary, and E. U. Knapp Treasurer. Henry S. Clubb, S. L. Fuller, and L. S. Scranton, were appointed a committee to draft articles of association, and Jacob Ganzhorn, Wm. Voorhis, and James Hamilton were appointed an Executive Committee.

THE FIRST REGULAR MEETING.

An adjourned meeting of the Society was held in Luce's Hall, on Saturday, February 26, 1870, S. L. Fuller in the chair. At this meeting, Articles of Association were introduced by H. S. Clubb, which were discussed, amended, and adopted:

ARTICLES OF ASSOCIATION OF THE STATE POMOLOGICAL SOCIETY OF MICHIGAN.

The undersigned, at the city of Grand Rapids, this 26th day of February, 1870, hereby associate themselves under the name and style of the State Pomological Society of Michigan, and agree to be regulated by the following Articles of Association until proper legislation is obtained for a legal organization:

ARTICLE I.—The object of the Society is to develop facts and promulgate information as to the best varieties of fruit for cultivation in the fruit regions of the State of Michigan, and the best methods of cultivation.

ARTICLE II.—The officers of the Society shall consist of a President, Treasurer, and Secretary, who, together, shall constitute an Executive Committee, with full power to call meetings and transact business under the direction of the Society. [NOTE.—This Article was amended at the December meeting.]

ARTICLE III.—The office of the Society shall be in the city of Grand Rapids.

ARTICLE IV.—The annual meeting for the election of officers shall be on the first Tuesday in December, in each year, the officers elected at such meeting to commence service on the first of January following.

ARTICLE V.—The officers shall remain and perform their respective duties until their successors are appointed or elected by the Society; but the regular term of office shall expire on the 31st of December in each year.

ARTICLE VI.—The Society shall hold a meeting on the first Tuesday of every month, at such place as the Executive Committee shall designate, under the direction of the Society.

ARTICLE VII.—Every person who subscribes, or who may subscribe, to these Articles, and pay to the Treasurer the sum of one dollar per annum, shall be entitled to membership, unless otherwise voted at a regular meeting of the Society.

ARTICLE VIII.—No money shall be disbursed except on an order signed by the Secretary, and by direction of the Executive Committee.

ARTICLE IX.—These articles may be amended at any regular meeting of the Society by a majority vote of such meeting.

ARTICLE X.—By-laws may be passed at any regular meeting.

ARTICLE XI.—The Executive Committee shall require of the Treasurer such security as they may deem necessary for the safe keeping and proper disbursement of the funds of the Society in his hands.

These articles of association were signed by a goodly number of gentlemen present, who congratulated themselves that the Society was thus formally organized.

The following gentlemen were elected honorary members: Wm. Adair of Detroit; J. G. Ramsdall of Grand Traverse; Townsend E. Gidley of Grand Haven; Daniel Upton of Black Lake.

On motion of Mr. Jacob Ganzhorn of Spring Lake, the

following officers were elected by ballot: President, H. G. Saunders of Grand Rapids; Treasurer, S. L. Fuller; Secretary, A. T. Linderman.

A Corresponding Committee was elected, composed of Henry S. Clubb, James Hamilton, and Daniel Upton.

THE APRIL MEETING.

The first display of fruit made at a meeting of this Society, was on Tuesday, April 5th. President Saunders brought in a basket of beautiful fruit, among others very fine and rich specimens of the Russet, and large, bright-looking Baldwins. Mr. Holt of Cascade offered fine samples of the Swaar and Peck's Pleasant. Mr. Houghtaling of Grand Rapids town, exhibited large, healthy, brown-looking Baldwins, and a few genuine Roxbury Russets. Mr. Erastus Hall of Grand Rapids, sent in a basket of bright, red-looking Baldwins. Mr. J. H. Ford of Paris, brought a basket of brotherly-looking Jonathans and some hardy English Russets. Rev. H. E. Waring of Grand Rapids town sent specimens of the Baldwin, Roxbury Russet, and Talman Sweeting. Mr. Noah P. Husted of Lowell presented a basket of splendid Wageners, attractive to the eye and delicious to the taste.

THE IMPORTANCE OF THE SOCIETY—THE STATE OUGHT TO AID IT.

At the afternoon session, a letter addressed to the Treasurer, from Mr. George Parmelee, of Grand Traverse county, was read:

OLD MISSION, Mich., March 28, 1870.

DEAR SIR—From a notice in the *Western Rural*, of March 10th, I learn that a State Pomological Society has been organized at Grand Rapids, and not knowing the postoffice address of the President, I send a line to you as a resident of the city, and likely to be present at the next meeting. But for the bad traveling just as our sleighing is leaving us, I would be present at the meeting on the first Tuesday in April.

If the first "Articles of Association," copied in the notice referred to, set forth mainly the objects of the Society, I will, if this reaches you in season and meets your views, ask you to suggest to the Society at its next meeting, the propriety of including among its objects the taking of measures to secure the aid of the State, more effectually than is now given, to the interests of Pomology. While we feel, and take pleasure in acknowledging, that the State has, through its Board of Agriculture, done, and well done much in the interests of knowledge in several branches, it must be apparent to all who understand the resources of the Peninsular State, that Pomology, as a branch of agriculture, has not had that prominence given to it which its value and importance, both present and prospective, in justice require. While anything that may add to our taxes will and should be closely scrutinized, the intelligence of our citizens is certainly equal to appreciating the value and economy of reasonable expenditures in that direction, if practically applied. Just what steps will best accomplish the desired ends, the united wisdom of the Society can best point out.

The selection of Grand Rapids as a point to initiate a State organization, seems eminently fit, for, while being nearly in the center of that interest, the more distant southeastern portions of the State enjoy facilities which render a few miles more of travel of little objection; and we in this other extreme section recognize the fact that we could not expect a selection that would be more convenient.

Yours truly,

GEORGE PARMELEE.

REMARKS ON THE CULTIVATION AND VARIETIES OF APPLES.

Members were solicited to describe their specimens of fruit in brief speeches. Mr. Ford said he had brought scions of English and Golden Russets. These apples were often confounded, when they were entirely different, as any one could see by looking at the scions. The Golden Russet limb is slim and light colored. Mr. F. also presented scions of the English Russet, which has a green and russet color. His English Russets were very hardy,—had kept them a year and a half. The English and Golden Russets were as different as the Baldwin and Spitzenburg. The English Russet grows upright and spreads; his soil was lightish. The value of the Russets was in their hardy keeping qualities. His Jonathans kept well; the chief value of this apple was that it was an annual and abundant bearer; there were always apples where there were Jonathan apple trees.

Rev. H. E. Waring made a statement of his fruit experience. He resides two miles east of the city, on elevated table land, possessing the naturally drained loamy elements most favorable to successful orcharding, and makes fruit a prominent feature of farming. Mr. W. ranks Steele's Red Winter, and Rhode Island Greening, among his most profitable sorts in full bearing, but has a more extensive collection of varieties that have not yet borne in sufficient quantity to judge of their comparative merits. He also makes peaches a specialty, and says he has not failed of a crop for fifteen years, although there have been a few seasons when the yield was not more than one-half or third of a full one. He places the early and late Crawford, and Barnard, at the head. Finds yellow peaches sought after when the white-fleshed are a drug. Of the latter, the Large Early York, Stump the World, and Crockett's White, are favorites,—the latter was brought from New Jersey and matures the latest of the three. He has also paid some attention to both standard and dwarf pears. Among the latter, he says, the Louise Bonne de Jersey has paid twice as much as any other.

THE BALDWIN.

Mr. Thompson suggested that the difference in the color of the Baldwins upon the table was occasioned by a difference in soil.

Mr. Husted said the distinction was important. A variation

in soil produces not only a difference in color, but also a corresponding variation in size, quality, and flavor. The Baldwin is one of the peculiar varieties which are materially changed by the variations of soil. There are low places that are rich and black, very cold in winter and very hot in summer; this soil produces a very inferior Baldwin. This apple does best on elevated situations, where the temperature is mostly uniform. He had watched the Baldwin for years in this State. There was an unusual variety of soil in Michigan, and the Baldwin succeeds remarkably well in some places, while in others it has been discarded. Hence a New England man coming here is often disappointed in the Baldwin.

Mr. Holt said the Baldwin did well with him until the severe winter of '56–'57, when they were all killed, though other varieties survived. Since then he had had no luck in raising young trees, but had done well in grafting the Baldwin upon old stock. He had concluded that the Baldwin was not a hardy tree in all parts of Michigan. The Baldwin grafted on mature stocks will be the hardiest.

Mr. Congdon of Lowell asked wherein the superiority of the Baldwin was so apparent? A member suggested that it was in its size, color, good keeping and cooking qualities; it would always sell.

THE SWAAR.

Mr. Holt said the Swaar was a good keeper and good producer. He had always cultivated his orchard, and raised crops therefrom. He thought the Swaar did well on a gravelly soil.

Mr. Husted said his experience was that the Swaar was well adapted to gravelly and sandy soil; but on a clay soil he would discard it. On dry, light soil the Swaar will rival the Baldwin in beauty. The variation in soil produces a corresponding variation in flavor and keeping qualities of apples.

THE WAGENER.

Mr. Holt had the Wagener in his orchard, and was in favor

34

of that apple. They bore at an early age, and he considered them one of the best of apples.

Mr. Husted confessed that he was enthusiastic on the subject of the Wagener. It had been suggested that it lacked constitution,—that it was an overbearer, and would soon wear out or use itself up. On the contrary, he knew of trees twelve or fourteen years old, and they bear annually, and good crops; there were no signs of decay; considered them a marvel of hardihood. He considered its cooking qualities as good as those of the Rhode Island Greening, and in this part of Michigan it is a hardier tree than the Baldwin or Greening.

THE SPITZENBURG AND RUSSET.

President Saunders said he had Spitzenburg trees, and they had prospered, and he considered it a very fine fruit,—not often excelled. His soil was a clay-loam. The specimens of Russet which he exhibited, and which were so generally admired, were taken from trees bought for the Roxbury Russet.

Mr. Husted did not recommend the Spitzenburg for general culture, because, after bearing a few years, it fails to perfect the fruit on most of the soils of Michigan.

Mr. Holt said his Spitzenburg apples bore heavily every other year, and scattering alternate years, for the past twelve years. He thought well of them.

Mr. Houghtaling then proceeded to detail his experience in relation to the various fruits, as follows:

PEACHES.

In regard to raising peaches, I depend mainly upon seedlings, and have shipped from $500 to $700 worth in one season, from about 200 trees. The latter sum was realized two years ago. The last year, although the crop was good, there were too many in market, and the profits were very small. I have a good many varieties, and some that are excellent in quality. I have none as early as the Crawford, nor as showy as the

Melancthon, but they are a nice yellow peach, and as good for all useful purposes. They are far more hardy than most budded peaches, and will yield many more to the tree as a rule.

PEARS.

Pear trees, with me, have proved a dead failure, having planted over one hundred and lost nearly all of them with the pear blight, and those trees that live do not grow nor flourish very well.

APPLES.

In reference to the best variety of apples, the speaker regarded the Baldwin as one of the very best, both as a good bearer, good keeper, and salable in market. The Red Canada and Spitzenburg come next. The Jonathan is also a very choice apple, but is rather small in size; but it is a good keeper, very showy, and of an excellent quality, as you will see by the specimens shown to-day. In regard to the keeping of apples, I find that burying them out, like potatoes, is a good plan. Have about ten bushels this year buried out-doors that seem to be as fresh as when they came from the trees.

QUINCES.

Of quinces I have but few, and they do first rate for the chance they have. They bore well last year, and the fruit was large and nice. I give them a mulching of half-rotten straw occasionally, and once a year a good dose of salt, or old brine, about the roots; this keeps them healthy and free from blight, and makes the fruit large and sound. I think it is a very profitable fruit. I shipped one barrel last year over the lake and realized $10 for it.

Henry S. Clubb inquired—Was the soil in which your pears failed rich or poor? Do you manure your pear trees?

Mr. Houghtaling replied—It was good strong, heavy soil, such as wheat and corn would grow well in, and pretty well manured.

Mr. Holt said that was the cause of the blight.

PRUNING.

Mr. Houghtaling said, May is the worst time for pruning. March is the best month, or June is good. The month of April is a good month in late seasons.

Mr. Holt liked pruning in March best. Mr. Houghtaling said, wax or gum shellac should be used to prevent bleeding.

The fruit on exhibition was referred to a committee, to report thereupon at the next meeting. The meeting then adjourned till the first Monday in May.

ADDRESS OF THE CORRESPONDING COMMITTEE.

GRAND HAVEN, April 9, 1870.

To the Pomologists of Michigan:

The State Pomological Society has been organized with a view to promote your interests.

In order to accomplish all its beneficent objects, the co-operation with the Society of fruit-growers in all parts of the State, either by personal attendance at its monthly meetings, or by correspondence, is absolutely necessary.

The Society desire to collect and publish such information in reference to every locality in the State, as may give to the outside world a correct idea of the extent and importance of the fruit interest, as well as a scientific view of the effects of locality, aspect, soil, water, and protection, on the varieties of fruit in improving or deteriorating certain kinds. For instance, at the favored locality of Spring Lake, in Ottawa county, it is believed that the Delaware grape is grown to greater perfection than in the Eastern States, being larger in berry and of superior flavor. There may be locations where the Baldwin apple attains greater perfection than at others, and other locations where it is inferior. Such information, carefully collected and published, will be of immense value to Pomology, and furnish scientific men of other States an opportunity of judging of the relative merits of Michigan as a fruit-producing State.

Fruit-growers of Michigan, co-operation and union will accomplish your interests far more than can ever be achieved by isolation. We have a Peninsula which stands unrivalled by any State in the Union for the production of choice, hardy fruits to the greatest perfection. Not even California can compete with Michigan, in the quality of those fruits suited to this climate. We have, therefore, the basis of success, and if we take steps to avail ourselves of the advantages within our reach, Michigan, even when stripped of her wealth of lumber, will become a hundred fold more wealthy in her production of fruit.

In order to develop the fruit resources of our State, the first thing to do is to make known what has been accomplished. The experience of those who have been pioneers in fruit-growing is of the most importance, and to secure the publication of the results of that experience is one of the principal objects of the Society. The Committee, therefore, respectfully but earnestly invite the co-operation of every fruit-grower in the State.

Signed by the Corresponding Committee.

HENRY S. CLUBB, Grand Haven.
JAMES HAMILTON, Newaygo.
DANIEL UPTON, Muskegon.

MAY MEETING.

An important and interesting meeting of the Society was held on Tuesday, May 3, at the Circuit Court Room, at Grand Rapids. A valuable list of apples was adopted and recommended for the use and cultivation of orchardists and pomologists.

THE DISPLAY OF FRUIT.

Considering the advanced season of the year, the exhibition of apples was very creditable. Mr. Walter G. Sinclair of Spring Lake, Ottawa county, presented some bright specimens of the Baldwins. They were placed on the shelves of his cellar when gathered, with all his other varieties, and no other care was bestowed upon them. Large, brown, sound-looking Baldwins were presented by Mr. Joshua Bradish of Grand Rapids town, grown on clay soil. Ionia county was represented by bright-looking Baldwins, sent by George N. Jackson of the township of Keene, raised on gravelly loam. When gathered they were buried under straw and covered with soil. The town of Cascade, Kent county, was represented by bright and red Baldwins, from the farm of President Saunders. Other specimens were presented by H. Holt of Cascade. A plate of large, fresh-looking Rhode Island Greenings was exhibited by Noah P. Husted of Lowell. The Russet family were gathered together in friendly accord and rivalry. The smallest was the American Golden Russet, or Bullock's Pippin, sometimes called the Sheepnose; the largest and finest were those presented by Dr. Saunders; samples of the Golden Russet, the Roxbury Russet, and the English Russet were presented by Mr. Husted, Mr. Van Buren, Mr. Holt, and others. Mr. Holt's samples of the Swaar were very large, handsome and sound. His Wagener, Holt's Seedling, Esopus Spitzenburg, Jonathan, Yellow Bellflower, Green Winter Sweet, and Rambo, all attracted attention.

The Corresponding Secretary read a number of letters from gentlemen residing in different parts of the State,—among others was one from Mr. Goodwin of Ionia:

LETTER FROM H. H. GOODWIN OF IONIA.

IONIA, April 30, 1870.

DEAR SIR:—Feeling a deep interest in the fruit culture of our great fruit State, I inclose the membership fee, and instruct you to put my name to the constitution of the Society. I have an orchard of 400 apple trees just coming into bearing; the soil is mostly of clayey loam, and the trees are promising well, but the Wagener has gone beyond the promising point, and for two years has been paying down for all it received, and canceling the old score of purchase price and care which I had charged against it. From the twenty-five Wageners that I have, I expect to pick more fruit for the next five years, than from all the other varieties in the orchard. The long season of use, and the perfect satisfaction rendered when used either as a dessert fruit or a cooking apple, make it the most desirable of any with which I am acquainted.

Success to the Society, and continued advancement to the great interests that called it into being.

H. H. GOODWIN.

GRAFTING.

Mr. Houghtaling said the subject of grafting is a thing very little understood, and very many people know nothing at all how it should be done. There are thousands of trees in the country that need to be grafted over to make them worth the ground they occupy or profitable to their owners. It is just as easy to raise the best fruit as it is the poorest. I have found by experience that it is very easy to put a new top on an apple tree even after it is fifteen or twenty years old. I have some in my orchard that were large enough to bear ten bushels of apples before they were grafted, and now have a new top and as handsome as any tree in the orchard. There is a right way to do it; it is very easy to spoil the shape and beauty of the tree. For this reason I have brought along a few specimens for ilustration, that I may show you something of the right and wrong way to success in the art of grafting.

First. We should cut the limbs out as far from the tree as we can, to keep the top open and well spread out. *Second.* It should all be done at one time, that the top may make an even growth and be well balanced. *Third.* They should be watched and attended to, keeping away all the suckers that take the growth away from the graft, and sometimes kill them out entirely, as here shown by specimen.

Mr. Houghtaling here exhibited the *modus operandi* of grafting, in a very interesting manner.

REPORT OF THE COMMITTEE ON APPLES.

The committee to whom was referred the apples on exhibition at the April meeting reported by their chairman, William Voorhies of Frankfort, Grand Traverse County:

Mr. President and Gentlemen:

We will make two classes of apples, those planted for *individual use* and those planted *for market* or shipping purposes. For the grower's own use we recommend the following varieties:

For Summer Use—Early Harvest, Early Strawberry, Williams' Favorite, Red Astrachan.

For Autumn—Porter, Jersey Sweet, Maiden's Blush, Gravenstein, Rambo.

For Winter—Baldwin, Steele's Red Winter, Wagener, Rhode Island Greening, Swaar, Esopus Spitzenburg, Peck's Pleasant, Tallman Sweeting.

For Market Purposes—Steele's Red, Baldwin, Rhode Island Greening, and Wagener, the red predominating, as they generally sell the best, though the Greening sells full as well. The Wagener is a comparatively new apple, and is not very well known, but receives high commendation from all quarters. The best apple for market should combine the following qualities: The tree should be hardy,—do well on any moderately fertile soil, and should be long-lived, free from disease, and an annual bearer. The fruit should be of good size, not too large or too small, skin thick (to bear handling), red or bright, lively color, flesh fine-grained, tender, crisp, sub-acid in flavor. The apple should hang well on the tree, bear early, fruit without spots. Fruit should be in good condition for eating from December until May. An apple with all the above qualifications can hardly be found. Get one as nearly like it as possible. Such an apple will pay the best.

In regard to the Wagener apple presented by Mr. Husted, the sample before us is light red, indistinctly striped and splashed with dark red, stem inserted in a deep cavity, calyx nearly closed and set in a rather shallow basin, juicy, sound to the core (April 12th), flesh white, somewhat tough, skin tough, mild, sub-acid flavor. Judging by this sample, the Wagener is certainly a good keeper.

Holt's Seedling, presented by Mr. Holt, very much resembles the Swaar in appearance, is mild, sub-acid bordering on sweetness; rather dry, and very fine-grained; at this season (April 12th), is in a partial state of decay at the core. On account of the flavor, we do not anticipate this seedling will be valuable for cooking purposes; yet its hardiness and prolific bearing are valuable characteristics, and would undoubtedly be conveyed to the next generation of seedlings, which, if some could be obtained that had a more decided flavor combined with the good qualities already possessed by the parent, would undoubtedly prove an acquisition. Will Mr. Holt try again?

DISCUSSION OF THE REPORT.

The report was accepted, and the question was on its adoption. Mr. Holt, one of the committee, said he did not agree with all there was in the report; thought the Swaar should be placed on the list of shipping apples,—it always sells well. Mr. Linderman, another of the committee, thought the report recommended too many varieties.

SUMMER APPLES.

Mr. Husted agreed that there were too many varieties, and he had often felt the annoyance and vexation in filling the long lists of customers. Mr. Holt could not help but recommend the Williams' Favorite as a summer variety,—it was the earliest, and very thrifty. That and the Red Astrachan could not be ignored. A member suggested that the Early Harvest be added. Rev. Mr. Hamilton said the Early Harvest would scab and crack. The Red Astrachan was his favorite summer variety the world over.

Mr. Husted would add the Duchess of Oldenberg as the coming summer variety. Its superiority was unquestioned. It was tough, hardy, and is coming on at a great rate. By

35

vote it was added to the list. So was the Sweet Bough. Mr. Holt thought the Red Astrachan was a little tender, and the Duchess of Oldenberg was an autumn apple. Mr. Husted said they covered the summer season,—were good for all purposes, for desert, cooking, and market. It was now moved that all summer varieties recommended be stricken from the list except the Red Astrachan, Duchess of Oldenberg, and Sweet Bough. Carried.

AUTUMN APPLES.

Rev. Mr. Hamilton of Newaygo highly commended the Maiden's Blush; it could not be dispensed with. It headed the list for autumn, as the Red Astrachan did for summer. He spoke highly of the Gravenstein.

Mr. Husted said the Maiden's Blush was a beautiful apple, always sold readily, saw it retailed last season, in Chicago, at five cents an apple. The Gravenstein was beautiful, a fine grower, but in its place he would substitute the Cayuga Red Streak. It was a splendid apple. Mr. Holt had the Maiden's Blush in his orchard; it was a great producer.

Mr. Linderman felt a little delicate about recommending the Snow; he had known it to crack.

Mr. Schermerhorn suggested that the Fall Pippin be added to the list. It was an old variety and very popular. People would be surprised to see it left out.

Major Light of Greenville said he would strongly indorse the Snow; had known it from his youth; in was a great success in Ionia, Kent, and Montcalm counties. Never knew a man to reject it. One good characteristic of the Maiden's Blush was that it would dry well; so would the Snow; this is a valuable qualification for a fall variety. Every list should have a sweet apple; the Jersey Sweet was not excelled. The Cayuga Red Streak, Maiden's Blush, and Snow would keep to spring.

Mr. Holt didn't know how to leave out the Porter. On motion, all the varieties recommended by the committee were stricken out, and the following substituted for autumn varieties: Maiden's Blush, Snow, Jersey Sweet, Cayuga Red Streak, Fall Pippin.

WINTER APPLES.

Mr. Holt said that Peck's Pleasant, in his experience, was not a long keeper. Mr. Linderman said it was his favorite.

Mr. Hamilton said he had the Northern Spy in all its perfection.

Major Light said the Tompkins County King does not succeed in a sandy-loam soil; on a clay loam it was a success and a splendid apple.

Mr. Carrier said his Baldwins had not ripened the last season.

Mr. Hamilton—To throw out the Baldwin would put the State in an uproar.

Mr. Husted—The Baldwin and Greening cannot be dropped.

Mr. Hamilton—As this was a list for general cultivation as well as for the market, they had better enlarge it. He suggested the Golden Russet, the Baldwin, Greening, Wagener, Hubbardston Nonsuch, Northern Spy, and Tompkins County King.

Mr. Light—I have a list of seven apples. The Baldwin cannot be stricken out; the Wagener is a coming apple,—uniform good bearer and a valuable variety. The Greening was one of the best cooking apples. The Northern Spy does not mature early, but when it comes it sells, and is a great bearer. The Hubbardston Nonsuch was a great favorite. We have a great variety of soil in Michigan. He recognized Steele's Red as valble in certain localities; nothing could be said against the fruit, but the tree was hard to start. We can substitute the Wagener in its place and get a better apple. Some very fine things could be said in favor of the Swaar, but it is not a constant bearer. The Spitzenburg of fine flavor was liable to fail; Peck's Pleasant, so fine, was too large, but a good grower; the Jonathan was too small. On his motion, this list was adopted for winter varieties; Baldwin, Wagener, Rhode Island Greening, Golden Russet, Tallman Sweeting, Northern Spy, and Hubbardston Nonsuch.

The report of the committee, as amended, was then adopted.

THE APPLES RECOMMENDED BY THE SOCIETY.

It will thus be seen that the Society, by their action, recommended the following varieties:

For Summer:—The Red Astrachan, Duchess of Oldenberg, and Sweet Bough.

For Autumn:—The Maiden's Blush, Snow or Fameuse, Jersey Sweet, Cayuga Red Streak, and Fall Pippin.

For Winter:—The Baldwin, Wagener, Rhode Island Greening, Golden Russet, Tallman Sweeting, Northern Spy, and Hubbardston Nonsuch.

After listening to an address by J. P. Thompson, the Society adjourned.

OUR ORCHARD SYSTEM.

THE NEED OF CONCENTRATION—WANTED, FEWER KINDS AND SORTS—THE LEATHER-JACKET SYSTEM.

ADDRESS BY J. P. THOMPSON.

Mr. President and Gentlemen of the State Pomological Society:

In contemplating the Apple Culture of the famous Fruit Belt of Western Michigan, the characteristics are so well defined that they will not escape the attention of the most casual observer, and these are:

1. The total absence of *system*;
2. The want of a correct and authorative *nomenclature;* and,
3. The lack of a well defined *purpose.*

The system appears to be to plant a great variety of trees

without regard to soil or climate, or intended use; the nomenclature comes from provincial traditions, and is derived from Northern, Southern, Eastern, and Western associations; and the purpose seems to be undefined and vague, partly experimental, but mainly to raise as many as possible of all kinds and sorts. Every apple culturist, in a certain sense, appears to be an amateur; he goes into the business hap-hazard to suit himself, without due regard to the demands of the market, and with no well conceived calculations about profit or utility. Now *amateurs* are well enough in their places; they serve oftentimes a valuable purpose; are useful as well as ornamental; but to be a successful amateur requires leisure and capital, as well as taste and cultivation. Men who settle a new country and lay deep the foundations of a profitable and intelligent fruit husbandry, have little time to waste in experiments; they need most to come directly to the utility and profit of the business in hand.

Calculate, if you can, the value of the time and the amount of money foolishly expended in tree culture, for the last twenty years, in Western Michigan! It would amount to tens of thousands of dollars, and where, to-day, we have few valuable orchards, we ought to have the land covered with productive, fruit-bearing varieties, sufficient to support every man, woman, and child in the section alluded to. There has been a great lack of skill and knowledge—a want of a well matured purpose. But let us not be too harsh in our judgments.

WHY THIS GREAT DIVERSITY?

It is very easy to account for this great diversity and indirectness in our fruit culture. Men have come from Northern and Southern Ohio, totally unlike our latitude; many have come from the hardy clime of Canada; a large swarm are here from New England, who, true to their nature, brought with them all their New England prejudices; a great mass are from Northern and Southern New York; Pennsylvania and New

Jersey have sent their share, and other sections theirs; so that we are a mixed multitude, and every man has sought to bring his old favorite apple tree with him! Every one has fondly desired to transplant from his old homestead the apple tree of his boyhood days, and has endeavored to perpetuate those sorts and kinds that gave a glow and cheer to the hearthstone of his sires. It is not a weakness, and is no disgrace, to love the fruits, the trees, the flowers, the birds, that blossomed, bloomed, and sung on the old homesteads. It could not be expected that men would know in advance; it was all they could do to bring the knowledge of their sections with them. Certain it is, however, that every clime, every latitude, has its own peculiar favorite fruit; certain it is that Northern and Southern Ohio are unlike in their productions; but few of the apples of New England flourish in New Jersey or Pennsylvania; while it is also certain that there are species in Western New York, under the lee of the lakes, which succeed well in Western Michigan under the protective influences of our inland seas.

TOO MANY VARIETIES.

So that it is true, as I have said, that the misfortune with which we are most afflicted is the misfortune of too many kinds. We are smothered, overburdened, crushed, with apples, such as they are! Practical men need to go to pruning with the axe. Let the amateur have all he wants, but let the pomologists, who have homes, schools, and churches to provide for, get down to business! With peaches we are doing better,—we have a *repertoire*, a system, and a nomenclature.

There are three kinds of peaches,—the Early Hale and the two Crawfords,—that are bound to add thousands of dollars to the productive wealth of the fruit belt. What if, ten or fifteen years ago, our peach orchards could have been planted with those kinds? Men who are toiling could now have lived at their ease with their Crawfords. "Yes," says this and that neighbor, "it would have been dollars in our pockets, and wealth to these counties and towns." But six hundred different kinds of peaches are recommended, and we have only three leading varieties,—not over six at most,—and are doing well. Let the amateur have his six hundred kinds, and we will stick to the three kinds, and have orchards. We cannot all be amateurs. Here is the lesson the peach teaches the apple. We want this *system* introduced into our apple-culture, —we want our six kinds of best market varieties.

DECLINE OF ORCHARD CROPS IN OHIO.

In Ohio they are talking about the decline of orchard crops. The apple crops in Ohio and the adjoining States have very materially declined, in quantity and quality, for some years past, especially where the orchards are in considerable age. but the evil is not attributed to the age of the trees nor the exhaustion of the soil, so much as to the increase of injurious insects and diseases affecting fruits and fruit trees. This is a doleful sound to come from that grand pomological State.

Not to weary your patience, I wish to read from the report of their State Horticultural Society.

The Ohio Horticultural Society have given much attention to this matter for the past two years, and have called the attention of the State Legislature to the subject in the following memorial:

" *To the Legislature of Ohio:*

"The undersigned, Executive Committee of the Ohio Horticultural Society, respectfully invite the attention of the members of the Legislature to the following facts and considerations relating to the orchard crop of the State for 1868:

Bushels of apples produced	11,637,515
Bushels of peaches produced	599,499
Bushels of pears produced	66,712
Total	12,303,726

If the apples were of fair average quality, they might be estimated as worth 50 cents per bushel	$5,818,757
The peaches $1.50 per bushel	899,248
The pears same price	100,068
Total value of orchard crop	$6,818,073

"If we add a moderate estimate for the thousands of gardens and door-yards partly occupied with fruit trees and omitted by the assessors, it will swell the aggregate of orchards to nearly 500,000 acres, and the value of the products to over $7,000,000—saying nothing of the cherries, grapes, and smaller fruits.

"Large as this sum appears, we are convinced, by much observation and inquiry, that the amount and value of the orchard products of the State have diminished very greatly within the past ten or fifteen years.

"The report also shows that while part of the evil may be attributed to neglect or bad management, the chief causes are injurious insects and fungus diseases. Both of these have increased greatly in variety and extent within the past ten or fifteen years, and seem likely to increase still more in the future, as our orchards and gardens increase in age, unless measures are taken to prevent such result."

THE REASON OF THIS DECLINE.

My own opinion about this decline is that that they have too many kinds and sorts in Ohio. If they would thin out one half, the deterioration would itself decline. There is nothing like useless varieties to breed poisonous insects,—pestiferous bugs and destructive borers. A tree of but little value attracts but little attention, and it is very likely to draw on these pests and communicate them to more valuable kinds. Thin out your orchards; revise your catalogues; reduce the number of your kinds one-half; is our humble advice to the orchardists of Ohio.

This, it appears to us, is the lesson for Western Michigan. One-half of the trees in this section should be grafted. A useless tree cumbereth the ground. It costs just as much to feed it; it exhausts the earth and the air just as much as does a good tree. Time, precious time,—half a lifetime,—is lost. The loss is not only to the soil, but to the pockets of the producer. For what is an apple worth that won't sell?

THE BEST KINDS.

Now comes the question, what are the best market varieties for this section, for our soil and climate? The market gives this answer; the *market* affords the best test. If a bushel of Northern Spys will bring a dollar, that is a test of value. "Steele's Red" is quoted in the *Western Rural*, April 28, at $3 50@3 75 per barrel, in Detroit. That's a good criterion of its value in that market. The *Country Gentleman* reports the Rhode Island Greening worth $4 50 per barrel in Albany, and recently barrels of them sold quick at $2 50 in this city. Here it is, in the market we find the test of value. There are 2,500 varieties of apples,—wanted, six of the best varieties! That is the question, and this is the policy. In Western New York they have partly learned to do the business better than we do. Their orchards are not filled with unsalable varieties. There they have established the kinds most profitable for the markets, and they raise and sell those kinds—tens of thousands of bushels.

THE QUESTION IN A NUTSHELL.

To sum it up: In looking over this vast field spread before us by a powerful providence, it has seemed to me that the one great mistake that has been made by those who raise fruit for the market, is the planting of too many sorts. If this Pomological Society could give to this people a small and carefully considered list, sanctioned by trial and experience, it would confer a benefit that would well repay for its institution. In looking for a few of the best varieties my attention was drawn to the

RUSSET FAMILY.

I know, practically, little about the Russet Family, but from the numerous names of its members it must be a decidedly mixed family.

THE RUSSET NOMENCLATURE.

We have the Pumpkin Russet, which is distinct from the

Sweet Russet and more worthless; American Golden Russet, or Bullock's Pippin, or Sheepnose—too small to be popular or useful; English or Poughkeepsie Russet—a profuse bearer—will keep twelve months; Golden Russet, known favorably as the Golden Russet of Western New York—distinct from the American Golden Russet or Bullock's Pippin—valuable; Red Russet, somewhat resembling the Baldwin, described by C. Downing, but not recommended; the Roxbury Russet or Boston Russet—sometimes called the Putnan Russet in Ohio—keeps late in spring—valuable; Bagby Russet or Egyptian Russet—in use in Illinois but not in Michigan; Cheseborough Russet, an autumn variety of little value; Darlington Russet, very moderate quality; Fay's Russet, known in Vermont; Goble Russet, an autumn variety of no value; Howe's Russet, closely resembling the Roxbury; Hunt's Russet, another Massachusetts Russet; Kingsbury Russet, identical with the Cheseborough; Spice Russet, no account; Sweet Golden Russet, no account; Whitney's Russet, a Canada variety; Winn's Russet, a Maine sort; York Russet, similar to the Pumpkin Russet; the Perry Russet and the Pommegrise or Grey Russet—not recommended. Downing has an English Golden Russet, inferior. Of all these sorts, but three are worth talking about—the Golden, the English, and the Roxbury. If all others were discarded, the pomological world would not know the loss. It cannot be said that this family lacks for a nomenclature,—it has any quantity of it. But hereafter let us be particular and not talk about the English Golden Russet or the American Golden Russet, but let us call the apple we mean, the Golden Russet—its proper name.

THE RESPECTABLE RUSSETS.

The Roxbury Russet was a prominent apple before the Baldwin was known in New England. Roxbury, now a part of Boston, was its birthplace, and it is consequently also correctly known as the Boston Russet. President Wilder of Massachusetts says: "It is a singular fact, that the three most popular Eastern apples—the Baldwin, the Roxbury Russet, and the Rhode Island Greening—are the three most popular apples in the West, where they grow so many millions of bushels. At many of the Western fruit conventions, votes have been taken as to what were the best varieties, and they have always resulted in favor of the Baldwin, the Roxbury Russet, and the Rhode Island Greening."

President J. F. C. Hyde of Massachusetts says: "If we were asked to name varieties for market use, we should give the Williams, if a summer variety is desired; the Dutch Codlin for cooking, the Gravenstein, Hubbardston Nonsuch, Rhode Island Greening, Baldwin, and for late keeping. the Roxbury Russet, which, though not a great, yet is a regular or annual bearer."

It is a somewhat curious incident in pomological literature, that the late Mr. Downing, in his "Fruits and Fruit Trees of America,"—edition of 1852,—recommended, as particularly favorable, what he called the "Putnam Russet." He says: "For a knowledge of this celebrated Western apple we are indebted to that zealous pomologist, our friend, Professor Kirtland of Cleveland. It is considered decidedly the most valuable late-keeping apple in the West—not inferior to the Newtown Pippin, and the growth of the tree is very luxuriant. It originated at Marietta, Ohio, and is largely grown for the New Orleans and West India markets." This apple was recognized and ascertained to be identical with the Roxbury Russet, and in later editions Downing so refers to it. It came to be called the Putnam Russet from the fact that it was taken to the Ohio Valley in 1796, by Rufus Putnam, and from thence it was distributed over the Southwest.

A vote of the Fruit-growers' Society of Western New York gave the following as the best six varieties for that section: Baldwin, Tompkins County King, Golden Russet of Western New York, Roxbury Russet, Rhode Island Greening, and the Northern Spy. It will be seen that the Russet family occupies one-third of this vote. It will also be seen that this apple (the Roxbury) has the preference in the East, in Ohio, and in Western New York, as one of the best of long-keepers.

Gentlemen, I suppose you will expect a more minute description, and I do not wish to exhaust your patience. I submit the following, which has been well matured, but yet may fail in accuracy. The varieties are very similar. It requires a close observer, in this section, to distinguish them. Oftentimes it is only by the shoots that they can be classified. As has been said, the Roxbury Russet originated in New England, and has been extensively distributed throughout the West; it is a good annual bearer, best on heavy soils, and in many localities the most popular of this species, attributable to its thick skin, and keeping qualities, enabling it to be taken long distances, often to New Orleans and other Southern markets, but is hardly above second rate in quality. The tree is of spreading growth, with rather downy shoots. Fruit, medium to large, roundish, flattened, dull green, overspread with brownish, yellow russet, occasionally a faint blush on the sunny side; flesh, greenish white, moderately juicy, mild sub-acid, good for market from January to June. Has been known to keep a year.

The English Russet is supposed to be an American seedling, was first disseminated on the Hudson River, is equally as long a keeper as the above, of less acid and more pleasant flavor. The growth of the tree is upright, with shoots of a lively brown color. Fruit medium size, roundish, conical, or ovate, light greenish yellow, mostly overspread with brownish russet, in large exposed specimens wholly russeted, flesh yellowish, white fine texture, rather firm, aromatic. Its profuse bearing, and quality of keeping over a year, render it very profitable.

The Golden Russet, an apple of foreign origin, and is extensively planted in Western New York, and resembles the English Russet in size and general appearance, but is more tender and crisp, and is not so long a keeper. Growth, spreading and irregular, with many slender weeping branches; has light-colored, speckled shoots, by which it is easily known; fruit medium size, roundish, usually a little oblong, sometimes slightly flattened; surface at times wholly a thick russet, at others a thin broken russet on a greenish yellow skin, sometimes a tinge of red on the exposed side; flesh greenish, fine-grained, high-flavored, nearly "best" keepers through winter.

The English is distinguished from the Golden and Roxbury by its straight, upright shoots, and from the Roxbury by its less flat form, and less acid flavor, while the Golden may be most readily known from its peculiar light shoots and its brisk rich flavor.

CONCLUSION.

Gentlemen, I have thus given you my opinion about the apple orcharding in this section. I have not aimed to be sentimental, learned, or nice, but have endeavored to present a practical question in a practical manner. I am convinced that until some of these suggestions are adopted, our apple orcharding will not amount to much. We want concentration of varieties. I want to hear of and see an orchard of one thousand Wagener apples; another one of five hundred Baldwins; another one of five hundred Rhode Island Greenings; another one of two hundred Red Astrachans; one of two hundred Red Canada, or Steele's Red; one of three hundred Maiden's Blush, or an equal number of the Fall Pippin; another one of two hundred Tompkins' County King; another one of a thousand of the Roxbury, or Golden, or English Russet, and one of five hundred Snow. When such orchards are established, fruit buyers will be attracted here, and it will not be necessary for any man to "peddle" apples from his orchard. I would not advise that over ten varieties be cultivated in the Fruit Belt, and would not recommend any one man to cultivate more than two summer varieties, two fall or autumn kinds, and not over three winter sorts. Indeed, I am convinced that three varieties would be better for one orchard that a larger number. The demand is for marketable apples, apples that will bear transport, and apples that will keep. What is known as the Fruit Belt ought very soon to produce and market half a million bushels of apples per annum. It can be done by systematized orcharding.

POPULAR VARIETIES OF HARDY APPLES.

DESCRIPTION AND REVIEW OF THE FIFTEEN VARIETIES OF APPLES RECOMMENDED BY THE STATE POMOLOGICAL SOCIETY, BY W. L. WARING; READ BEFORE THE SOCIETY, JUNE 7TH, 1870.

In preparing the following descriptions of popular hardy apples, comprising the sorts recommended for general orchard culture in the State, at a late session of the State Pomological Society, the writer has been guided by personal experience, observation, and the suggestions and views of practical fruit-growers and intelligent pomologists. It is designed to furnish an outline of the most prominent distinguishing features of these leading market fruits, with brief notes that may be useful or interesting. A minute description of old, well-known kinds is not deemed necessary, as reference can always be made to standard works. The following terms used in describing apples, oblate—flat; ovate—egg-shaped; round—globular; conical, oblong, ribbed, etc., are such as are commonly accepted and generally understood.

SUMMER APPLES.

Red Astrachan.—Rather large, sometimes quite large, roundish-flattened, slightly approaching conical, nearly whole surface brilliant deep crimson, overspread with a thick, white bloom; flesh white, crisp, juicy, rather acid. Excellent for cooking, and, when fully ripe, an agreeable dessert fruit. Ripens from the 20th of July to the middle of August,—a few days after the early harvest,—and should be picked before fully mature. Shoots stout, clear reddish chestnut, with many white specks; leaves broad. The tree is a vigorous, upright, regular grower, forming a very handsome head, and apparently adapting itself to all soils and locations.

It comes early into bearing, and bears annually a fruit always fair, good for eating, and of excellent culinary qualities; so beautiful, and bearing carriage so well, as to make it worthy for general cultivation, and especially desirable for marketable purposes. Hardy north and west. First imported into England from Sweden in 1816.

Sweet Bough.—Large, roundish, remotely conical ovate; sometimes distinctly conical; skin smooth, pale greenish yellow, becoming yellow when fully ripe; flesh white, tender, crisp, sprightly, with an excellent sweet flavor. Ripens from the middle of July to the tenth of August. The Sweet Bough, or Yellow Bough as it is often called, is a native apple, and a popular sort everywhere, highly valued for dessert and much esteemed for baking. Shoots yellowish; tree hardy, a somewhat irregular, upright, spreading, moderate, compact grower, forming a good round head, and an abundant annual bearer. It is not as desirable for the kitchen as the Red Astrachan, but is generally greatly admired for the table; very valuable for market purposes, and should be in every collection.

Duchess of Oldenberg—Rather large, roundish, a little flattened at the ends, skin smooth, finely washed with light red in broad broken stripes and splashes on a golden or yellow ground, with a faint blue bloom; very handsome; flesh yellowish white, tender, juicy, sub-acid, with an excellent flavor; first-rate for cooking, and a pleasant dessert fruit.

Ripens early in September. Very hardy, and a young and abundant bearer. The strong growth of the tree, its early bearing, and endurance of severe winters, and the fair and showy appearance of fruit, render it one of the most valuable sorts for the West.

So much is it prized by many who have grown it, that they think it surpasses most all early autumn apples. Of Russian origin.

REMARKS.

These three varieties are all hardy, and, according to general experience, are among the best and most profitable for the Northwest, covering nearly the whole period of summer, keeping up a supply of good apples from about the time of wheat harvest to early September, and some practical fruit-growers plant nine of them to any other early sorts.

The Duchess of Oldenberg, however, is more properly an autumn variety, and I would suggest as a substitute the Carolina Red June as the most valuable. Very early market apple; above medium in size; ripens among the earliest in July, and keeps long after ripe; is hardy, vigorous, and a young, abundant, regular bearer.

The Williams, a rather large, handsome, good fruit, in season for several weeks late in summer, would be preferred but for its later ripening. Or the Keswick Codlin for cooking, and tolerable for eating, if quantity of fair fruit is more desired than a high standard of quality.

AUTUMN APPLES.

Maiden's Blush.—Rather large, oblate or flat, skin thin, smooth and fair, clear lemon yellow, with red cheeks, varying from a delicately tinted flush to a rich, brilliant crimson; remarkably beautiful; flesh white, fine grained, tender, sprightly, pleasant, rather sharp sub-acid unless fully ripened; begins to ripen at the end of August and until the last of October; will keep through winter.

It is much admired as a dessert fruit, and is also highly esteemed for cooking or drying.

The tree is a rapid grower, forming a fine round, spreading head, and bears annually large crops. Hardy, valuable at the West.

Although not highly flavored, it is greatly valued for its fair, tender, and beautiful fruit, and uniform productiveness. A native of New Jersey, first described by Coxe.

Snow Apple.—Medium in size, round, often oblate, or somewhat flattened; surface even, handsomely striped and blotched with fine deep red on whitish ground; where much exposed to the sun, becoming nearly a uniform rich dark crimson. Flesh very white, tender, juicy, sub-acid, slightly perfumed, delicious.

Late autumn, at the North in use from October to February. Not very rich, but much admired, as a dessert fruit, for its handsome appearance and exceedingly pleasant, refreshing flavor, and is regarded as excellent for drying. Shoots, dark, diverging. Tree a vigorous grower and regular bearer. A celebrated, hardy, productive apple, especially valuable far north.

Originated in Canada, and has its name, Pomme de Neige, or Snow Apple, from the snow-white color of its flesh.

Jersey Sweet.—Fruit medium size, round-ovate, often oblong-ovate, thickly washed and striped with fine red on greenish yellow, sometimes entirely covered with pale or deep red; flesh whitish, fine grained, exceedingly juicy, tender, sweet, and sprightly. Its good flavor and remarkably rich sweet make it popular for table use, and especially suited for baking. Owing to its saccharine quality, it is largely planted throughout the Middle States for the fattening of stock. Commences maturing the end of August, and continues ripening till mid-autumn. Young shoots, stout and short jointed; tree a strong, fine grower, and a profuse annual bearer; are incorrectly said to bear themselves out early. A very valuable sort, succeeding in all localities, and highly esteemed in almost all parts of the country, both for market and for uses about the homestead. First disseminated in New Jersey.

Cayuga Red Streak.—Very large, roundish, remotely conical, surface slightly uneven, sometimes smooth; splashed with stripes of rich purplish red, on greenish-yellow or yellowish-white ground; flesh coarse grained, with a sprightly, sub-acid, pleasant, but not rich flavor; of fair eating quality, and excellent for baking. Late autumn and early winter. An upright, compact grower; a fine, regular, annual bearer, growth in large trees becoming straggling. One of the best very large apples,—showy, fair, productive, and profitable.

Though not high flavored, its remarkably handsome appearance and large size render it one of the most popular fruits in market. Origin, Western New York.

Fall Pippin.—Fruit very large, roundish, approaching oblong-conical, flattened at the ends, sometimes with obscure ribs; color greenish, becoming a high, rich yellow when ripe, with often a tinge of brownish blush on one side when grown in good soil and well exposed to the sun. Flesh yellowish-white, quite tender and juicy, with a rich, sub-acid, aromatic flavor. Is esteemed everywhere as the first of autumn apples, from its beauty, large size, and delicious flavor, for the table or cooking. Season, October to January.

Shoots, dark; tree vigorous, spreading, becoming large. In strong clay loams it is one of the very best, hardy, and a good bearer. Fine in nearly all localities. Origin uncertain, supposed to be an American seedling raised from the Holland Pippin, from which it differs most strongly in its later keeping.

REMARKS.

The list of autumn apples embraces a sufficient variety that have been well tested and approved throughout the State for general good qualities, and such as are commonly admitted to rank among the most hardy and profitable.

It would be a larger number than necessary, if grown strictly for market, were it not for the fact that they not only supply the whole fall season, but a majority of them continue in use nearly through winter.

There are some who would substitute for one of these, the Lowell, here called Greasy Pippin, a rival of the Fall Pippin, which, although hardly as good in quality, excels that variety in its uniform fair surface, and great and early productiveness.

WINTER APPLES.

Baldwin—Large, roundish, tapering a little toward the eye, nearly covered and striped with crimson, red, and orange, on yellow ground; on light loams well exposed to the sun, a beautiful bright red; a few russet dots and russet streaks about the stem; flesh yellowish white, crisp, with a rich sub-acid flavor, not very fine grained. A first-rate winter apple in all respects. Ripens through winter, but keeps well later. Shoots, reddish, stout, slightly downy, long jointed, with white specks; tree a vigorous, rapid grower, with curved erect branches; forms a regular open head in the orchard; comes early to maturity and bears abundantly; in strong soils supplied with lime and potash, produces very even, perfect, and uniform-sized fruit. It is the most popular winter sort throughout New England, New York, and Michigan, and as a profitable market variety, stands among the very first. Originated in Massachusetts,

Wagener—Above medium to large, round-ovate, sometimes oblate, flattened at the ends, outline somewhat irregular, slightly ribbed, skin smooth, mostly overspread with indistinct stripes of two shades of red; full deep red in the sun, pale light red on the shady side, on warm yellow ground often streaked with russet; flesh whitish, fine grained, compact, mild sub-acid, vinous, aromatic, excellent. A celebrated variety, and one of the finest for dessert, cooking, or market. Ripens through winter, but a late keeper, retaining its freshness of flavor and appearance till May, and superior for culinary uses at the beginning of autumn.

Tree a stout, upright, rapid grower, becoming spreading; very hardy; exceedingly productive when quite young; a heavy annual bearer, and esteemed by many intelligent market-growers, especially in Western Michigan, as the most valuable winter apple yet introduced. In quality, as a table apple, marked as best. A native of Yates County, New York. First brought into notice in 1848.

Rhode Island Greening—Large, roundish, flattened, sometimes angular, always fair, green, yellowish when fully matured, a brownish blush on sun-grown specimens, many rough russet dots and patches; flesh yellowish, tender, slightly aromatic, with a lively acid juice; as excellent for dessert as for cooking, and its presence in orchard or garden, for kitchen or table use, cannot be dispensed with. Keeps well till March or April. Growth strong, of a broad-spreading habit, healthy, medium sized shoots, broad leaves; a great and constant annual bearer in nearly all soils and situations, and is more generally known and esteemed than any other winter sort; single trees yielding forty bushels of fair fruit in favorable years, and neglected orchards 200 bushels per acre. Like the Fall Pippin, which it resembles in wood and leaves, and all trees of the same vigorous habit, it is a gross feeder, and should be supplied with fertilizers by turning in heavy crops of clover or dressings of compost, where there is any deficiency in the soil.

Fine throughout the Northern States. Is said to be a native of New Jersey, once known as the Jersey Greening; was first widely distributed in Rhode Island and adjacent region.

Golden Russet.—Medium size, roundish-conical, a little oblong, flattened at stem end, nearly regular; skin rough and thick, sometimes wholly a dull russet, and at others a thin broken russet on a greenish-yellow ground, rarely with a tinge of red on the exposed side; flesh greenish, fine grained, firm, crisp, juicy and high-flavored, nearly "best." Season, November to May. Popular and extensively grown in the Genesee Valley, where it is known as the Golden Russet of Western New York, and considered here, from its productiveness and superior keeping and eating qualities, the most valuable of all the Russets.

Tree hardy, a fine grower, spreading and irregular, with many slender weeping branches; has light-colored, speckled shoots, by which it is easily known; bears large crops, and ranks among the first as a profitable market variety. Of English origin.

Northern Spy.—Large, roundish-conical, often flattened, sometimes ribbed, smooth, pale yellow mostly covered with glossy red, and distinct stripes of purplish crimson, some russet around the stem, and coated with a fine bloom; flesh yellowish white, very tender, crisp, juicy, sprightly, sub-acid. Keeps through winter and late into spring, preserving its flavor remarkably fresh. One of the largest, most beautiful, and excellent long-keeping apples yet known. Shoots, dark reddish, spotted, stout; the tree forms a very handsome upright head, requiring pretty severe pruning and thinning out until it comes into bearing, after which it requires little or no pruning.

The flowers open late and thus sometimes escape spring frosts; it does not come early into bearing, but once in bearing it is very productive. A fruit of the highest keeping and eating qualities, and with good culture and care in picking and packing, profitable for market. Hardy, succeeds throughout the north. Originated in Ontario county, New York, about 25 years ago.

Tallman Sweet.—Size above medium, roundish, slightly conical, pale, whitish-yellow, faintly tinged with red on one side, with a brownish line from stem to eye; flesh white, firm, rich and very sweet; hardly first-rate as a dessert fruit, but from its exceedingly rich sweet, and great productiveness, is especially valuable for winter baking, and as food for stock. Keeps late into spring. Quite hardy; has dark colored wood, of an upright, strong, rapid growth, becoming spreading; vigorous and healthy when young, and one of the handsomest in the orchard, where it makes a round, regular, open head, and forms a rather large tree, bearing annually great crops of fair, even-sized fruit.

A popular, profitable sort, and from its firm texture and late keeping suited for shipment to distant markets. Originated on the grounds of Mr. Tallman, in Maine.

Hubbardston Nonsuch.—Large, round-ovate, largest at the middle, nearly regular, smooth, glossy, rich yellow, nearly covered with deep, warm red, with small, broken stripes, and numerous dots of light crimson; russeted next the stem, and sparsely dotted on the surface with large russet specks. Flesh yellowish, very rich, slightly sub-acid, with a strong mixture of a rich, sweet flavor; excellent. A famous New England sort, that for ordinary family uses has few equals. Season, early winter; loses flavor by late keeping.

Wood, brownish chestnut, with whitish specks; annual shoots, slender, downy at ends; leaves large, deep green above, whitish beneath; tree very regular, upright, becoming spreading; more hardy than the Baldwin, when grown in very rich, heavy soils; a strong grower and great bearer.

Sells well in the market. Extensively planted and succeeds well throughout the Northwest. Originated in the town of Hubbardston, Mass.

CONCLUDING REMARKS.

So far as I can learn, the apples adopted in this list are those most generally popular throughout the State. There are a few prominent in other sections, not fully tested in Michigan, that may yet rank among the most profitable here, but the shrewder class of orchardists prefer to raise the few leading sorts of decided excellence, well known in the markets, rather than plant for the sake of variety,—such as produce well in most seasons, and especially those that are found to suit our soils and latitude.

The Northern Spy, although a tardy bearer, preserves late into spring, when there are but few good fruits, and therefore holds a high rank among market sorts.

There are many who would be unwilling to omit the Roxbury Russet, which, from its very long keeping properties, is in season after the others have gone by, and commands a high price. Or Steele's Red, a rather slow grower, but much esteemed as a very reliable sort in most parts of the State.

I should name Tompkins County King, in place of Hubbardston Nonsuch, as a more valuable variety. It is a splendid apple; a strong grower, good bearer, and keeps through winter.

The Ladies' Sweet, also, is larger and better than the Tallman Sweet, is one of the best and most profitable winter sweet apples; hardy, a fair grower, bears young and abundantly, keeps well, bears carriage to market well, and is one of the best for table, cooking, or stock uses.

A SMALL SELECTION.

For a small list, to be planted solely with a view of obtaining the largest income from 1,000 trees, it is believed that 50 each, Red Astrachan, Duchess of Oldenberg, Cayuga Red Streak, Maiden's Blush, and 200 each, Baldwin, Wagener, Golden Russet, and Rhode Island Greening, would make a selection that would be found the most productive, reliable, and profitable.

It is a matter of the first importance to those engaged in raising fruit, to be assured the sorts they cultivate are true to name, and they should look to them closely at the season of ripening, compare them with the descriptions, and better, when practicable, bring or send in specimens at the regular monthly meetings of this Society, for exhibition, examination, and testing.

JUNE MEETING.

STRAWBERRIES AND CHERRIES.

AN ADDRESS DELIVERED AT THE FIRST ANNUAL STRAWBERRY EXHIBITION OF THE MICHIGAN STATE POMOLOGICAL SOCIETY, GRAND RAPIDS, THURSDAY, JUNE 16, 1870, BY HENRY S. CLUBB.

On Thursday afternoon, June 16, there was a pleasant meeting of the Society, though not a large attendance. The exhibition of strawberries and cherries included the following standard and excellent varieties: A plate of Wilson's Albany, presented by O. R. Wilmarth, won general approbation; the samples of Hovey's Seedlings and Russell's Prolific, sent in by E. Carrier, were greatly admired; specimens of the Agriculturist and Green's Prolific, from the garden of S. S. Bailey, were superb; while the Boston Pine, from D. Schermerhorn, were amongst the finest on exhibition. Mr. C. C. Rood presented specimens of Black Tartarian Cherries, which were represented to be prolific bearers. Mr. A. T. Linderman brought samples of the White Ox-Heart Cherry. Mr. James D. Husted was tendered a vote of thanks for beautiful and large bouquets. The flowers and fruits were referred to a committee. After a discussion of the merits of the best varieties, Mr. H. S. Clubb, editor and publisher of the Grand Haven *Herald*, was invited to address the Society, and he proceeded as follows:

Mr. President, and Fellow Members of the Pomological Society—In compliance with your resolution, I appear before you to discuss a subject of my own selection. I have selected "Strawberries and Cherries," not only because these are the first fruits of our gardens and orchards, and are now in season, but because there are some considerations in regard to them that I have long desired to present to you. I therefore thank you for this opportunity, and hasten to improve it.

A UNIVERSAL FRUIT.

The strawberry is as nearly as possible a universal fruit. The question is not, Where does it grow naturally? but, Where does it not grow? In its wild, uncultivated form, the strawberry can be found in all parts of the globe we inhabit, where the foot of man has trod. In its cultivated and improved form, it has become the favorite fruit of every civilized country, so far as heard from. It is therefore a subject of vast extent and absorbing interest.

Whether regarded for its beauty or its use, its rich, luxurious flavor, its medicinal, health-producing qualities, the strawbery has no peer, and at this season of the year it has no competitor.

As soon as the snows of winter disappear from the surface of the earth, the beautiful, fresh, green eye of the strawberry exhibits its vitality, and every ray of sunshine seems to expand the tiny plant into opening leaves. The plant has not grown large before a cluster of buds appears, and a few weeks of sunshine develop these bunches of buds into flowers, rivaling the daisy in its modest beauty, and excelling it in delicacy and profuseness. Rapidly the season advances; the blossom, like all things delicate and beautiful, soon falls off, giving place to a small green berry. Sunshine and shower succeed, and in a very few weeks we have before us an object not only "exceeding pleasant to the sight," but "good for food."

Few, perhaps, of the thousands who purchase strawberries in our city markets ever contemplate this gradual unfolding, this process, step by step, which results in the perfection of this most perfect of all fruits; and yet it is a study replete with interest and fraught with delight at every stage, the climax of which is the enjoyment which the perfected fruit contributes to the sum of human pleasure.

> But who their virtues can declare? Who pierce
> With vision pure, into their secret stores
> Of health, and life, and joy?

See how eagerly the children watch the first ripe strawberry, and how industriously they fill the basket with this most beautiful of all the berry family. See how the burning fever is assuaged and the parched lips refreshed by this cool and luscious fruit, infusing as it does new vigor and new hope to the heart of the desponding invalid. Nor is the hope fallacious in many cases, for in the whole catalogue of simple, direct, and efficacious remedies, what is there more potent to temper and purify the blood and infuse healthful exhilaration? Linnæus was himself cured of the gout by this fruit, and how many have been cured since can never be told.

To produce strawberries in abundance, so that they can be enjoyed by all, is a philanthropic ambition worthy of a Howard, and we are sure it would be applauded by Florence Nightingale, with all the enthusiasm of her generous and sympathetic nature. The mission of the fruit-grower is as much one of blessing to mankind as if he confined his labors to fine-spun theories of metaphysics, or advanced the most abstruse theologies. And the man who masters the art of strawberry culture and practically overcomes the obstacles to its success, is worthy of high consideration, and will long be remembered by coming generations. The names of Linnæus, Downing, Wilson, Hovey, Longworth, Wilder, and Knox are already household words, and their hold on the gratitude of mankind is renewed every returning spring, when the strawberries which bear these names contribute their life-restoring nectar to millions of homes. These men, for their labors in perfecting seedlings and hybrids, cannot be overestimated. They have marked out a career which renders the road to fame and fortune much easier to their successors.

THE POOR MAN'S FRUIT.

The strawberry is emphatically the poor man's fruit. It does not require the new beginner in fruit culture to wait five, ten, or fifteen years for a crop that shall repay him for his outlay and labor. While the apple, peach, pear, and other prominent fruits are coming into bearing, the strawberry, with its rapid growth and early maturity, is furnishing the support necessary for a beginner with no other capital. Plants put into the ground this year become productive next, and with proper cultivation a large crop may be secured every season with as much certainty as can a crop of corn.

THE SOIL.

The soil for strawberries should be rich and well drained. Bottom lands, where the water can be got rid of quickly, answer well for the strawberry. Under-drainage is undoubtedly the best, but where this cannot be readily attained, the land should be laid as nearly as possible like a well graded street, highest in the centre, and very gradually sloping off to a ditch or gutter on each side. Summer fallow with repeated plowing is a good preparation of the soil. It should be thoroughly mixed and pulverized by cultivators or drags, and the condition known as "mellow" should be attained. All the weeds should, if possible, be destroyed. If fall planting be adopted, as is most convenient after a summer fallow, cool, moist days should, if possible, be secured for putting out the plants. Rather than wait too long, however, for such weather, where a large amount of planting is to be done, artificial watering should be resorted to. The roots of the plants should be kept moist, in thick mud, and sufficient water poured into the hole before putting in the plant to secure a firm hold in pressing the earth around the root.

PROCESS OF PLANTING.

I have sometimes seen elaborate instructions for spreading out the roots of the strawberry plant and placing it on a conical form of earth, covering the root with a trowel, etc. This may be very well where a small patch of a few yards only are to be planted, but is too slow a process where a large number of plants must be inserted in a short time. A dibble that will make a hole the size of a fifty-cent piece, is the best instrument to use in planting. If the soil be dry, some one should go ahead of the planter, making holes and filling them with water. The planter should follow with another dibble, putting in the plants and pressing the earth quickly around the root by another insertion of the dibble after the plant is in, care being taken not to cover the eye, which, in strawberries, is very near the root. None but strong young, plants should be used. My observation is that it is not the old root of the plant that grows. This is useful to hold the plant to the ground, but it is the new roots which strike out all around the plant soon after planting, which become the main support of the growing vine. This is why we claim the spreading of the old root unnecessary. Having the plants firmly set is of much more consequence than any theoretical arrangement of the old roots, and this can be secured most readily by one stroke of the dibble on one side of the plant, pressing the earth towards it. By this process a large number of plants can be put in in a day, and if the ground be moist and fine, as it should be, the new roots, which strike out just above the old ones, will spread in their natural form and sustain the plant in it's future growth. August planting can be made successful in this manner, no matter how dry the season, if careful hoeing and cultivation be attended to.

If, in extending a strawberry plantation, a sufficient number of plants cannot be secured, or if the beds from which the young plants are to be obtained are required to bear fruit, a large number of good plants can be raised by commencing a nursery bed for plants early in the season, say in May, in some shady place where water is at hand. The method is to prepare a small bed by digging, and then thoroughly drench it with water, or liquid manure, pulverizing it and making it level. Sometimes I have used boxes for this purpose. Little girls will soon learn to do the work which follows, and enjoy it very much. As you cut the runners from the fruit-bearing vines, reduce each eye to a convenient length and prick out in this nursery-bed, two inches apart, keeping the whole shady until the plants have well struck. These eyes will strike whether they have roots when pricked out or not. In this way we can obtain both plants and fruit from the same vines; the fruit will be larger and the plants more abundant than if you had allowed the runners to strike in their native beds, and the plants thus nursed will be strong and vigorous for planting as soon as the ground can be prepared,—some of them as early as July.

BEST TIME FOR PLANTING.

Of course, in speaking of summer or fall planting, I do not recommend it as the best. But when the ground has to be got ready the same season, it is better to plant in the fall, in ground that has been well prepared by summer-fallowing, than it is to plant in the spring, on ground rank with grass and weeds, it being much less labor to destroy these before planting than after.

Mr. Peabody prefers the first of July for planting strawberries, and with the method of having nursery-beds as just described, the operation of transplanting can be performed at any convenient time, commencing when the runners first pricked out of the nursery-bed become good strong plants. They can, and if the weather be very dry when transplanted, had better be removed with the soil about them from the nursery-bed to the rows, without checking their growth, making planting in summer almost as good as in the spring, so far as the next year's crop is concerned. The convenience of this method consists in having a ready and abundant supply of plants whenever opportunity occurs to use them, and without damaging the bearing vines by digging around them for new plants.

CULTURE.

If the ground was well prepared, the work of keeping the growing plants clean will be comparatively easy. Hoeing should be carefully performed, and if any of the plants become loose in the ground, as is frequently the case in light soils, the earth should be pressed around them. The benefit of early planting will be seen in the vigorous character of each plant for standing the trials of winter.

MULCHING.

Mulching is greatly recommended, and if done carefully with some light material, is beneficial, but in a breezy country like Western Michigan, it is not easy to find a light mulch that will remain where it is placed. A heavy mulch of barnyard manure is more likely to destroy the plant than is frost, and of the two I prefer to risk the severest winter, depending on the natural covering of snow, to any mulching I have tried. Last winter, while one of my neighbors destroyed a magnificent bed of Agriculturists, planted last spring, by mulching with manure, my young plants of Wilson's Albany, planted in August and September, were uninjured, and are now producing a crop of berries, of which a sample is now before you.

BEST VARIETIES—WILSON'S ALBANY.

With regard to the varieties to be cultivated, I regard this as a matter of taste for amateur culture. For profit, as a market berry, the Wilson's Albany has long stood at the head of the list, and maintains that position throughout the entire western country. As a bearer it is very prolific. It stands transportation well, and it is a good selling color. Its strong acid flavor is considered by some a strong objection to it. I think this its chief recommendation. It gives it character. For what is the lemon so highly esteemed in all cases of fever? It certainly is not its sweetness. It is its fine acidity. Your sweet strawberry, to my taste, lacks character and is insipid.

Not so with the Wilson's Albany. Whatever may be said of it, no berry has done so much to establish strawberry culture as a profitable business pursuit, as this much-abused variety, and no berry is so universally cultivated for market. On suitable soil and with good culture, Wilson's Albany can be relied on for a good crop every season. It is hardy, early, prolific, and when ripe, a delicious berry. It is sometimes gathered before it is ripe, and this has, in most cases, given it a bad reputation for flavor, which it does not deserve.

The Triomphe de Gand is a favorite variety for amateurs, and so is the Agriculturist, but neither are as hardy as the Wilson. The more recent varieties, although possessing many good qualities, have not shown themselves remunerative market berries, in the West, and have been generally discarded by growers. The varieties will, however, be a proper point for discussion.

For market purposes, even, I would not use the horse-cultivator. The small hand-cultivator is much better adapted to the purpose, and can be so adjusted as not to injure the roots, while keeping the ground stirred during the growing season will prevent injury from drought. The rows need not be more than thirty inches apart for using this implement, and the plants can be a foot apart in the rows. Any closer than this would be detrimental.

EXTENT OF CULTURE.

With regard to the extent of strawberry culture in the United States, some idea can be arrived at from the fact that on the second of this month a train on the Illinois Central railroad arrived in Chicago with twenty-three cars, all filled with strawberries, the aggregate weight of which was one hundred tons! This was only one train on one railroad! The idea that the business can be overdone here is a great mistake. Steamers from the mouth of Grand River could just as well take a similar cargo every day from the middle of June to the middle of July, as to take the moderate shipments now exported. Instead of here and there a patch of two or three acres along the banks of Grand River, at Bass River, at Eastmanville, at Lamont and Grandville, every landing should have hundreds of acres from which strawberries are gathered. Our crop here arrives in Chicago and Milwaukee in much better condition than the crop of South Illinois, because our water transportation is so much better for this delicate fruit; and being later, the strawberry season would be greatly prolonged in the cities across the lake, by the more extensive culture which the future will undoubtedly develop. Small shipments are always unprofitable; large shipments are alone productive to both grower and dealer.

As a society, we should endeavor to so extend the culture of each fruit adapted to our climate as will give Michigan prominence, not only in the quality of the fruit produced, but likewise in the quantity, so that the eastern shore of our noble lake shall become the resort of ambitious buyers, who will find it profitable to come hither to purchase our products. To this end a large quantity as well as a good quality of fruit must be produced. A small shipment attracts no attention, and can scarcely find a purchaser. A large cargo commands the market, and secures the competition of buyers for the prize.

Much has been said about the profits of strawberry growing. When well attended to, the strawberry interest is a good one, and no business is profitable that is not looked after.

My aim has not been to give full directions in strawberry culture,—this is done in many useful little manuals,—but to throw out such suggestions as have occurred in the course of a few years' experience, and to awaken an ambition among our fruit-growers to excel in the production of what Mr. A. J. Downing describes as "Arcadian dainties," * * "the most wholesome of all fruits."

CONCLUDING SUGGESTIONS.

The principal objection to the very extensive production of strawberries,—the perishable character of the fruit,—seems likely soon to be overcome. At a recent meeting of the Horticultural Society of Black Lake, Captain Walker stated that by enclosing strawberries in a dry, closed box, and placing the box in an ice-house, he succeeded, last year, in keeping strawberries fresh three weeks, and he had good reason to believe the same plan would have kept them good twelve months. This plan should be fully tested, as ripe, fresh strawberries, at seasons of the year not now provided with them, would be a luxury for which many would be willing to pay handsomely. And with proper facilities provided, strawberry-dealers need never crowd the fruit into the city markets so as to bring down the price below what is profitable, to the producer and dealer. Safes will undoubtedly be constructed for this purpose, and our favorite berry may become an article of every day consumption, instead of a luxury of a few summer weeks. Should this be accomplished, the production of strawberries will become as extensive as that of any staple article, and much more profitable.

I have only time to say that the cherry (the other fruit on our programme) is gradually coming into favor among growers. I know of one fruit-grower who is planting 3,000 cherry trees, and others who are intending to increase their plantations. Although the improved varieties of cherries are very fine, as a general thing they are not good bearers, whereas the common Duke Morrello—the sour red cherry—are very prolific bearers. For throat diseases, the cherry appears to be a specific of great value. The sour cherry is but little liable to the ravages of either birds or insects, and will bear, year after year, most abundantly. As a canned fruit, the sour cherry preserves its natural flavor and color better than any other fruit, and dried cherries (stoned as they can be by a new patent invention) always command a very high price in market,—much higher than our best imported fruits. The Morrello cherry will grow vigorously and bear abundantly on the north side of a hill, or in many positions where other trees would fail and become unproductive. For these reasons I think we should be safe in recommending the extensive culture of the Morrello cherry.

ADDRESS OF HON. FLAVIUS J. LITTLEJOHN,

BEFORE THE MICHIGAN STATE POMOLOGICAL AND THE KENT COUNTY AGRICULTURAL SOCIETIES, AT THE FAIR GROUNDS, GRAND RAPIDS, SEPTEMBER 29TH, 1870.

LADIES AND GENTLEMEN:—If I shall presume to deviate somewhat from the beaten track of addresses at annual fairs, I trust that you will pardon the digression.

The spontaneous productions of the earth in their primal condition are seldom adapted to meet the varied wants of man. In the system of providential economy, both mental and physical labor are essential prerequisites for the attainment of many things alike desirable and necessary for human use. The seasons may wheel their accustomed rounds—the sun may pour forth his effulgent beams—fertilizing rains may descend upon the lap of the earth—genial breezes may continue to fan the luxuriant verdure mantling the hills and carpeting the valleys, and yet *man,* without a constant draft upon his own resources, would perish from alternate heat and cold, or starve in the vestibule of nature's granary. The stern decree, "In the sweat of thy face shalt thou eat bread," still, rests in pristine vigor upon the whole brotherhood of humanity. The fearful lesson has been written a thousand times upon the page of human suffering,—meagre famine induced by want of proper forecast, by indolence, or by crime,—has often exacted the most appalling tribute of life.

Even mental and physical activity, strenuous and continued, cannot alone purchase immunity from want. The noblest conceptions of genius have been penciled on canvas and chiseled on marble, whilst the pangs of hunger were consuming the artist. At the very time when myriads of husbandmen were compelled to labor in rearing useless pyramids, men died of starvation within their shadow. Impelled by ambition or revenge, intellect has schemed, and men have strenuously labored to make desolate the fairest portions of the globe, trampling to dust the bounties of nature garnered up by care and prudent forethought, and crushing out the image of God from the face of humanity. Generally, as a direct result, the recoil of pinching famine has proved more fearful than the sword.

Intellect, then, must operate in useful channels, and labor must be skillfully directed and diligently applied to the legitimate pursuits of industry.

In the midst of gratulations for the eminent success which has so fully crowned your past efforts, it is natural for you to indulge in comparison and retrospection. You instinctively glance over other communities, nations, and countries, for the proper measure of your own attainments and true position in the scale of civilized being. The result to you must be highly gratifying, when tried by every rational standard of morality, of social happiness, general intelligence, or public prosperity. As it regards matters of practical utility, you may safely conclude that in mechanic arts, inventions, and scientific discovery, our country has no superior.

In the line of agricultural production, whether of stock or cereals, grains, grasses, vegetables, or fruits, with proper allowances for soil and climate, you have abundant cause for an honest pride in your community efforts.

Upon occasions like the present, we are also inclined to glance away backward over past ages, and to mark the progress made, at successive eras, in human affairs. We become curious in observing by what advancing steps the physical wants of man have pushed his intellect into investigation, research, invention, and discovery, until the actual circle of human power over elemental nature has become immensely enlarged.

Ill-shapen and rude as must have been the first tools of trade, still Tubal Cain, the primal artificer in brass and iron, unquestionably furnished the crude models from which successive improvements have finally wrought out the innumerable variety of useful, elegant, and ornamental implements and utensils. Men were first clad in skins, and then with garments made of more flexible materials, elaborated by intellect for the occasion. But the twirling distaff and spindle, with the spool and shuttle of ancient matrons, have been entirely eclipsed by the modern spinning-jenny and power loom, performing the labor of myriads of hands, and daily throwing off incredible amounts in textile fabrics of surpassing beauty and of gossamer fineness.

Nor are these the only changes compassed by science and skill. Who, whilst viewng the glorious models of naval architecture, in this our age, would dream of tracing out their archetypes in the dug-out or frail canoe once creeping along the coast of Tyre, or in the more recent but clumsy Trireme of Grecian pirates in the Levant? Who, as he now in fancy contemplates the rude cabins and huts erected by Adam and his sons, would think it possible that genius, from such a starting point, could ever reach the proportion, style, and finish, much less the grand architectural design of St. Paul's, the Pantheon, or of St. Peter's at Rome?

What modern astronomer, standing in yonder observatory, as he traces with matchless accuracy field after field of the starry vault, measuring the planets in their orbits, and assigning to each its law of motion, does not forget to remember that his first lessons were derived from the nightly watchings of shepherds on the plains of Chaldea?

Who now detects in the mystic stonewrought hieroglyphics,

of some coptic priest of Egypt, the germ of the printer's art, or the embryo of the power press, daily throwing off to millions the free-born thoughts of cultivated minds? Or, stranger still, deems it true, that in such mystic groupings and durable stamping of thought are shadowed forth captive lightning, laden with telegrams, and racing with light?

These are by no means the only inroads made by intellect upon the silent domain of nature. Geology, mineralogy, and chemistry are daily presenting us with astounding developments. I have time simply to point your attention to the steam engine, flying in mighty power, with its lenghtened train, across kingdoms and continents, or driving leviathan ships, with sure direction and resistless force, over the storm-crested billows of the ocean.

Time and space have thus, for the purpose of travel and transit, been practically annihilated. Differences in soil and climate have been adjusted on a scale more nearly approaching equality. By a rational division of labor, far greater excellence has been attained in the several branches of industrial pursuits. The facilities of land and water communication now open the world as a market for the producer, whilst the products of every climate may be easily reached by the consumer.

Measured thus by the standard of present attainment and progressive improvement in the various arts, we are fain to conclude that civilization is now in advance of any former period.

We have thus reached a point of view whence we may still more closely inspect the three grand departments of human industry: Agriculture, Manufactures, and Commerce. In this general division, agriculture includes all the products of the soil resulting from skill and labor, together with the avails of flocks and herds. By manufactures we understand all the products of the mechanic arts, whatever material may be used or means employed by the artificer, whilst by commerce we mean the export and import, the sale, exchange, or other domestic disposition of all marketable commodities.

As thus defined, each of these general departments has challenged the attention of our race from its earliest foothold on earth; and yet, although each originated in the necessity or convenience of men, still there has ever existed not only a perceptible, but marked difference in their public estimation and progress. Despotic power, national pride, and aspirations for personal renown, have contributed to swell the amount of the distinctive difference thus created.

Commercial enterprise has ever been cherished, both as a source of profit to the merchant, and for the articles of taste and comfort it has been wont to furnish, alike to palate and person. We now speak of the extent and variety of traffic, and not of the facilities for land and water transit. Beasts of burden overland, and clumsy craft for the water, were the principal means employed for ages. As recently as the time when Venice, through her merchant princes, controlled the commerce of Southern Europe and the Mediterranean, very slight improvements had been made in naval architecture, always excepting the high-beaked, elegant, scull-driven gondolas of her own canals.

Mechanic art, in its extended sense, was early pressed to its utmost tension for supplying, not only the wants and conveniences of men, but to meet the large demands of taste, whether ostentatious, voluptuous, or refined. Genius and talent were alike subsidized, whilst invention, discovery, and improvement were, at times, rewarded with the contents of public treasuries, to be replenished again by cruel exactions from agricultural classes.

When viewed with an understanding eye, a popular error seems to have prevailed in regard to the genius and attainments of the old Roman world. As a race, they could imitate, but not invent. They could furnish hand-pattern artisans, but they borrowed from Greece, Egypt, and from Assyria, their exquisite designs and models, and imported their principal architects. With a few noble exceptions, they were neither a literary nor scientific people. They were servile copyists. Even their system of jurisprudence, down to their Emperors, was derived from Greece, and compiled by an Ephesian. Their refinements were ostentatious, their pleasures were sensual, and their amusements often brutal. Their genius was for war, and their principal progress was in human slaughter.

Pardon this digression. It was wrung from me by reading upon their gilded palaces, their majestic temples, their triumphal columns, and their eighty thousand seated amphitheaters, the thrilling record of blood and famine in their impoverished provinces. Why should it then be deemed incredible that the Etrurian plow of Cincinnatus, composed of a wooden crotch, and harnessed by thongs of rawhide to the horns of the oxen, should have continued the Roman plow for many ages? Why should their uncouth reaping-hook, with its still more crooked left-hand accompaniments, have been superseded by a more improved utensil? Why should agriculturists have taken delight in extended fields of waving grain; in sleek, well-fed flocks and herds; or in spacious barns and well-filled granaries? Such possessions would have proved the certain signal for rapine and plunder; for personal violence, and perchance for murder. No! their only shield was apparent destitution; their only granaries were hidden excavations in the earth, whilst the forest or the mountain-steeps furnished a hiding place for their sheep and cattle. The imperial tax upon all their known possessions, in itself onerous, was usually doubled by the rapacity of their governors and their numerous officials.

With slight modification, agriculture has labored under similar embarrassments the world over, until modern times. Wherever the relation of lord and vassal exists, whether as serfs of the crown, or by any of the forms of feudal tenure, there you will never find either agricultural prosperity or progress. True, small parcels of land may be farmed out by the lord of the manor, to his retainers and dependents, and be, by them, kept in productiveness by hand-trenching tillage.

But this is not what we, in America, call farming. When we speak of a farmer, we mean both the owner and cultivator of broad acres. Five hundred farms in portions of France would hardly equal in extent the possessions and enclosures of individual farms in our Northwestern States.

We have thus endeavored, in a cursory manner, to point out the wide difference in attainment and progress heretofore existing between the three great departments of industry. We have also endeavored to exhibit the leading causes for this essential difference.

Agriculture is, emphatically, a peaceful occupation. It demands social order, with effective and stable laws for protection from aggressive inroads. It requires science to direct, intelligence to guide, industry to accomplish, and a fee simple in the soil for its successful management. And, then, in the wide range of human effort, there is no field of human enterprise more useful, more honorable, or more promising in results, than American farming. The first settlers of the Union were mainly agriculturalists, and land tillage has all along been nominally regarded as taking a high rank in the pursuits of our citizens. And yet, until within a few years, what marked improvements were introduced into this system? What useful inventions to facilitate cultivation and to lessen the burden of labor for man and beast, had been regarded by our husbandmen? What discoveries of science had been generally adopted? What associations for an interchange of views and a comparison of products were in active operation? What newspaper, sheet or periodical, laden with the gleanings of experience and scattering agricultural intelligence, was ever either circulated or read? What careful examination into the properties or different soils, and their adaptation to the various kinds of grasses and cereals, had ever been prosecuted to any satisfactory results? And what intelligent system of rotating crops and recuperating the exhausted energies of the soil by rest and fertilizing ingredients, was received as the basis of operations by any considerable number of our farmers?

The truth is, that agriculture, for more than a century, was conducted negligently, shiftlessly, without system, and without the requisite amount of intelligence, in every part of our country. The unavoidable result was everywhere experienced. Farmers became impoverished; the average yield was lessened in quantity, and depreciated in quality; labor was but poorly requited, and the vocation (for it was not then deemed a profession), fell into disrepute. What wonder, then, that the son, forced to toil in the treadmill routine of his father's unthrift, should have acquired a thorough disrelish of all that pertained to the business, and have sought for himself some more congenial employment? But, thanks to the persevering energy, skill, and science of a few large-hearted and philanthropic individuals, agriculture has been rescued from opprobrium, and resurrected into activity and productiveness throughout the country. The series of spectacles like the one of to-day, now actually exhibited in every part of the land, are evidences of progress made and of triumphs achieved, and are also the harbingers of still higher attainments.

The hitherto dormant energies of our agricultural classes have been successfully aroused, and the broad and deep furrow will not only render earth's surface beautiful and productive, but will in turn penetrate far downward into the hitherto sterile domain of uncultivated mind. There are but few branches of human learning that will not ere long be drafted into the ranks, and render effective aid in the field operation of practical, scientific farming. Men will no longer, as the poet writes, follow the plow, "whistling for want of thought."

Competition will force into use all the valuable improved implements in husbandry. As the rich furrow yields to the patent share, and takes its place with but trivial effort, the physical science and mental faculties of the operator will have a wide margin for studying the broad volume of nature thus being opened up before him. The experimental tests previously established by combination, comparison, and analysis, will be diligently applied. Geological indicata will be observed; chemical affinities detected, and the elemental ingredients, and fertilizing properties of the upturned soil will be fully arranged for the class-table of seeds. Skillful manuring will go far towards supplying deficiencies and preserving exhaustion.

Judicious management in rotating and maturing crops will relieve the soil by skillful drafts of gaseous nutriment from the atmosphere. The aftermath of dry meadows will be kept sacred from scythe, hoof, or tooth. To tillable land a season's rest in clover will be frequently awarded.

A corresponding improvement will be visible in the condition, breed, and blood of all domestic animals, whether reared for home use or market abroad. Color, shape, size, attitude, activity, muscular strength, beauty in proportion, and graceful movement, weight, early maturing in taking on fatty matter and fleshy fibers,—all these will be severally considered, and the respective properties wanted will be made the basis of judicious selection.

No gaunt, long-nosed, apron-eared, large-jointed, crooked-back specimen of swine will longer offend the eye in the street, or at two years of age, with a clear live weight of 125 pounds, be thrust into a sty, to wallow in filth, and be reduced in six weeks to killing order, upon flint corn in the ear.

Bald-headed, hairy-limbed, light-quartered sheep, with a scanty covering of coarse wool upon back and sides, will be entirely displaced by Leicesters, Southdowns, Saxons, and Merinos.

Neat cattle of high breeds and pure blood will range in green pastures, well fenced, shaded, and watered in summer, and in winter will no longer shiver over their scanty allowance of refuse fodder in open yards, or wend their way two hundred rods along slippery paths in search of water, and then, like a Mahomedan at his devotions, be compelled to take it kneeling.

Horses of approved varieties, by close breeding, keep, and proper training in position, step, gait, and general movement, will become what nature designed them to be,—the most active, useful, and beautiful of all domestic animals. We shall no longer behold them, with head and tail projecting toward the ground at an angle of forty-five degrees, creeping along with a ludicrous mixture of all kinds of gait, upon a couple of legs and a pair of setting-poles, galled upon back and breast, troubled with fistula, glanders, and heaves, and limping with pin-hip, ringbone, curb, and spavin.

Should the truth be honestly spoken, there can be no doubt that our domestic animals have too often been stinted by cruelty and dwarfed by neglect, until all beauty, size, and proportion have nearly disappeared. Let the sin and shame rest where they belong.

The change of which we have thus far spoken is already extending itself to the conveniences and comforts of rural life. Taste, neatness, and methodical arrangement are already evinced in the buildings, fixtures, fences, orchards, and gardens of the farmer. Shade and ornamental trees are taking their proper position in the yards, along the highways, around the dwellings, and at judicious points in every field. Flowering shrubs and fruit-bearing vines are now being clustered and trained over trellises, delighting the eye and gratifying the taste. The choicest varieties of fruit from vine, shrub, and tree, each in its season, now add to the pleasures of the family, and give a zest to toil itself.

At this precise point of my address, I am reminded by the public call for this meeting, by the splendid specimen varieties of fruit put on exhibition here, and by the attendance of representative gentlemen from various parts of the State, that a distinct branch of our main theme now presses itself upon our attention, and demands more than a passing notice. The Michigan State Pomological Society are entitled to have their specialties considerately treated.

Fruit culture, in proper localities, is a pleasant field for human care and effort, full of taste and beauty, and not only richly remunerating labor, but largely contributing to the comfort and luxury of living. It is a field of operation where a good degree of intelligence, sound skill, observation, and experience must be combined to achieve excellence or to insure success.

In prosecuting fruit culture, men must have reference to the elemental condition of the proximate natural world. They must not ignore the immutable laws which govern locality and latitude. They must ascertain and recognize the normal condition and aptitudes of soil and climate; the appropriate degree of heat and moisture, and the length and general adaptability of the seasons requisite for the proper development and maturity of any given variety of fruit.

Our observation and human experience alike confirm the fact, that, by the economy of natural law, a zone has been assigned to different species of fowls, fish, and animals. To each, in its appropriate sphere, all the appliances are at hand for its highest development.

But, as in the animal, so in the vegetable and fruit kingdoms, many varieties of the latter are susceptible of being transplanted, and reared to fair perfection, in zones and climates, and even soils, where they are not indigenous. This process, however, demands additional care, skill, and knowledge of all their peculiar properties and requirements on the part of the grower.

In glancing over the map of earth's indigenous productions, we readily discover that ours is not a favorable portion of the globe for ravishing displays of nature in the floral kingdom, and that we are also debarred the pleasure of witnessing the expanding loveliness or maturing richness of the spices and luscious fruits of the lower latitudes. Our conception of equatorial landscapes, where Nature is draped in her heaviest verdure and her must gorgeous floral attire, must fall almost infinitely short of the beautiful reality. So also of the fruits, of plants, shrubs, and trees of those regions, exotics here, and never reaching us without being denuded of their freshness, and bloom, and tint, their exquisite odor, and their luscious taste.

But whilst frankly admitting our inferiority to the tropical climes in the particulars thus alluded to, we may readily ascertain that we enjoy countervailing advantages incident to our position, and fully compensating us for the deprivation of those exquisite though simple pleasures of taste. We may also have fruits of our own planting, handling, and maturing, in great abundance and variety, and fully adequate to supply our substantial wants, and at the same time both satisfy and gratify the palate of the fastidious epicure.

Whilst fruit-growing in Michigan, generally, may be pronounced a success, yet we, of the western portion of the State, shall not be deemed arrogant if we affirm that the peculiar fruit zone of the Northwest is limited to a belt of country ranging in breadth from fifteen to forty miles, and stretching along the Eastern shore of Lake Michigan. Outside of this special region, however, various kinds of fruit may be grown and matured in different sections of our Lower Peninsula. The various small fruits indigenous here, and even growing spontaneously, are the strawberry, the raspberry, the currant, the gooseberry, the blue and whortleberry. You all recognize the fruits. All these are susceptible of improvement by skillful cultivation, and may be made a source of large profit as basket fruits, as well as by conversion into jellies, wines, canned fruits, etc. Of the large fruits growing on climbing vines and trees, we have a large and increasing variety, many of them taking as kindly to our soil and climate as if indigenous to this region.

Through the economy of natural laws, as well as by our judicious selection and training, our distinct varieties may be again subdivided into classes.

As general standard varieties, we would designate apples, pears, peaches, plums, apricots, nectarines, quinces, Siberian crab, cherries, and grapes. Of these, as sub-varieties, which may be classed to mature in gradation, and thus adapted to meet the wants of successive seasons of the year, and possessing excellent properties, and sufficiently prolific to warrant cultivation, we name, of apples, the Astrachan, Sweet Bough, Duchess of Oldenburg, Strawberry, Fall Pippin, Cayuga Co. Red Streak, Snow, Jersey Sweet, and Maiden's Blush; and then the Baldwin, Wagener, Golden Russet, Rhode Island Greening, Wine Sap, Tallman Sweet, Hubbardston Nonsuch, and Northern Spy. Of peaches, Early and Late Crawford, Early Barnard, Smock Free, Stump the World, and Hill's Chili. Of pears, Bartlett, Flemish Beauty, Louise Bonne de Jersey, Duchesse d'Angouleme, and White Doyenne; and of grapes, the Concord, Delaware, Clinton, Iona, Isabella, Diana, Northern Muscatine, Ives' Seedling, and Rogers' Hybrid. Other equally excellent sub-varieties may have been overlooked, or are unknown to the public.

From the remarks already submitted, we must all admit that fruit culture, as a science and a branch of human industry, is entitled to high rank, not only as a source of profit to the producer, but as a regulator of the blood and stomach, largely contributing to the health and comfort of the consumer. Without its presence, in some of the myriad forms of its preparation for our use, what refreshment table would be pronounced fully set?

But to achieve success in fruit culture requires unwearied oversight, skill, and persistent effort. The ground must be selected with due regard to the requisite amount of light, heat, and air. Then come mulching at the root, staking, training, pruning, foreshortening, and, for vines, trellising and cutting back. Preventives against blight and mildew must be sought for, and a constant warfare waged against the army of fruit and tree destroying insects. The borer, grub, caterpiller, and the root-worm, will assault root, bark, body, branch, and foliage, whilst the curculio, with its numerous allies, will sting and plant their fatal larvæ in the tender germs and incipient fruit.

It thus becomes apparent that protracted care and labor are imperiously demanded for successful fruit culture. A detailed statement of the mode and manner of practically applying this labor, we must decline to enter upon, and for two cogent reasons. First, because our masters in the art are present before us; and second, because our personal knowledge and experience have been far more exclusively devoted to the consumption than to the growing of fruit.

In concluding our remarks upon this special topic, it is peculiarly gratifying to be able to say that the measure of your past success in fruit culture is full of hope and promise for the future. For these auspicious results furnish a sure guaranty that intelligence, skill, and energy have been already embarked in the enterprise, and that the Michigan State Pomological Society will make this, their first anniversary, a bow of promise, spanning our whole Peninsula, flooding it with brilliant beams from its particolored groundwork, and pledging to us a renewal of the Garden of Eden on the eastern shore of Lake Michigan.

From the slight glances we have been able to take of the various fields of labor in the great department of agriculture, we may glean the great lesson of the hour, that both excellence and progress must depend upon intelligence and effort, abiding in and springing from a well developed manhood.

That this lesson should have a peculiar force and significance with our rural classes, may be demonstrated in a few words. First, because the precious deposit of our national prosperity, glory, and honor, the prevalence of intelligence and virtue, the maintenance of social order, the perpetuity of our free institutions, and the hopes of our race are largely in the keeping of our agricultural classes.

And second, there is far less of trust and stability in other departments. With them wealth is more emphatically power, and that power is liable to be abused. In our larger towns and cities, with their mixed population and heavy influx of foreign elements, a wide departure from the republican simplicity of our forefathers is clearly perceptible.

In the earlier history of our country, and down to a recent period, it has been an acknowledged fact that our people have stood foremost in the leading qualities and essential elements of a well developed manhood. The reason is obvious. In the rapid opening and settlement of the country; in the activity, thrift, and economy necessary with many to keep want from the fireside; in the surging tide of advancing trade and commerce, and the bustle and activity at business centers, and in the ceaseless struggle to furnish marketable products, amply supplying the demand, the constant strain on body, brain, and muscle was calculated to round out and ripen up in full development our American manhood.

But now, in the widening and shifting fields of human effort, and the changing aspect of business affairs, new motives operate, and new incentives are brought to bear. In the ambitious yearnings for power and place, and the selfish cupidity for pecuniary accumulation and family aggrandizement; in the ostentatious and fashionable displays of opulent *parvenus*, and the graspings and revelings of those below to attain some higher grade or more exclusive circle, the tax and strain upon the brain and muscle, with the forced and unnatural direction of these new competing efforts have become intense and exhaustive, whilst the average moral nature of society is being warped and coerced by design, or dwarfed by neglect.

Who then shall longer maintain the doctrines of our ancient republican simplicity, or practice the stern virtues of our forefathers, if the agricultural classes fail us? Where shall we look for mature judgment, for enlarged views, and integrity of heart and life, if not to them? In the hour of the nation's peril, where shall we search for the man with head to plan, with heart to brave, and hand to execute, if not to them? How immensely important, then, that the youth, in all rural districts, should be physically, mentally, and morally trained and schooled into a round and ripe development!

But grant that the flag of our Union shall continue to wave over the length and breadth of our country, and that our institutions shall continue to afford protection to all beneath the shelter of that flag, until the valley of the Mississippi, the steeps of the mountains, and the far off shores of the Pacific, shall teem alike with a dense population. The vast preponderance of commercial wealth and power will still unquestionably be centered in the cities, along the seaboard East, and West, and South; manufacturing enterprise is wont to seek out particular localities, and its wealth and power might remain concentrated in one extremity of the Union.

Where then shall be found an all-pervading interest, permeating the grand masses of society, and imparting freshness and vigor to the arterial tides of life, pulsating from the great heart of the Republic to its extremities? Agriculture, with its varied productions and diffusive interests, must solve the problem.

Aye! Practically, scientific agriculture will ever be found adequate to supply the wants of all, still binding the extremities to the center, and that center to freedom, as anatomical ligatures unite the bones in healthful articulation.

Like as among the Aborigines, the belt of wampum, passing from tribe to tribe, was the precursor and herald of the pipe of peace, so shall the sheaf of wheat remain from ocean to ocean, and from the Lakes to the Gulf, the symbol of amity and the token of brotherhood.

AMONG THE ORCHARDS AND VINEYARDS.

The following report, made by Messrs. Clubb and Fasset, Committee on Orchards, September 29th, 1870, addressed and submitted to the President of the Western Michigan Agricultural and Horticultural Society, which has its headquarters at Spring Lake, in Ottawa County, is inserted in this place to give information respecting the system of orcharding now being introduced and established in that county. It is especially the object of the State Pomological Society to encourage and promote the growth of orchards and vineyards, and demonstrate the feasibility and profitableness of broad and extended culture. The officers of local Agricultural and Horticultural Societies throughout the State are earnestly invited to furnish information respecting the orchards and vineyards of their several counties; and information and statistics of this kind from reliable and intelligent sources, will be given publication and wide circulation.

The committee to whom was intrusted the inspection of such orchards and vineyards as were entered for competition, respectfully report that they have performed the duty assigned them, and recommend the following

AWARDS.

The best three-year-old apple orchard—First premium, Jas. B. Soule; second premium, Charles E. Soule.

Best one-year-old pear orchard—First premium, Mr. Job Sessions.

Best three-year-old quince orchard—First premium, Rev. H. Beckwith.

Best bearing peach orchard—First premium, Chas. E. Soule; second premium, Geo. Seagrove.

Best four-year-old peach orchard—First premium, Thomas Petty.

Best three-year-old peach orchard—First premium, Walter G. Sinclair; second premium, Ambrose L. Soule.

Best two-year-old peach orchard—First premium, E. L. Treadway; second premium, Thomas Petty.

Best one-year-old peach orchard—First premium, Frank Hall; second premium, Orrin Douglass.

Best bearing vineyard—First premium, Messrs. Cutler & Savidge; second premium, Thomas Petty.

Best five-year-old vineyard—First premium, Jas. B. Soule; second premium, A. L. Soule.

Best four-year-old vineyard—First premium, Thomas Petty.

Best three-year-old vineyard—First premium, Thomas Petty.

Best two-year-old vineyard—First premium, Chas. E. Soule.

Best one-year-old vineyard—First premium, Timothy Hall; second premium, Job Sessions.

REMARKS.

As is customary in a report of this character, your committee also submit remarks in reference to the several orchards and vineyards entered for competition, some of which will indicate the reasons which actuated the committee in recommending the above awards.

Mr. James B. Soule's three-year-old apple orchard has made good promising limbs. Mr. Soule's system of pruning encourages several good branches, which are led out in different directions so as to form a well balanced tree. We prefer this method to heading in short and producing numerous shoots.

Mr. Chas. E. Soule's apple orchard of 350 trees, planted in 1867. These trees, which have not been cut back, but simply thinned out, have made good limbs, and are doing finely. Those which have been headed in make bushy growth of small wood, which we deem objectionable. The orchard is thrifty and promising.

Mr. Job Sessions enters his pear orchard, planted with 50

pear trees this spring. His pears appear to be all alive and doing remarkably well for the first season. This orchard well deserves the first premium awarded it.

Rev. Hiram Beckwith, residing between Spring Lake and Crockery, has the best quince orchard we have seen, and being the only one entered, is properly awarded the first premium. It contains 200 quince trees, planted three years ago. Those pruned by Mr. Ganzhorn are doing well, the plan being to encourage the growth of several limbs rather than numerous slender shoots.

Mr. Chas. E. Soule's peach-bearing orchard, planted in 1863, has done admirably. It contains 300 trees. His Barnard trees averaged him $10 a tree this year; his Old Mixon, $5; his Seedlings, $2; his Early York, $1, and his Crawfords, nothing. This little orchard yielded $700 in 1868; $100 in 1869, and $400 last year. It is a miscellaneous orchard with regard to varieties, but its bearing qualities must be regarded as very superior.

Mr. George Seagrove has sold a considerable quantity of peaches from his orchard, even this year, but so far as we could ascertain it was less than the amount sold by Mr. C. E. Soule. Both are good bearing orchards.

Mr. Thomas Petty's peach orchard, planted in 1867, contains about 700 trees. It is located on the south side of a hill, and its fine, even rows of trees contrast with the white sand upon which they grow so luxuriantly, presenting a scene from the lake of surpassing beauty, being washed by the water below and tipped with clear blue sky above. The trees have perfected their growth, and the wood is ripened to the tips. The growth has been ample, and the show of double buds indicates a heavy crop for next season.

Mr. Walter Sinclair, the zealous Secretary of the Western Michigan Horticultural and Agricultural Society, enters his peach orchard, planted in 1868. For healthy growth and uniformity, except where trees were replanted, it is difficult to find the superior of this three-year-old orchard. It is on the north side of the railroad, and with a slightly northern slope toward Spring Lake,—a very favorable aspect for peaches, especially with ample water protection as this has.

Mr. A. L. Soule's peach orchard, planted in 1868, contains 1,000 very handsome trees, with a long and gradual slope to the southwest. It is one of the most promising orchards on Spring Lake, and cannot be easily surpassed of its age.

Mr. Charles E. Soule's peach orchard of 300 trees, planted in 1868, is on a fine elevation, and located on a peninsula, giving it water protection in almost every direction. The land is rolling, and the growth this year has been all that is desirable. The wood is ripening rapidly to the tips.

The peach orchard of Mr. J. B. Soule contains 2,000 trees,—800 Early Crawfords, and the balance miscellaneous. It has made as much growth as is safe, even on this favored location, and begins to look like a bearing orchard. Show for fruit next year, excellent.

Mr. Elmore A. Treadway enters his splendid peach orchard, planted in 1869. The growth of these trees is surprising, and yet it is ripening up rapidly, and the terminal buds have formed. The double buds are quite numerous, and the prospect of a crop, even in 1870, is good.

Mr. Thomas Petty's peach orchard, planted in 1869, contains 500 trees. For regularity of growth, and uniformity of size and trimming, this orchard stands high in our estimation. Its aspect—sloping to the south—gives it a fine appearance, although it is probable a northern aspect would be better for the fruit. The fine elevation of most of the trees, however, renders their chance for avoiding frost excellent.

Mr. Frank Hall has a peach orchard of 500 trees, planted last spring. Not one of the 500 failed to grow, and all appear to have made a healthy and vigorous growth. The long rows of uniform trees give high promise of future productiveness. It can scarcely fail to become one of the best orchards in the township. The varieties planted are 160 Mountain Rose, 75 Smock, 25 Crawford's Late, 25 Crawford's Early, 25 Hales, and 75 Old Mixon.

Mr. Orrin Douglass, Stahl's Bay, near Fruitport, enters a peach orchard planted last spring. It is at present surrounded by timber, being a new clearing. It contains 1,200 trees, mostly branching very low. They are very vigorous trees. Buckwheat sown between grew almost as high as the trees. This, probably, was a good precaution, to prevent an over-growth. The trees have now completed their terminal buds, and the wood is ripened to the tips, although the leaves remain perfectly green. The varieties are good: 600 Early Crawford, 250 Early Barnard, 250 Smock, 50 Early Hale, 100 Stump the World, etc. Mr. Douglass, with but little help, has cleared 18 acres this season, and has raised an abundance of corn and buckwheat, besides planting the orchard. Such industry and perseverance deserve the highest commendation, and but for the fact that 300 out of 1,500 trees died, and the newness and consequent imperfect preparation of the land, we should have accorded this the first premium. Soil, sandy, with clay subsoil.

Messrs Cutler & Savidge and Mr. George Seagrove entered their orchards, but when your committee called Mr. Seagrove was absent at the State Pomological Exhibition, so that we could not ascertain the particulars as to the amount borne by their orchards and vineyards this year, and must therefore pass them by with the remark that these are the best cultivated orchards on Spring Lake. Especially is this the case with those of Messrs Cutler & Savidge, Mr. Seagrove evidently devoting his chief attention to them, somewhat at the expense of his own.

Mr. Jas. B. Soule's Concord vineyard was planted in 1866 It has borne a good crop this year, and has made very long growth, one vine measuring 19 feet.

Mr. Ambrose L. Soule's is a fine Delaware vineyard of about

300 vines, planted in 1866–7–8. These vines have made immense growth from the roots this year, and bid fair to bear a large crop next year.

The vineyard of Mr. Thomas Petty has produced this, its fourth season, 3,500 pounds of grapes. It contains 1,100 Concords, 100 Catawbas, 50 Isabellas, and 300 Delawares. It is on a hill rising some 60 feet above the lake. The trellis is formed by posts with light rails top and below, with lath about 10 inches apart, running perpendicularly, like pickets. The vines, although bearing the above crop, have made a good growth this season, and the prospect is favorable for a heavy crop next year. Sandy soil.

Mr. Thomas Petty's vineyard of 500 Ives' Seedlings, is of three seasons' growth. It has similar advantages of aspect and elevation of the other vineyards, has made a good growth, and indicates that the Ives' Seedling will be a successful grape in this region. Soil sandy.

Mr. Chas. Soule's Delaware vineyard of 100 vines was planted in 1869. It has made surprising growth for the Delaware in its early stages.

Mr. Timothy Hall enters his vineyard of 500 vines, mostly Concord, planted this spring. It has made good substantial growth, although the soil looks like a pure fine gravel. His plan is to keep the ground stirring while the plants are growing, as preferable to manuring. The health by growth is a good recommendation of this theory. When the season of warm weather is protracted as this year, there is probably no danger from over cultivation in a soil of this character. Mr. Hall's vineyard has this advantage over that of Mr. Session: the land was prepared the year before planting and the soil is naturally warmer on account of its gravelly character.

Job Sessions' vineyard of 200 vines was planted this spring. It contains 60 Delaware and 140 Concord vines. The growth is good, especially as this land is new, and Mr. S. only commenced this year. Mr. Session is a careful and zealous cultivator. His place is a credit both to his judgment and industry.

Your committee conclude with the assurance that the present condition of all the orchards visited is highly satisfactory and promising. Never was wood better ripened or buds more promising of an abundance of peaches next year, and nothing but the most excessively severe winter can affect the buds or wood produced this year.

The vineyards have also done well, both in bearing fruit and maturing wood for next season, and although the grape crop is large this year, the prospect is good for a much larger produce next year, on account of numerous young vineyards coming into bearing that have not hitherto borne fruit.

DECEMBER MEETING.

The Society met at its rooms in Fuller's Bank, Grand Rapids, Tuesday, December 6, being the annual meeting for the election of officers, etc. In the absence of the President, Mr. J. P. Thompson was, on motion of Henry S. Clubb, elected President *pro tem.*

Mr. A. T. Linderman, the Secretary, stated that he had received communications from Dr. E. Ware Sylvester, of Lyons, N. Y., promising valuable contributions of varieties for test by the Society.

Also, a letter from Mr. Geo. E. Waring, Jr., of Ogden Farm, inclosing Trophy tomato seed for that purpose.

SEEDLING APPLES.

Prof. Whitney of Muskegon presented samples of seedling apples from trees twenty years old, in Macomb county. They have been grafted from, and are extensively known as the Whitney apple. They have been kept till February, although this year they are like all other apples, not keeping as well as usual. The apple is a pale yellow, with a slightly strawberry flavor.

Mr. Pearsoll of Alpine also presented some fine seedling apples raised from seed sown by Mr. J. L. Tuxbury of Casenovia, some eighteen years since. It is a large rich apple, similar to the Spitzenburg. This apple was very highly spoken of by the members present.

OLD VARIETIES.

Mr. Holt presented several varieties of apples and pears, including Fall Pippin, Cayuga County Red Streak, Peck's Pleasant, Esopus Spitzenburg, Jonathan, Buerre Diel pear All of these have kept well for this season.

ROTTING OF APPLES THIS SEASON—CANNING APPLES.

Prof. Whitney said, in reply to a question by the President *pro tem.*, that it was conceded that the cause of the rotting of apples was the ripening of the fruit much earlier than usual. Another reason is, the fruit ripened under a dense mass of foliage this year, and had consequently been brought on by a sort of hot-house maturity. Another reason is, the fruit was picked too late.

The best way to remedy the evil is to take the choice apples, such as the Northern Spy, cut them up as for sauce, and can them. The cans, being emptied now of other fruit, could be refilled with apples with decided advantage. It could be done either with or without sugar.

Mr. Pearsoll said: For keeping winter apples in the cellar, I would rather not pick until the first frost; but for packing in barrels, I would pick a little earlier. If you pick winter apples too early they will wilt.

BARRELING APPLES.

Mr. Whitney said: East we always barrel, even apples for our own use, and they keep better than in any other way. They are kept in a cool cellar, and, being well barreled, are kept in the dark and from the air. It is like canning fruit.

Mr. Pearsoll: I do not like moving and jarring apples. It injures them worse than freezing.

FROZEN APPLES.

Mr. Knapp said, with frozen apples it is best to let them remain till the frost draws out of them, and then you would not know they had been frozen. If you move them and make them thaw suddenly, they will rot. One reason the fruit rots this year is the fact that apples are more wormy than usual.

Mr. Pearsoll—When gathered wet?

Mr. Knapp—Yes. If you gather apples when they are wet, they will not keep so well.

Mr. G. H. Linderman—The best way to keep apples is, after they are picked, heap them up to sweat, then pack them in buckwheat, chaff, or bran.

Mr. Fuller was always in the habit of picking apples on the 10th of October. It is not natural for an apple to rot. It is in the careless handling of apples that the apples are made to rot. The apples should be gathered in a small basket, and then carefully packed in barrels, and headed in. In this way I never lost any apples. But I never could find a man that would handle the fruit with sufficient care.

Mr. Holt: A person may not drop an apple, and yet the apples may be injured by pulling out the stem. I believe with Prof. Whitney, that we ought to gather apples earlier. You take a juicy apple like the Spy, and it is much harder to keep than the Russet, which is dryer. The difficulty of apples wintering by being picked green will be overcome by packing well in barrels.

Mr. Knapp mentioned an instance where the earliest gathered apples rotted worse than those gathered later.

It was stated that the Commissioner of Agriculture had just received from the Imperial Botanical Gardens of St. Petersburg, a collection of Russian apples, embracing about 100 varieties. These have come in perfect condition, and are well provided with grafts, of which the Society will receive a few specimens. It will be recollected that the Russian apples flourish exceedingly well in Michigan, to prove which it is only necessary to mention the Astrachan and Duchess of Oldenburg.

A. T. Linderman then read a paper upon the subject of establishing a Test Garden in connection with the Society, which was ordered printed.

On motion of Henry S. Clubb,

Resolved, That the Society hereby heartily indorse the action of the Secretary in issuing the circular to disseminators.

Resolved, That the Secretary be and is hereby further instructed to procure the services of competent horticulturists to conduct the tests of such new varieties as may be sent for that purpose.

Resolved, That a committee of five be appointed to prepare a petition to the Legislature for a special charter, under which the Society can be incorporated.

Resolved, That Article II. of the Constitution be so amended as to read as follows:

II. The officers of the Society shall consist of a President, eight Vice Presidents, a Treasurer, a Secretary, and such local Secretaries as may be elected by the Society.

On motion of S. L. Fuller,

Resolved, That the Society now proceed to the election of officers.

The following persons were then elected officers of the Society for the ensuing year:

President—Jonathan P. Thompson, Grand Rapids.

Vice Presidents—Henry S. Clubb, Grand Haven; George

Parmelee, Old Mission, Traverse Co.; Henry Holt, Cascade, Kent Co.; T. T. Lyon, Plymouth, Wayne Co.; George Taylor, Kalamazoo; William Bort, Niles; Payne K. Leach, Utica, Macomb Co.; S. B. Peck, Muskegon.

Secretary—A. T. Linderman, Grand Rapids.

Treasurer—S. L. Fuller, Grand Rapids.

Executive Committee—J. P. Thompson, A. T. Linderman, N. P. Husted, C. L. Whitney, George H. Linderman.

Local Secretaries—B. Hathaway, Little Prairie Ronde; Thomas Archer, St. Joseph; Joseph Chapel, Eastmanville; E. Bradfield, Ada; James Hamilton, Big Rapids; Judge Ramsdell, Grand Traverse.

On motion of Henry S. Clubb,

Resolved, That the Executive Committee be and is hereby instructed to add to the list of Local Secretaries such names as it shall select.

Resolved, That the thanks of the Society be and are hereby tendered to President Saunders, Secretary A. T. Linderman, and Treasurer Fuller, for their efficient services during the past year.

Resolved, That a special committee of three be appointed to compile the proceedings of the Society during the past year, for publication in pamphlet form.

On motion of C. L. Whitney, J. P. Thompson, H. S. Clubb, and A. T. Linderman were appointed said committee, to report at the next meeting. The Society then adjourned to meet on the first Tuesday in January, 1871.

The subject for discussion will be, "Does the stock upon which a graft is set influence the fruit?" This question arises from the statement made by Prof. Whitney, relative to the Wagener,—facts which he has learned from M. S. Wagener of Muskegon, and the grandson of the progenitor of that apple. Mr. Wagener brought the scions from Philadelphia, about the year 1800, and grafted them upon a wild apple stock on his farm, in Penn Yan, Ontario County, New York. From this tree, which ten years ago was still vigorous and fruit-bearing in Penn Yan, has been disseminated this popular apple. Prof. Whitney and other members of the Society hold that the original stock has a vital influence over the future character of the graft. Mr. Pearsoll and others believe the strictly opposite. The Society expect to hear from its Local Secretaries on this subject.

A STATE TEST GARDEN.

A PAPER READ BY A. T. LINDERMAN, SECRETARY, AT THE DECEMBER MEETING.

Mr. President and Gentlemen of the Society:

Whether or not it is practicable or possible to establish a Test Garden in connection with the Michigan State Pomological Society, for the purpose of assisting it to furnish to its members more reliable and authentic information than could otherwise be easily obtained, is a question, which to say the least, has two sides; and to impartially present these two sides to your notice is my object, resting assured that the intelligence of those present to-day, as well as those absent,—into whose hands this little waif may fall,—will furnish the correct answer.

THE BENEFITS.

The benefits of a Test Garden are many; in fact, so numerous that to attempt a notice of but a small portion is consistent with your time or my space. The first thing to be considered when a person has decided to plant fruit of any kind for sale, is what variety pays the most money, and to the intelligent fruit-grower this is a matter of no little importance, for right well he knows that it costs just as much land, labor, and money to grow a second, third, or fourth grade fruit, as it does to grow a first grade; and an error committed in planting other than the best kinds is a mistake that can never be repaired, and one that costs him a loss every year of the difference between the grade he plants and the best.

LACK OF INFORMATION.

A very large proportion of those planting orchards have not experience enough to enable them to decide which sorts belong to the first grade, and which to the lower grades. Knowing this, they are naturally led to apply for this information to some source which they deem authentic. There are but few experienced fruit-growers to whom a person in need of such information might apply, but that would have a favorite list to recommend. There are, perhaps, as few who would furnish a list that might correspond with another list from another source. In consequence, the matter, to say the least, becomes somewhat mixed in the mind of the planter, who oftentimes, in despair, makes a list composed of each kind that has been recommended, feeling reasonably sure that some of it will be right at all events, and sends this order to the nurseryman to fill. The damage arising from this method of commencing—which is not by any means unusual—is much greater than would appear, at the first glance, possible, and an evil, gentlemen, which needs your earnest efforts to overcome; and I shall occupy a brief space upon the magnitude of this point, as I believe it has great force in this connection.

Allowing that four-fifths of the list that is furnished the nurseryman to fill, is composed of the second, third, or fourth grade fruits; then we find, when the orchard bears, that the owner has a great variety of fruit which never attracts the market-man, in consequence of which his fruit obliges him to find a home market by peddling it out from his wagon by the bushel or peck, or in some equally perplexing manner dispose of it for the best price offered, until, instead of the anticipated pleasure of handling the products of his orchard, it becomes a dread and a perplexity. What wonder is it, then, that so many orchards are neglected, and, by that neglect, what wonder that so many trees are diseased, and become the rendezvous of the thousand and one beetles, bugs, and borers? But not here does the result of this wrong commencement stop. The nurseryman, compelled by this demand for variety, must propagate many kinds, must keep a great variety of sorts, or shut up shop; and so long as the nurserymen of Michigan are obliged, in order to supply the popular demand, to grow fifty varieties of apples, forty of which are of second quality, just so long will the fruit-growers of Michigan plant fifty varieties, forty of which are of second grade, to say the least; for the nurserymen do not grow their trees to throw them away,—somebody plants them. To strike, then, at the root of this evil, the purchaser must know that such and such are the best and only kinds to plant. This knowledge must be widespread, and accepted by unanimous consent. Such kinds only will then be propagated to any extent. To make any knowledge widespread and unanimous, it must be convincing. To make people generally convinced that a list of the kinds of fruit to be planted—furnished by this Society, for instance—is correct in every particular, is not so easy a matter, where that list is made up from members' opinions; for it is classed by many as simply an opinion, which is not necessarily any more nearly correct than their own, and consequently loses a great degree of its force. But if, in giving to the world our opinion as to the best and most prolific varieties to plant, that opinion is backed by the figures obtained from our tests which show that such is the case, the matter assumes a different aspect at once; and although this Society possibly might furnish a list to-day that the test of five years might not change in any particular, yet if it was not unanimously accepted it would lose a great share of its value, as I think has been clearly shown.

But to proceed. New varieties of all kinds of fruit are constantly being produced by people all over the States, some of them, without doubt, excelling in quality, productiveness, etc., anything heretofore known, while many, very many, are worse than useless perhaps, while others, again, are of medium worth. Now we must test them all, or a greater portion of them, before we can know what claims each one has to our favor. It certainly needs no demonstration to show that it is much cheaper and better to unite together in making this test do for all, especially where the subject to be tested will cost us, when united, nothing, but when procured for an individual trial will cost a large price. It may not be out of place in this connection to state, that, under your direction, I have recently issued a circular to those who were sending out varieties,—as far as I have been able to secure their names,—stating that this Society would like samples of any new variety they were disseminating, for the purpose of testing the same, and promising a full report of the hardiness, quality, and probable value of the subject for this section. In reply to which I have received, and am still receiving, prompt responses from a large number, who state that when the proper time arrives,—next spring,—they will gladly forward samples for trial. From some, on the other hand, I receive no response, which, although not conclusive, is an intimation that they do not care to have a test of their introductions made public; and there is no doubt in my mind, that if this Society had a Test Garden, and it became well known throughout the States that such was the case, that every new variety of worth in the country would be furnished for testing; and as the great object of the disseminator would be to embody the report of this Society—should it be favorable—in the first offer of his plant to the public, it would be necessary for him to furnish the sample for test of the first stock, in order that it might have time to fruit and be reported upon by the time his stock was sufficiently large to offer.

This arrangement will not only be useful to the producer of new sorts, but it will also be of great benefit to the purchaser, for he will be kept informed of the value of all the new varieties offered, and in purchasing can rest assured that he is not getting a worthless article. And, in order that nothing should in the least hinder the full working of this plan, all scions or cuttings are reserved subject to the order of the donator of the plant. The end of this must be, if followed up, that any one offering, in future years, any new variety of fruit, without the favorable report of this test connected therewith, will give reasonable cause, especially to the members of this body, to beware, for it will be reasonable to suppose that, knowing the failings of their new sort, they refused to submit it to our test, or, having subjected it to the test, it was found wanting. But, aside from the uses before alluded to, there should be a department devoted to growing seedlings, fruiting only those which, as seedlings, possess marked characteristics. Another department should be devoted to hybridization. All this can be carried on at the same time, and with but little extra expense, and, beyond doubt, will result in producing new varieties of which the Society and the State at large may well feel proud.

So much for a rough outline of the general benefits to be derived from such an institution. On the other hand, it might be said that a test in one portion of the State would not be reliable for another portion. This might have some bearing in testing new kinds, for, owing to climatic differences, a variety which would prove hardy at one point might not be sufficiently so in some other section. In consequence, it has been deemed necessary to establish four points, in as many different sections, at which to conduct the tests already begun, but it does not become essentially necessary that any other part of the benefits arising from a test garden would be deprived of their usefulness on account of the locality at which it was placed, provided that location was central.

The last and great difficulty to be overcome, is to secure the means sufficient to carry the work to a successful issue, a portion of which must be furnished by the members of the Society. For, although there are several towns which will undoubtedly bid liberally to secure the location of this garden, with its monthly gatherings and its annual crowds, yet there will need to be some funds raised by the Society to erect buildings, etc. It is thought by many that the best method to adopt in this case to supply the funds, would be to divide the amount needed into shares of stock, which would at once give to the purchaser an equivalent for his money, and at the same time be the means of founding what to me seems to be one of the enterprises of the age. The shares would probably be about ten dollars each, and, as it is very desirable to obtain the expression of fruit-growers throughout the State upon this subject, I would respectfully request all who may read this article to notify the Secretary of this Society, as soon as practicable, whether the enterprise meets with their approbation or not, and if so, whether they will assist, more or less, if a move is made to establish a State Test Garden.

FRUIT LANDS OF WESTERN MICHIGAN.

From the Report of the U. S. Commissioner of Agriculture, 1869.

The "Michigan Fruit Region," popularly so called, is now known to extend the whole length of the eastern shore of Lake Michigan. The peach belt may be said to vary from five to twenty-five miles in width, and its length is about two hundred and twenty-five miles. With an average width of ten miles, the area comprised amounts to two thousand two hundred and fifty square miles. It is estimated that the proportion of this area actually suited by circumstances of elevation, etc., to the successful culture of the peach and the grape, amounts to one-third, or 480,000 acres, of which about 10,000 acres are already planted in peaches, and probably 2,000 acres in grapes, only a small percentage being yet in full bearing condition. Throughout the entire fruit region the rise of real estate has been remarkable, especially in the neighborhood of the principal shipping points and harbors. There are also numerous small piers and second-rate harbors, where fruit is shipped to some extent; and in the vicinity of these points, land which formerly was considered valuable only for ties and wood is now salable at $10 to $50 an acre, when all the valuable timber is stripped off, and before it is cleared for cultivation. The increase in the price of real estate has kept even pace with the confidence which each succeeding season inspires in the success of peach culture. The location of Grand Haven being central with regard to north and south, perhaps the increase here may be regarded as a fair average of the whole region. It is greater at St. Joseph, on account of the longer time the business has been in operation there; while the increase north is proportionate to the time since it was discovered to be practicable to grow peaches at Manistee and Grand Traverse.

Planting peaches and grapes for market commenced in the vicinity of Grand Haven and Spring Lake in 1859-60. At that time only four or five orchards were commenced. They were planted on land which had been stripped of the pine, and was regarded as of very little value, commanding not more than ten or fifteen dollars per acre, notwithstanding the favorable locality. In 1867 one of these orchards, then containing sixteen acres planted to fruit, together with twenty-five acres of scrubby oak land, full of "grubs," sold for $10,000 cash. A portion of another orchard, at Ferrysburg, sold in

1869 at $500 per acre. The wild land in the vicinity of these orchards now varies in price from $50 to $200 per acre, as shown by actual sales. There is an abundance of land, however, equally good for the production of peaches, situated two and three miles from navigation, which can be bought at $5 to $10 per acre. This land could not be sold at any price two or three years ago, and on account of taxes was considered a burden to owners.

LIST OF PREMIUMS, AND RULES AND REGULATIONS

OF THE

MICHIGAN STATE POMOLOGICAL SOCIETY,

FOR ITS FIRST ANNUAL FAIR, HELD AT GRAND RAPIDS (IN CONNECTION WITH THE KENT COUNTY AGRICULTURAL SOCIETY), TUESDAY, WEDNESDAY, THURSDAY, AND FRIDAY, SEPT. 27, 28, 29, AND 30, 1870.

RULES AND REGULATIONS.

This fair will be held upon the grounds of the Kent County Agricultural Society, and on the same days that the fair of that Society is held. One admission ticket will admit the purchaser to all the exhibitions of both fairs. There will be but one ticket office, and tickets must be purchased at that office. Prices of admission: Single admission, adults, 25 cents; children under 12 years of age, 10 cents; teams, 25 cents; saddle-horse, 15 cents.

Individuals who wish to join the Society can do so by the payment of one dollar. Life memberships, ten dollars. It is desirable that all persons interested in its objects should join the Society, and it is expected that all exhibitors of articles will become members, though this rule will not be enforced with persons under 21 years of age.

Fruit-growers, and persons interested in fruit culture, are earnestly invited to become exhibitors, and use their influence to establish the Society upon a sound and prosperous basis.

All entries of all articles for exhibition must be made at the office of the Secretary, at Pomological Hall, on the Fair Ground, and should be made on the first day of the fair, or by 12 o'clock M. of the second day.

Entries may be made for the exhibition without competition; and awarding committees in the several classes may notice such as they shall deem worthy, in their reports; but all such entries must be made by a member of the Society.

There will be a Superintendent of the Pomological Department, who will have the general charge and arrangement of the fruits exhibited, and to whom, with the fruits, must be delivered correct lists of the specimens and varieties entered by each exhibitor.

The several specimens and varieties shown by any exhibitor should be labeled with the name by which they are known to such person.

All doubtful cases which any special viewing committee either may not or cannot decide, may be referred to the Pomological Committee for final adjudication.

Exhibitors entitled to first premiums will be allowed to take in place thereof the diploma of the Society.

Members of awarding committees are requested to inform the Secretary of their acceptance as soon as they are notified of their appointment. Upon their arrival at the fair grounds, they will report to the Secretary on or before 12 o'clock M. on the second day. A vacancy on any committee shall be reported to the Executive Committee, who shall fill such vacancies in the usual manner.

Members of awarding committees are requested to report promptly for duty on Wednesday, at 1 o'clock P. M., and their reports, in writing, must be handed to the Secretary by 12 o'clock M., on Thursday.

No person who is an exhibitor can act as a judge on the class in which he exhibits.

Exhibitors, when requested, are expected to make written or verbal statements respecting their contributions.

As one great object of the Society is to collect valuable

information upon pomology, the several committees are requested to gather all the information possible from the exhibitors in their classes, and to make their reports as full as time and circumstances will permit.

In case there should be deficiency in the funds of the Society, the premiums awarded will be liable to a *pro rata* reduction.

When articles are not deemed worthy of a premium, the judges will, in all cases, withhold it.

Any article entered for exhibition in one class shall not compete for premiums in any other.

Under no circumstances will the name of exhibitor appear on the entry card.

When the judges have made their decisions, premium badges will be attached to the fruit. First premium, a blue ribbon; second premium, a red ribbon.

Fruit will be marked with cards furnished by the Secretary, designating the class and number of entry; and during the exhibition all articles must be placed entirely under the management of the officers of the Society.

All articles entered for exhibition will be required to remain on the grounds during the days of exhibition, under penalty of forfeiture of the right to premium, unless permitted by the Superintendent to take them off the ground.

When a majority of the Viewing Committee are present, they shall constitute a quorum, and be authorized to award premiums; and the first on the list of those present shall be chairman.

No person will be allowed to sell the articles they have on exhibition until special permission is granted by the Superintendent.

LIST OF PREMIUMS.

DIVISION A—COLLECTION OF FRUITS.

Committee—Hon. Flavius J. Littlejohn, Allegan; Martin Walsh, Spring Lake; J. M. Harwood, Jackson; A. S. Stannard, South Boston; J. P. Thompson, city.

	Best.	2d Best.
For collection of fruits from any one township in the State	$20 00	$10 00

RULE.—These collections shall not include fruits shown individually in any other class, and the committee of judges shall be composed of not more than one member from any one township.

For the best collection of the varieties recommended by the Michigan State Pomological Society, for general cultivation	$10 00	$5 00

NOTE.—The varieties recommended by this Society, referred to above, include the following: For summer use, the Red Astrachan, Sweet Bough, and Duchess of Oldenburg; for autumn use, Fall Pippin, Cayuga County Red Streak, Snow, Jersey Sweet, Maiden's Blush; for winter use, Baldwin, Wagener, Golden Russet, Rhode Island Greening, Tallman Sweet, Hubbardston Nonsuch, Northern Spy.

Exhibitors, to obtain the above premiums, must exhibit at least twelve of the above varieties.

For the best collection of fruits exhibited by any individual	$5 00
For the second best	2 00
For the third best	1 00

DIVISION B—APPLES.

Committee—Rev. James Hamilton, Newaygo; H. E. Light, Greenville; Noah P. Husted, Lowell; Jacob Ganzhorn, Spring Lake; E. U. Knapp, Grand Rapids.

	Best.	2d Best.
For peck, any one variety summer apples	$2 00	$1 00
For peck, any one variety autumn apples	2 00	1 00
For peck, any one variety winter apples	2 00	1 00
For the best single variety of summer apples, not less than six spec.	1 00	50
For the best single variety of autumn apples, not less than six spec.	1 00	50
For the best single variety of winter apples, not less than six spec.	1 00	50
For the best peck Sweet Bough	1 00	50
For the best peck Red Astrachan	1 00	50
For the best peck Duchess of Oldenburg	1 00	50
For the best peck Fall Pippin	1 00	50
For the best peck Cayuga Co. Red Streak	1 00	50
For the best peck Maiden's Blush	1 00	50
For the best peck Snow	1 00	50
For the best peck Jersey Sweet	1 00	50
For the best peck Wagener	1 00	50
For the best peck Baldwin	1 00	50
For the best peck Greening	1 00	50
For the best peck Northern Spy	1 00	50

	Best.	2d Best.
For the best peck Golden Russet	$1 00	$0 50
For the best peck Hubbardston Nonsuch	1 00	50
For the best peck Tallman Sweet	1 00	50
For the best peck of any other variety	1 00	50
For best collection of apples grown by exhibitor	5 00	2 00
For best collection of Siberian Crab Apples	2 00	1 00
For best single variety Siberian Crab, not less than 20 specimens	1 00	50
For twenty specimens Transcendent Crab	1 00	50
For twenty specimens large Red Crab	1 00	50

DIVISION C—PEARS.

Committee—Judge J. G. Ramsdell, Grand Traverse, T. T. Lyon, Plymouth; Dr. A. Peck, Lowell; Wm. Davis, Kalamazoo; Asa W. Slayton, Grattan.

	Best.	2d Best.
For collection of pears, not less than six varieties	$2 00	$1 00
For peck summer pears, one variety	1 00	50
For peck autumn pears, one variety	1 00	50
For peck winter pears, one variety	1 00	50
For single variety summer pears, not less than six specimens	1 00	50
For single variety autumn pears, not less than six specimens	1 00	50
For single variety winter pears, not less than six specimens	1 00	50
Plate Bartletts	50	25
Plate Flemish Beauty	50	25
Plate Louise Bonne de Jersey	50	25
Plate White Doyenne	50	25
Plate Duchess d' Augouleme	50	25

DIVISION D—PEACHES.

Committee—Hon. Louis S. Lovell, Ionia; Hon. Henry Penoyer, Crockery; S. B. Peck, Muskegon; Hon. Stephen Rossman, Greenville; Jacob Barnes, Grand Rapids.

	Best.	2d Best.
For collection of peaches, not less than six varieties	$2 00	$1 00
For half peck of early peaches	1 00	50
For half peck of late peaches	1 00	50
For half peck of clingstones	1 00	50
For dish single variety of peaches, not less than six specimens	1 00	50
For plate early Crawford	1 00	50
For plate Early Barnard	1 00	50
For plate Smock Free	1 00	50
For plate Late Crawford	1 00	50
For plate Stump the World	1 00	50
For plate Hill's Chili	1 00	50

DIVISION E—GRAPES.

Committee—Jacob Quintus, Grand Rapids; Wm. Bort, Niles; George Taylor, Kalamazoo; George Seagrove, Spring Lake; E. Bradfield, Ada.

	Best,	2d Best.
For collection of native grapes, not less than four varieties	$2 00	$1 00
For ten pounds of native varieties	1 00	50
For five pounds of Concord	1 00	50
For five pounds of Clinton	1 00	50

	Best.	2d Best.
For five pounds of Isabella	$1 00	$0 50
For five pounds of Delaware	1 00	50
For five pounds of Iona	1 00	50
For five pounds of Diana	1 00	50
For five pounds of Ives' Seedling	1 00	50
For six clusters of Rogers' Hybrids, Nos. 3, 14, 19, and 33	1 00	50
For plate of any valuable variety, not mentioned above	1 00	50

DIVISION F—PLUMS, APRICOTS, AND NECTARINES.

Committee—T. J. Ramsdell, Manistee; H. H. Goodwin, Ionia; W. O. Houghtaling, Grand Rapids; Joseph Bray, Middleville; J. P. Chapel, Owosso.

	Best.	2d Best.
For collection of plums	$2 00	$1 00
For half peck of plums, single variety	1 00	50
For collection of apricots	2 00	1 00
For half peck of apricots, single variety	1 00	50
For collection of nectarines	2 00	1 00
For half peck, single variety	1 00	50

DIVISION G—SEEDLING FRUITS.

Committee—A. T. Linderman, City; J. N. Keeler, Middleville; Fletcher Fowler, Black Lake; W. H. Gregory, Pine Grove; Hunter Savage, Spring Lake.

	Best.	2d Best.
For seedling apple	$1 00	$0 50
For seedling peach	1 00	50
For seedling pear	1 00	50
For seedling plum	1 00	50
For seedling grape	1 00	50
For seedling quince	1 00	50
For seedling Siberian Crab	1 00	50

NOTE.—Specimens of the above seedling fruits, it is expected, will be accompanied with history, origin, and such other information as may be useful to characterize the fruit.

Premiums for seedlings may be awarded to others besides the original producer, when it is shown that the originator does not compete.

Exhibitors of seedlings must not expect a premium because the fruit exhibited is a "seedling." It must have merit equal to well-known varieties, the object being not to encourage inferior sorts.

DIVISION H—QUINCES, CRANBERRIES, AND SMALL FRUITS.

Committee—H. S. Clubb, Grand Haven; G. S. Linderman, Grand Rapids township; Warren Hale, North Brownville; Thomas J. Slayton, Lowell; Allen Thompson, Otisco.

	Best.	2d Best.
For collection of quinces	$2 00	$1 00
For peck, single varieties	1 00	50
For peck cranberries	1 00	50

NOTE.—Discretionary premiums will be allowed on all small fruit exhibited.

For one quart of American chestnuts	$1 00	50
For one quart of Spanish chestnuts	50	25
For half peck butternuts	50	50
For half peck walnuts	50	—

DIVISION I—DRIED FRUITS AND PICKLED FRUITS.

Committee—Mrs. Hunter Savidge, Spring Lake; Mrs. E. L. Craw, Fruitport; Prof. C. L. Whitney, Muskegon; Mr. and Mrs. O. R. Wilmarth, City.

	Best.	2d Best.
For half peck dried apples	$1 00	$0 50
For quart dried pears	1 00	50
For quart dried peaches	1 00	50
For quart dried cherries	1 00	50
For quart dried currants	1 00	50
For quart dried raspberries	1 00	50
For quart dried blackberries	1 00	50
For quart dried quinces	1 00	50
For quart dried strawberries	1 00	50
For quart dried whortleberries	1 00	50
For collection of pickled fruit	2 00	1 00
For sample of pickled pears	1 00	50
For sample of pickled peaches	1 00	50
For sample of pickled apples	1 00	50

DIVISION K—CANNED AND PRESERVED FRUITS.

Committee—Mrs. S. L. Fuller, city; Mrs. Henry Holt, Cascade; Mrs. N. P. Husted, Lowell; Mrs. Solomon L. Whitney, city; Hon. P. R. L. Pierce, city.

	Best.	2d Best.
For collection canned fruits	$2 00	$1 00
For sample canned apples	1 00	50
For sample canned pears	1 00	50
For sample canned peaches	1 00	50
For sample canned plums	1 00	50
Best sample canned cherries	1 00	50
Best sample canned Siberian apples	1 00	50
Best sample canned strawberries	1 00	50
Best sample canned raspberries	1 00	50
Best sample canned blackberries	1 00	50
Best sample canned whortleberries	1 00	50
Best sample canned quinces	1 00	50
Best collection preserved fruits	2 00	1 00
Best sample cider apple sauce	1 00	50
Best sample preserved pears	1 00	50
Best sample preserved peaches	1 00	50
Best sample preserved plums	1 00	50
Best sample preserved cherries	1 00	50
Best sample preserved strawberries	1 00	50
Best sample preserved raspberries	1 00	50
Best sample preserved blackberries	1 00	50
Best sample preserved whortleberries	1 00	50
Best sample preserved quinces	1 00	50

DIVISION L—WINES, BOILED CIDER, AND CIDER VINEGAR.

Committee—Dr. Charles J. Hempel, city; L. H. Randall, city; John W. Pierce, city; E. P. Fuller, city; Dr. G. K. Johnson, city.

	Best.	2d Best.
Best collection domestic wines	$2 00	$1 00
Best sample currant wine	1 00	50

	Best.	2d Best.
Best sample blackberry wine	$1 00	$0 50
Best sample grape wine	1 00	50
Best sample Clinton wine	1 00	50
Best sample Concord wine	1 00	50
Best sample Ives' Seedling wine	1 00	50
Best sample Delaware wine	1 00	50
Sample native grape wine of any other variety	1 00	50
Gallon cider	1 00	50
Specimen bottled cider, not less than six bottles	1 00	
Gallon boiled cider	1 00	
Gallon cider vinegar	1 00	

DIVISION M—JELLIES.

Committee—Mrs. J. Morgan Smith, city; Mrs. James Hamilton, Newaygo; Mrs. W. W. Hatch, Lowell; Mrs. J. Mason Reynolds, Plainfield; Mrs. Spencer L. Shaw, Saranac.

	Best.	2d Best.
Collection of jellies	$2 00	$1 00
Specimen currant jelly	1 00	50
Specimen apple jelly	1 00	50
Specimen Siberian crab jelly	1 00	50
Specimen grape jelly	1 00	50
Specimen raspberry jelly	1 00	50
Best specimen blackberry jelly	1 00	50
Best specimen any other variety	1 00	50

THE PRESS.

For the convenience of editors and reporters of the press, accommodations will be provided, and every facility will be afforded them to obtain and transmit intelligence. A committee of reception from the press will receive their brethren from abroad, on the field, and further the purposes of their coming. They are requested to announce themselves on arrival, and to present their names or credentials at the Secretary's office on the grounds, when they will be furnished with cards of admission.

Committee on Reception of Representatives of the Press—A. B. Turner, city; C. B. Smith, city; M. M. Clark, city; E. F. Harrington, city; C. C. Sexton, city.

Committee on Reception of Invited Guests—Moses V. Aldrich, city; Wilder D. Foster, city; S. L. Fuller, city; Ransom E. Wood, city; Charles H. Taylor, city.

POMOLOGICAL COMMITTEE.

There will be a Pomological Committee appointed, to whom will be given the charge of the nomenclature of the fruits exhibited, and before any examination shall be made by the Viewing Committees, the Pomological Committee will examine all specimens and correct the names of varieties which may be wrongly named by the exhibitors, affixing a doubtful mark in cases where the name of the variety shown may be uncertain.

Pomological Committee—Henry Holt, Cascade; Townsend E. Gidley, Grand Haven; J. D. Husted, Lowell; A. Parmelee, Old Mission; Thomas Archer, St. Joseph.

SUPERINTENDENTS OF POMOLOGICAL HALL.

Prof. C. H. Whitney, Muskegon; Geo. S. Linderman, city.

THE ADDRESS.

An address will be delivered before the Society on Thursday afternoon, by Hon. Flavius J. Littlejohn of Allegan.

H. G. SAUNDERS, *President.*

A. T. LINDERMAN, *Secretary.*

THE FIRST ANNUAL FAIR.

After much consideration, it was decided to hold the first fair of the Pomological Society on the fair grounds of the Kent County Agricultural Society, and in union and connection with that Society. Consequently, a premium list was prepared and issued, and all the necessary preliminary steps were taken preparatory to an exhibition of fruit. The old "Agricultural Hall" was enlarged and inclosed for the occasion, and the second story of the building, 100 feet in length by 30 feet in width, semicircular in form, was devoted to the pomologists. This first effort proved a perfect success, and established the fact that the fruit-growers intend to make the State Pomological Society a permanent and successful institution, and the Society will undoubtedly provide, next year, the largest building that can be obtained, for the purpose of giving all a good chance to exhibit their productions under the most favorable circumstances.

THE BEST TOWNSHIP COLLECTION.

We intend only to note a few of the most prominent features of this most interesting occasion. There was but one entry for the premium ($20) for the best collection of fruits from any one township in the State, and this was made by one of the best fruit towns in Kent county, the township of Grattan. The committee, consisting of Judge Littlejohn, Judge Tracy, A. S. Stannard, and J. P. Thompson, had but little difficulty in making their award. The collection embraced sixty-four varieties of apples, five of pears, six of grapes, and six of peaches, and the following were some of the varieties:

APPLES.

Eustis, Disharoon, Fall Orange, Jersey Sweet, Gabriel, Green Pippin, Winter Sweet, Seedling, Green Seek-no-further, Red Canada, Twenty ounce, Rambo, Black Detroit, Rhode Island Greening, Snow, King, Peck's Pleasant, Baldwin, Black Gilliflower, Compo Sweet, Tinifbate, Green Spitzenburg, Slug Sweet, Winter Pippin, Sweet Bough, Flushing Spitzenburg, Blue Pearmain, Pennock's Red Winter, Fall Sweet, Dwarf Bearer, Esopus Spitzenburg, Evening Party, Fall Jenneting, Winter Pippin of Vermont, Autumn Pearmain, Wagener, Ridge Pippin, Autumn Swaar, Jeffries, Westfield Seek-no-further, Holland Pippin, Duchess of Oldenburg, Tallman

Sweet, Mother, Summer Sweet, Paradise, Western Red Streak, Fall Pippin, Maiden's Blush, Surprise, Black Apple, Hill's Pie Apple, Northern Spy, Golden Sweet, Red Siberian Crab, Giant Apple (seedling).

PEARS.

Flemish Beauty, Seckel, White Doyenne, Winter Nellis, Beurre Bosc.

PEACHES.

Purdy's Seedling, Lemon, Late Yellow, Jersey Cling, Duga's Seedling.

GRAPES.

Concord, Black Cluster, Clinton, Delaware, Isabella, and Catawba.

The two best autumn varieties in this collection were the Maiden's Blush and Fall Pippin, apples which have no superior for culinary and dessert uses, and which are always welcome in the markets. The leading winter sorts of the collection were the Rhode Island Greening, Golden Russet, Wagener, and Northern Spy, and these were prime, sound, healthy, and gave every indication of good keeping qualities. Very many of the apples in this collection were partially useless when compared with the few standard sorts, and the committee did not award the premium to encourage the growing of a large number of varieties.

GRATTAN,

The township which won this sweepstakes, is twenty-two miles northeast of Grand Rapids—north of Vergennes—and joins Ionia county on the east. Its soil is a heavy loam, excellent for fruit and wheat. The orchards of the town are just coming forward and beginning to bear. Whenever there is fruit anywhere it can be found in Grattan, and one resident of the town has had peaches on his farm for eighteen years past without a failure. The inhabitants are intelligent and forehanded, and the soil, water, rolling land, all combine to make it a town desirable for settlement. The committee urged upon the people of that town to be cautious in their selections of fruit; make a few good selections, and set out larger orchards, and in this way buyers will come to the town and take the fruit from the trees. Messrs. Slayton and Duga, the gentlemen who had the collection in charge, are entitled to an honorable mention for their enterprise and labors.

THE BEST INDIVIDUAL COLLECTION.

One of the largest and finest exhibitions of fruit ever made in the State, was presented by Noah P. Husted, from his nursery at Lowell, but it was not entered for a premium. This collection embraced the following varieties, and those marked by a * are esteemed the best and most valuable sorts by Mr. Husted:

WINTER APPLES.

*Baldwin, *Wagener (among the very best), *Northern Spy, *Hubbardston Nonsuch, *R. I. Greening, *Golden Russet, *Tallman Sweet, *Peck's Pleasant, Seek-no-further, *Red Canada (for top graft), Swaar, Spitzenburg, Rambo, King of Tompkins Co., Roxbury Russet, Yellow Bellflower, Jonathan (good, but slow grower), Wine Sap, Winter Pearmain, Pennock, American Pippin, Ben Davis, White Pippin, Minister, Twenty-ounce Pippin, Old King, Neverfail, Pomme Grise, Black Gilliflower, Ladies' Sweet.

AUTUMN APPLES.

*Maiden's Blush, *Snow, *Cayuga Red Streak, *Autumn Strawberry, Tallow Pippin, *Hawley, *Jersey Sweet, Fall Orange, Ladies' Blush, *Fall Pippin, *Gravenstein, Cabbasha, Gloria Mundi, Black Detroit, Scolloped Gilliflower, Fall Jenneting, Pumpkin Russet.

SUMMER APPLES.

*Duchess of Oldenburg, Summer Belle.

CRAB APPLES.

* Transcendent Crab, best; Montreal Beauty, excellent for dessert; Hyslop, season, November; Soulard, long keeper; Cherry, small size.

PEARS.

Louise Bonne de Jersey, Duchess d' Angouleme, White Doyenne, Winkfield, Deurre Diel.

GRAPES.

* Concord, Isabella, Ives' Seedling, Rogers' Hybrids Nos. 15 and 4, Allen's Hybrid.

CANNED FRUIT.

*Transcendent Crab, Montreal Beauty, Concord Grape, Hartford Grape, Cherry, Gov. Wood, Whortleberries.

JELLIES.

Made from Concord Grape, Hartford Grape, Oporto Grape, Clinton Grape, Delaware Grape, Common Crab, Transcendent Crab, Cherry Crab, Hyslop Crab, Montreal Beauty.

Also, Orange Quince, English Sage, Sweet Chestnut.

This collection received complimentary notices from all the committees. The Wagener apple appeared in all its perfection; five Cayuga County Red Streaks weighed five pounds and two ounces; six samples of the Maiden's Blush averaged twelve ounces each; while the Hubbardston Nonsuch and Northern Spy were beautiful beyond comparison. A peck of beautiful Orange Quinces, a neglected but most valuable fruit, attracted universal attention, while the samples of the American Sweet Chestnut, a tree invaluable for its timber, shade, ornament, fruit, and beauty, were much admired. The show of Crab Apples, and of canned fruit and jellies, was large and very creditable.

THE MOST VARIED COLLECTION OF GRAPES.

Although the exhibition was overflowing with apples, it was apparent, also, that the grape was to fill a large space.

Mr. Edward Bradfield of Ada, a veteran grape amateur, made a remarkable exhibition, but not entered for a premium, of twenty-seven varieties of grapes.

Mr. Bradfield gave valuable information to the people on his favorite subject of grapes. He had great confidence in the Iona, and his really fine specimens of complete bunches of that delicious grape indicated that he has the secret of success in its cultivation. His Isabellas were also very fine. A seedling originated by his brother, known as "Bradfield's Seedling," is a very early grape, a little larger than the Delaware and similar in flavor; although it blossoms as late as the middle of June, it ripens in the Middle of August.

The following list contains the names of the twenty-seven varieties of grapes exhibited by Mr. Bradfield for the general advancement of the interest in grape cultivation, with his opinion and experience of their quality and fruitfulness, and the time of their ripening this season:

NAME OF FRUIT.	Quality.	Fruitfulness.	Time of Ripening.
Iona	1	1	September 10
Delaware	1	1	September 1
Crevling	2	2	September 1
Israella	2	2	September 1
Diana	2	2	September 20
Adirondac	1	2	August 20
Union Village	3	1	September 25
Catawba	2	2	September 25
Clinton	3	2	September 25
Early July	4	3	August 25
Blood's Black	3	3	August 25
Anna	2	2	October 1
Allen's Hybrid	2	2	September 10
Rogers' do. No. 19	3	2	September 10
Lenoir	3	3	September 20
Lincoln	3	4	Sepcember 20
Alvey	2	2	September 20
Rebecca	2	3	September 20
Pauline	0	0	September 20
Hartford Prolific	4	1	August 24
N. Muscadine	4	2	August 25
Concord	4	2	September 10
Isabella	4	2	September 25
Black Cluster	3	2	September 1
Elsingburg	0	0	September 25
Taylor's Bullitt	0	0	August 25
Bradfield's Seedling—flowers June 15	....	....	August 15

Note by Compiler—This list is undoubtedly correct with Mr. Bradfield's location, but it does not correspond with the experience of Ottawa county vine-growers, who would put the Concord, Isabella, and Hartford Prolific, and even the Clinton, much higher on the list in regard to quality and fruitfulness, while the Iona and Israella would have a much lower position than that assigned them by Mr. Bradfield. This list, however, is very encouraging in regard to new and important varieties, as indicating what careful culture will do for them.

THE SPRING LAKE GRAPES.

Messrs. Savidge, Seagrove, and Petty desired the world should learn that Spring Lake had other attractions besides her magnetic waters, and that the day was near at hand when the banks of their lovely lake would be literally vine-clad, and that the grape is to be as common there as on the islands of Lake Erie. Their exhibition consisted of about twenty boxes of different varieties, the whole weighing about two hundred pounds, and it was a tempting and luscious show. All were pleased to notice, by the blue ribbon, that this collection had won the sweepstakes premium.

The Committee on Grapes—Mr. Quintus of Grand Rapids, Mr. Taylor of Kalamazoo, and Mr. Bradfield of Ada, made the following awards: Best collection of grapes, first premium, Hunter Savidge of Spring Lake; best Delaware, first premium, G. W. Dickinson of Grand Rapids town; best Clinton, first premium, Charles Alford of Ottawa county; second best, Henry Allen of Paris; best Concord, first premium, Hunter Savidge, Spring Lake; second best, President Griggs of Paris; best Isabella, first premium, G. W. Dickinson; second best, Thomas Petty of Spring Lake; best Rogers' Hybrid, George Seagrove of Spring Lake; best foreign variety,—the Black Hamburg,—Geo. Kendall, Grand Rapids; best Catawba, W. I. Blakely, Grand Rapids; second best G. W. Dickinson. Mr. Quintus, in his report, made a valuable suggestion, that hereafter all exhibitors of grapes should be required to place their samples upon plates, in order that they might be examined the more closely.

OTHER EXHIBITORS.

One of the largest exhibitors was Mr. Charles Alford, of Talmadge, who has an orchard on the highest solid land of Ottawa county, containing over 100 varieties of apples, over sixty of which he had on exhibition at the fair. They were a very attractive feature, and deserve special mention. His Cayuga County Red Streak apples were a marvel of size for apples of their superior quality. He has, on his farm, a tree on which sixteen varieties of apples are growing. Mr. Alford received a number of premiums, among others, one for his Maiden's Blush, being the best single variety of autumn apples.

Another gentleman who took a good share of premiums was G. W. Dickinson of Grand Rapids town. He had twenty varieties of apples, and Catawba, Delaware, and Isabella grapes; while Mrs. Dickinson was represented by dried peaches, currants, raspberries, whortleberries, and canned peaches and crab-apple jelly.

Mr. Henry Allen of Paris, exhibited twenty-four varieties of apples, making a very beautiful display. They consisted principally of the varieties recommended by the Society, and fully sustained the recommendation by their appearance. He received the premium for the best collection grown by exhibitors; also, for the best peck of Fall Pippins.

C. J. Dietrich of Grand Rapids town had twenty-three varieties of apples, among which was the Minister, which Mr. D. esteems to be his best for culinary purposes, but inferior for dessert. The Scollop Gilliflower is another fine-tasting apple, but a scraggy grower. Mr. D. brought his apples to the fair in order to have them baptized with names,—a commendable object, and one worthy of imitation by his brother orchardists. His Greenings were honored with the blue ribbon.

Geo. Kendall, Grand Rapids, exhibited a small but superior collection of grapes, all of foreign varieties, including two bunches of the Black Hamburg, each bunch weighing nearly two pounds; two bunches of White Fontainebleau, the market grape of Paris; two bunches of the Dutch Sweetwater, very delicate and choice; two bunches of the Chasselas Musque, and one bunch of Child's Superb. These varieties only thrive in this country under hot-house treatment.

Henry Holt & Son of Cascade entered twenty-eight varieties of apples. Their Flemish Beauty pear received a premium as the best autumn pear; best winter pear, the Vicar of Wakefield; among other varieties of pears was the Buffum, the Passe Colmar, Buerre Diel, and Louise Bonne de Jersey. The Messrs Holt had splendid Swaar apples. The Porter they esteem as one of the best autumn varieties; the Jonathan and Snow excell, while the Roxbury Russet looked as if it would keep a year.

Rev. H. C. Waring, Grand Rapids town, received premiums for Late Crawford and Smock Free peaches. His thirteen varieties of apples embraced the most worthy and useful sorts.

A. W. Slayton of Grattan made a splendid exhibition of apples, pears, peaches, etc., and won several premiums. J. A. Duga, also of Grattan, had the best plate of Clingstone peaches.

S. Pierce, city, sent in the best single variety of winter apple,—the Northern Spy; second best,—Steele's Red Winter,—was shown by Charles Waterman.

The Wagener apple was well represented from the orchards of Jacob and David Yeiter of Lowell; also, of H. E. Light, Greenville, and Prof. Whitney of Muskegon, and this apple secured the premium as the best winter apple. The Northern Spy, exhibited by Charles Waterman, was favored by the second premium.

F. M. Rosenkrans, Cascade, brought twenty-six varieties of apples. Quite a curiosity was a sprig from a Hubbardston Nonsuch tree, bearing a cluster of six samples, weighing over three pounds. Miss Ellen D. Rosenkrans received premiums for a large variety of dried and preserved fruits and jellies.

Mrs. Wm. Gunn, city, had excellent jellies, and Mrs. C. C. Rood, city, won a premium with her Clinton grape wine.

John Ashley of Oakfield made an entry of thirty-three varieties of apples, among which were summer sorts that were not excelled. The Duchess of Oldenburg was noticeable, also the Sweet Bough, and a plate of Golden Sweet.

John W. Newhall of Wyoming brought eight varieties of apples, embracing fine samples of the Northern Spy, Greening, Twenty Ounce, Baldwin, Tallman Sweet, Ramsdell's Red Sweet, and Seek-no-further.

Mr. Wm. Rowe of Walker entered samples of Wagener and Spitzenburg apples.

G. W. Griggs, President of the Agricultural Society, sent fine Concord grapes, and took a premium.

John Suttle, florist, city, sent bouquets and plants of various kinds, to decorate Pomological Hall.

J. R. Renwick, city, florist, also sent flowers and plants.

Mrs. T. R. Williams, Paris, exhibited a splendid plate of Duchess d' Angouleme pears.

Mrs. D. Schermerhorn, Walker, had dried currants, apples, peaches, cherries, as all good housewives should have.

Among other exhibitors who were honored with premiums, were Mrs. J. J. Watson, Mrs. A. F. Linderman, Mrs. G. S. Linderman, Mrs. Elihu Smith, D. K. Emans, S. Pelton, Chas. Blaine, A. C. Barkey, O. W. Blaine, A. C. White, J. S. Davis, Geo. Van Nest, W. I. Blakely, C. L. Shomaker, J. H. Ford, R. J. Stowe, E. U. Knapp, N. L. Crocker, S. Pierce.

Asa W. Slayton was awarded a premium for the best seedling apple, which was named "Grattan;" Henry Holt & Son the second premium for a seedling apple, named "Cascade;" Chas. Alford a premium for a seedling peach, named "Alford's Late Yellow."

Among the active chairmen of committees, were Judge

Lovell of Ionia, Judge Ramsdell of Grand Traverse, T. J. Ramsdell of Manistee, Hon. P. R. L. Peirce of Grand Rapids, George Parmelee of Old Mission, Jacob Quintus of Grand Rapids, Geo. Taylor of Kalamazoo, Rev. Mr. Hamilton of Newaygo.

PROF. WHITNEY'S FLOWERS.

One of the most attractive features of the Fair was the exhibition of flowers by Prof. C. L. Whitney of Muskegon, who made an entry of two hundred varieties, the product of his garden and hot-house. The inconvenience of bringing flowers in pots was obviated in this case, by bringing them out in baskets and placing each one in a small vial of water. In this way they were kept fresh during the entire exhibition. The arrangement of two hundred varieties, embracing some fifteen genera of plants, in this way, was exceedingly tasteful and displayed the flowers to a great advantage.

The foliage plants were peculiarly attractive, owing to their vegetation and beautiful tints.

The Japan pinks, the pansies, the verbenas, the dwarf chrysanthemums, or asters, the zenias, the phloxes, the gladiolis, and the odoriferous tuberose, all manifested respective charms to the senses, and the ladies were especially interested in this part of the display.

Nothing is so refining and ennobling as the love of these most beautiful developments of Nature, and all felt grateful to Prof. Whitney for the opportunity he afforded of seeing so many perfect specimens.

PLENTY OF FRUIT.

One feature of this exhibition was worthy of commendation. It was not a thin show, and there was no effort to make a few samples cover a good deal of space, but the fruit was abundant, and lay in heaps around the Hall. There must have been one hundred and fifty varieties of apples, of all sizes, shapes, color, and quality, and each variety was represented by duplicates and triplicates, while the more popular sorts occupied as many as twenty plates each. Gentlemen who have attended like fairs in this and other States were free to acknowledge that the display of apples surpassed anything they ever saw before, and not a voice spoke otherwise than in terms of praise and surprise. Old residents were especially gratified, and the old settlers who have borne the toil and broke the ground, were thankful to see, this day, the fruition of their hopes. Here was the result of their labor, and here was the pledge of the bounty for the future. It was estimated that twenty thousand persons entered the grounds during the exhibition.

THE ADDRESSES.

Judge Littlejohn delivered a very elaborate address at the fair on Thursday, September 29th, and was followed by Gov. Baldwin, who expressed his delight at the progress made in Western Michigan as follows:

GOV. BALDWIN'S ADDRESS.

LADIES AND GENTLEMEN—I did not come here either for the purpose or with the expectation of being called upon to say a single word; and if I had come here with that purpose, I should feel that it was necessary to forego it for the reason, as you have already perceived, I am suffering from a severe hoarseness and cold, which renders it utterly impossible for me to speak on this occasion. But, ladies and gentlemen, while I shall not inflict upon you or detain you here with any extended remarks of mine, I feel it a pleasure that I have the privilege of being with you on an occasion of this kind, in this lovely portion of our beautiful and fertile State. [Cheers.] I congratulate you, my fellow citizens, on the improvement manifested, and the increasing improvement felt and witnessed throughout the length and breadth of our land in every thing pertaining to agricultural pursuits. [Applause.]

Fifty years ago, or a little more, what did we know, what was known, and, less still, what was thought of the doctrine and practice of rotation in crops? What improvements, what wonderful improvements have been made in agricultural implements! It has been said—I have somewhere read it—that, fifty years ago, a good strong man could carry upon his shoulders the entire implements of a farm, except the cart and that old-fashioned, clumsy harrow. To-day we see not only improvements in agricultural implements, but in the cultivation of the soil, as a consequence, and in the quantity and quality of everything that is grown upon the farm. While we live in a day of progress, in no other branch, in no other pursuit of life, has this progress been more marked, more telling, than in that great and noble calling, the cultivation of the soil.

The knowledge we possess,—the larger part of the knowledge we possess, or that is practical by us all,—has been obtained in our intercourse with one another. Now, fellow citizens, in every branch of agriculture progress has marked the day, but in no one branch has this improvement, perhaps, been more marked than in that of the cultivation of fruit.

If I remember aright, Mr. President, it is about twenty years since the formation of the American Pomological Society. But look to-day, throughout the length and breadth of our land, at the agricultural, horticultural, and pomological societies existing from North to South, and from East to West. The object of these societies is to bring together the people, especially those who are engaged—I will not say who are interested, because I trust every man and every woman within the sound of my voice is interested, whether they are pursuing the calling of agriculture or not,—but it is to bring together the people and those who are so engaged, that they may compare together the practice of farm, orchard, and garden, for the purpose of showing improvements.

If there is any one calling, if there is any pursuit intended by the Creator of all things for the special benefit of mankind; if there is any one pursuit more ennobling to man than any other, that pursuit is the cultivation of the soil. [Applause.]

Now, fellow-citizens, I shall not detain you or weary you with any remarks of mine, but I do most heartily congratulate you upon the evidence I have seen, to-day, of the improvement of this portion of our noble Peninsula State. I have never seen, in any part of our noble State,—I have never seen anywhere,—any more beautiful or more creditable exhibition of apples than I have seen here to-day. I say I congratulate you upon this, your first exhibition of the Pomological Society. I congratulate you, my fellow-citizens in this part of the State, upon these evidences of thought; these evidences of mind; these evidences that you intend to make use of thought, of mind, of intelligence; that you intend these to bear upon this noble calling. If there is any calling that requires thought, or any calling that is deserving of mind and thought, of education, of intelligence, it is that of the agriculturist. [Cheers.]

Again, fellow-citizens, I congratulate you upon this exhibition. I congratulate you upon the growing importance of this section of our beautiful State, not only in this branch, but in everything that is for the well-being of society. [Loud applause.]

Judge Ramsdell of Grand Traverse spoke earnestly in support of the Pomological Society, urging the people to subscribe liberally, and sustain a society which is doing so much to promote the vital interests of the State.

Rev. James Hamilton of Newaygo made an eloquent address, showing that the pursuits of horticulture are identified with the highest and best interests of mankind, and promotive of moral and religious improvement.

RETURNS FROM COUNTY SOCIETIES.

BARRY COUNTY.

The Secretary of the Barry County Agricultural Society reports: The eighteenth annual exhibition of the Society was held at Hastings October 5th and 6th, 1870. The exhibition of stock was very good, showing an improvement on previous years; while of fruits and vegetables the specimens were exceedingly fine. Membership tickets sold, 393; Number of entries, 560. The premiums awarded amounted to $304.50.

The following is a financial statement for the year:

RECEIPTS.

Cash on hand	$109 64
Cash from the county	100 00
Cash from rent of grounds	36 00
Cash from sale of membership tickets	393 00
Cash from gate fees	336 57
Cash from rent of dining-hall	50 00
	$1,025 21

EXPENDITURES.

Insurance	$7 00
Old premium orders	11 03
Express on State Reports	3 65
Repairing fence	93 86
Expenses of dining-hall	155 07
Expenses of exhibition	97 50
Expenses of printing	72 70
Secretary's salary	50 00
Premiums to date	295 00
Cash on hand	239 40
	$1,025 41

OFFICERS FOR 1871.—S. J. Bedleman, President; Richard Jones, Vice President; M. L. Williams of Hastings, Secretary; Wm. H. Powers, Treasurer; Peter Cramer, Marshal; T. Alteft, D. W. Ellis, W. P. Bristol, Wm. M. Garrett, John Keagle, Directors.

The Colorado Potato-bug first appeared in June, 1869. The damage in 1870, in this county, was about one-third of the crop. Some localities lost nearly all their potatoes, while others were uninjured.

BENZIE COUNTY.

The Secretary of Benzie County Agricultural Society, Charles F. Travis of Benzonia, writes: That they have no means of knowing the yield of crops in the county this year, but hope that arrangements will be made for compiling reliable statistics another year.

The Colorado potato beetle has not made its appearance yet, and some farmers thought they should not be troubled for some years. Apples grown here keep longer than those grown further south, the trees maturing much slower. Plums seem to be perfectly free from the ravages of insects, and yield largely. Fifty plums have been picked from the section of a limb twelve inches long. Peaches did not do well. Many of the trees were injured by the winter setting in before the leaves had changed color.

The capabilities of the soil have not been thoroughly tested as yet, but the results of the past season demonstrate the fact that this region will compare favorably in variety and quality of its productions with places further south, that have not the moderating influence of large bodies of water.

Crops of oats, clover, and timothy were cut in this county the past season that would have been a credit to the finest farming portion of Ohio. Three tons of timothy were cut from an acre, in several instances.

Potatoes seldom rot, if properly cared for when dug. Large quantities of potatoes are shipped to Chicago, where their quality is considered very excellent.

The Agricultural Society has been in operation three years. The annual meeting for the election of officers was held the first Monday in June. The fair was held at Benzonia September 28th and 29th, on grounds owned by the Society. The fair was characterized by a fine display of grains, fruits and vegetables. The display of wheat was not equal to former years; it was probably injured by the warm, moist days in June. Almost every variety of flint corn was exhibited in perfection. Vegetables in great variety were shown, fully ripened. The quality of apples, pears, and plums was most excellent.

CALHOUN COUNTY.

The twenty-second annual fair of the Calhoun County Agricultural Society was held on the 4th, 5th, and 6th of October, 1870. There was a full average attendance, and a healthy interest in the success of the fair, with about the usual number of entries.

The show of thorough-bred cattle, sheep, and swine was never excelled. The Short-horns were the favorites, and in goodly numbers. The sheep comprised fine and long-wooled, and grades, all of very superior quality. Of hogs, there were all the choice kinds usually found in the country. There was a decided improvement in good, square, substantial horses for all work, and roadsters.

The display of fruits and vegetables, and in domestic manufactures and implements of husbandry, was very fine and large. An elegant assortment of wagons and carriages made in the county was displayed.

The premium list offered about $2,000 for premiums; $900 were awarded.

Receipts and expenditures are as follows:

RECEIPTS.

Cash on hand	$43 48
Receipts at fair	1,444 68
	$1,488 16

EXPENDITURES.

For printing	$63 75
improvements	195 31
expenses of fair	222 20
forage	92 15
premiums paid	746 28
Premiums due	123 00
Cash on hand	45 47
	$1,488 16

OFFICERS FOR 1871.—Nathan Robinson, President; Frank Beach, Secretary; A. O. Hyde, Treasurer. Executive Committee, George H. Barber, Nathan Chilson, Isaac Hewett, T. Hoaglin, Wm. Harris, and a Vice-President for each township.

Calhoun county is twenty-four miles north and south, by thirty east and west, and is about midway between Detroit river and Lake Michigan. There are found within its limits all the grades of land and kinds of timber natural to Michigan, except, pine, hemlock, and cedar, and the soil usually indicated by such timber. We have the rich prairie, the heavy-timbered, clay soil, originally covered with maple, beech, oak of various kinds, white and basswood, white and black ash, hickory, walnut, butternut, elms, etc., with abundant tamarack in the low ground. We have the bur-oak plains, scarcely less fertile than the prairie; the timbered, opening, and rolling white-oak land, running through all grades of loam down to sandy.

This county is watered by the Kalamazoo and St. Joseph rivers and their tributaries, rich with numerous lakes, affording abundant and convenient water for stock, and water-power for milling and manufacturing purposes. The climate is healthful. When the Northern Central and the Marshall & Coldwater roads are completed, its market facilities will equal the most favored localities of the State. The soil and climate are admirably adapted for farming purposes, when well cultivated producing very fine crops of wheat, clover, and timothy hay, corn, oats, and potatoes, and, as a natural consequence, abounds in flocks of fine and long-wooled sheep, herds of cattle, including some very fine thorough-breds, and a stock of horses which for size, style, symmetry, and action is seldom excelled. There are already two or three cheese and butter factories, and this branch of husbandry is rapidly increasing in importance. Our reputation as a fruit-producing county is second to only two or three counties in the State.

Wheat may still be considered the leading crop, but the naked fallow has in a great measure disappeared under a system of rotation, in which the barley, oats, corn-stubble, or clover meadow takes its place. The crop of 1870 was of fine quality, but many pieces were badly winter-killed in consequence of ice and sleet in the early part of winter; and unfavorable weather in March and April made the average less than usual. Corn was more largely planted than usual, and was a very successful crop, many farmers harvesting from 80 to 100 bushels per acre, in fields of fifteen to forty acres. Oats and barley were fine crops, but the berry of the barley was in many cases impaired by excessive rain.

The farmers of this county are now pursuing a system of rotation of crops, including sheep and stock raising, which is far more remunerative than the former manner of farming. They are also paying more attention to fruit, including small fruits and grapes, with apparent promise of success.

It is hardly necessary to add that the potato-bug has

appeared and produced the usual destruction of the crop. They appeared about the 10th of June, and increased in geometric ratio until, like the plagues of Egypt, they covered the land. My impression is that the crop was lessened forty or fifty per cent by the beetle. It is believed, however, that general prosperity prevails in the farming communities.

CASS COUNTY.

The Cass County Agricultural Society was organized in 1851. The following year the whole number of articles and animals exhibited was two hundred and seven. Premiums awarded, ninety-eight. At the fair for 1870, the number of articles and animals exceeded one thousand five hundred, and the receipts for the year were $1,746 09.

The Society has been contending against difficulties, having to move its grounds twice, owing to the passage of railroads through its grounds. The Society is at present, however, in good financial condition, and a good prospect of greatly increased interest in the annual exhibition, from the increased railroad facilities. In the ladies' department there was abundant evidence of that refinement of taste which adds so much to the beauty and pleasure of the fair. The ingenuity, talent and skill shown by the ladies in the productions of their pencils, their needles, and their fingers, was creditable in the highest degree. It affords much gratification to find this branch of the exhibition so well attended to, showing that while the community is advancing satisfactorily in the pursuit of the substantial blessings of life, it is keeping equal pace in the culture of its graces. On the whole, the progress of the Society is onward, and is destined to become of very general interest.

Officers for 1871.—President, John P. Coulter, Howard;

Secretary, D. M. Howell, Cassopolis; Treasurer, W. W. Peck, Cassopolis; Directors, James Rivers, Calvin; G. B. Turner, Jefferson; Hiram Mecchum, Porter; M. J. Gard, Volinia; W. H. Doane, Howard; J. Haine, Jr., La Grange; A. Huyck, Marcellus; A. Knapp, Dowagiac.

The following items concering the county at large have also been kindly furnished by Mr. James Rivers, Secretary *pro tem.*

The county embraces almost every variety of soil and timber found in the State. It is well watered by numerous clear lakes and running streams. Owing to the natural fertility of the soil, the farmers have not paid the attention to the preparation and application of manures which their true interests demand, and to which they must attend if they expect to keep up the productiveness of their farms.

The county produces all kinds of grains and grasses common to this climate. For quality of fruit it challenges competition, while vegetables of all kinds are raised in abundance, and large quantities of both fruit and vegetables are exported.

Stock-raising is not followed largely for profit. The young stock is bought up by parties from other places. Oxen, on improved farms, are seldom seen, and cows are kept for home use only. Devons and Short-horns (Durhams) are the most prominent improved breeds. Considerable attention has been paid to the breeding of horses, the light and fleet roadsters being superseded by draft-horses and horses for all work. Sheep-raising is profitable, but it has not the attention the age demands.

Pork forms the most valuable export from the county, and improvements in breeding are being made by practical farmers. The method of feeding on clover through the season is largely practiced, it being generally conceded that hogs running in clover through the season double their weight by fall without further expense.

The potato crop is the only one that has failed for two years, and this was owing to the bugs.

EATON COUNTY.

Mr. T. D. Green, secretary of the Eaton County Agricultural Society, writes: The past year has been a very prosperous one for our Society, both financially and in the interest shown by the members, in the number of entries made. Our new grounds are well fenced, we have good buildings, and one of the best half-mile tracks in the State. From the organization of our Society, in 1855, up to the present year, we have been compelled to hold our fair in a seven-acre lot, where there was no chance to exhibit half the entries. We now have thirty-one acres, finely located, and the last fair proved that the farmers can appreciate nice grounds and buildings. Our receipts were very much in excess of former years. After paying all expenses and premiums, we had something left with which to make improvements another year.

Officers for 1871.—President, E. T. Church; Treasurer, E. Shepherd; Secretary, J. V. Johnson.

The potato-bug appeared early in the season, and early potatoes were more injured than late. Some pieces were very little injured, others nearly destroyed. No remedy has been found that farmers value.

GENESEE COUNTY.

In accordance with annual custom, and pursuant to the provisions of our By-Laws, the Executive Committee submit their twenty-first annual report of the Genesee County Agricultural Society, for the year 1870.

In contrast to the spring of 1869, which was cold, wet, and backward, the spring of 1870, after a severe winter, opened early, and was genial and favorable for agricultural operations. Taking advantage of this condition, a large amount of spring crops were put in by our farmers generally. The summer proved rainy to an uncommon degree, the wet being accompanied all through with much warmth, and a portion of the season was intensely hot. The autumn was remarkably fine, without frosts, and the mild weather extended late toward the close of the year. The effect of the wet and warm weather of summer was to produce a rank upright growth of straw in wheat and oats, and to largely obstruct the cultivation of corn, while promoting the vegetation of weeds.

FIELD CROPS.

Such of the wheat as was early sown in the fall of 1869, so as to attain a good growth to give it strength and vitality to withstand the sudden and severe winter of that year, did well. That which was sown late fared worse, for want of time to acquire the stamina necessary to be up against extreme cold, in consequence of which a large proportion of it was killed by ice and exposure. The combined warmth and wet of the growing season had an injurious effect upon the fruitfulness of the crop, by causing the young plant to spindle up with too rapid growth, instead of stooling. The wet and warmth were likewise unfavorable for early harvesting operations, and caused a little sprouting in the ear of the bald varieties of wheat in some situations. Notwithstanding all disadvantages, the general quality of the crop was excellent; but we estimate that the average yield for the county at large was reduced, by the conditions referred to, to twelve and a half bushels per acre, although some of our best and most thorough farmers were favored with heavy crops.

In the early part of the sason, oats looked uncommonly promising. But as the summer advanced they, like wheat, were injuriously affected by the heat and wet, causing the general yield to be below an average,—say thirty-five bushels to the acre.

Corn was planted in good season, and under favorable circumstances of weather, but the rains which followed affected injuriously its growth, and their frequency and long continuance prevented the usual amount of cultivation. However, the very fine fall weather favored its coming to maturity, and a good crop of superior quality was secured. The average yield is estimated at thirty bushels per acre.

Barley is not very extensively cultivated in this county. The yield in 1870 was short,—perhaps not to exceed on the average thirty bushels to the acre, owing to the same causes which affected the crop of oats.

The meadows yielded about one and a quarter tons of hay to the acre, with a decided improvement over the quality of the crop of 1869. The early haying weather was not very favorable, but it improved as the time advanced; and it was mostly secured in good order, excepting clover hay, much of which was considerably damaged by wet. For the maturing of clover-seed, however, the season was good, and a fine crop was realized. The pastures, also, have been excellent.

The season was favorable for the planting and growth of potatoes, but they were more or less injured by the operations of the Colorado beetle, the damage from this cause varying much in different localities. The yield is estimated at one hundred bushels to the acre, on the general average. Some portion of the crop has been lost by rot, after the potatoes had been dug and housed in apparently fine condition. This decay has been found especially to affect Peachblows, hitherto one of our most esteemed varieties.

The culture of hops in this county has been almost abandoned. We know of but two or three growers who deemed their vines worth picking this year. This was owing mainly to the depressed market price of the article, in consequence of which, many who have heretofore raised hops to some extent have plowed up their yards for other uses.

About the usual breadth of buckwheat was sown, but the returns were low, owing to the prevailing wet, hot weather. The average yield was about twelve bushels per acre.

Root crops were seriously damaged by the rank growth of weeds choking their progress. Aside from this, which was only an incident of the season, we regret to notice that less attention is being bestowed upon these valuable aids to the healthy support of stock than their worth would seem to demand. This neglect is not owing to any want of appreciation of the value of these crops, but is rather to be ascribed to the scarcity of labor, making it difficult or impracticable to take care of and keep them free from weeds at the most critical period of their growth; their season of sowing and cultivation, and when they most need attention, coming just at the time of year when the farmer's other work is heaviest and most pressing, and will not admit of postponement.

FRUIT.

The orchards of apples, at present the leading fruit of this county, produced a large crop. But owing to the peculiar atmospheric conditions of the season, the ymatured too rapidly and too early, which has been found to seriously impair their keeping properties. The fruitage of grapes was large and fine, and the success which has heretofore attended their cultivation is inducing an annual increase in the number and extent of the vineyards. Pears, plums, cherries, and most fruits raised here were abundant and good. Peaches, however, did not produce so well as in 1869; and the liability of the peach trees to be winter-killed prevents much attention being given to their increase.

HORNED CATTLE.

The display of blooded stock at the county fair shows that no diminution has taken place in the interest we have noted for several years back. The raising of stock for export is becoming an important item in the resources of this county, and doubtless our farmers find profit in improving their breeds.

HORSES.

As the wealth and population of the county grow apace, the number of fine horses kept for work or pleasure increases in proportion. Handsome carriage horses and roadsters which, a few years ago, were scarce enough to be remarkable, are now so numerous as to attract no special observation, while the class of horses used for farm work is constantly improving. A demand for heavy draft horses is introducing a greater number of that description, and directing attention to their propagation. The sire horses kept in the county for raising stock for farm work are numerous and valuable.

SHEEP.

To judge by the number of pelts offered for sale in the Flint market alone, the slaughter of sheep, alluded to in previous reports of this Society, must still be very large. While prudent and careful sheep farmers still adhere to their convictions that wool will be a profitable commodity to raise, and reserve their flocks accordingly, the reaction among those who embarked so incautiously in expensive sheep a few years ago it would appear has not yet quite subsided. The remarkably fine and extensive show of first-class sheep at the county fair in October afforded unmistakable evidence of the care and attention still bestowed upon this branch of stock-husbandry. The clip of wool has been rather light, and the average price, 40c., was a little higher than that obtained in 1869.

SWINE.

While the breed of hogs continues to receive marked attention, and their improvement is diligently sought, it is observable that the preference is being transferred from the large heavy breeds to those of smaller and more compact frames; the latter being easier to feed and cheaper to fatten. The favorite crosses seem to be the Suffolk with the Chester, and the Essex with the Chester. Either of these makes good marketable pork. The quantity killed this year has been large, and the price obtained satisfactory.

MANUFACTURES.

There is no falling off in the prosperity of the manufacturing interests of Genesee county. Additional mills continue to be erected for the manufacture of lumber, lath, shingles, cloths, cassimeres, flannels, shawls, yarns, etc. Flour, also, has become an article of large export to the Eastern States; and staves are manufactured very extensively. A number of cheese factories are also among the newly established evidences of industrial enterprise.

TRANSACTIONS.

The twentieth annual meeting of the Society was held on the 12th day of January, 1870, when the following persons were elected officers for that year:

President—Elijah W. Rising of Litchfield. *Vice Presidents*—John B. Cochran of Argentine; William Owen of Atlas; Henry Schram of Burton; Enos M. Miller of Clayton; Calvin Cartwright of Davison; Y. E. Benton of Fenton; Abner Randall of First Ward; John B. Hamiltou of Second Ward; Artemas Thayer of Third Ward; George W. Thayer of town of Flint; William Schram of Flushing; George W. Meriam of Forest; Zenas A. Gage of Gaines; William B. Wetherell of Genesee; John W. King of Grand Blanc; William Hulburd of Montrose; Samuel Greenly of Mount Morris; Walter Cross of Mundy; William Munger of Richfield; David Case of Thetford; Charles L. Cole of Vienna. *Secretary*—Francis H. Rankin of city of Flint. *Treasurer*—Oren Stone of city of Flint. *Executive Committee*—Jesse M. Davis of Genesee; Chandler H. Rockwood of Genesee; David Schram of Burton; George W. Thayer of Town of Flint; D. H. Stone of Grand Blanc; Josiah Pratt of city of Flint; Charles Pettis of Davison. *Auditors*—Alexander W. Davis of Grand Blanc; John L. Gage of Burton.

At that meeting, discussion was resumed on the subject of a change or enlargement of the fair grounds, which terminated in the adoption of the following resolution:

Resolved, That the question of removal or enlargement of the fair grounds be left to the discretion of the Executive Committee; and if, in their judgment, they deem it for the best interest of the Society to remove the same, that they are hereby authorized to sell the present grounds and purchase and locate elsewhere; or to purchase additional land adjoining the present grounds.

On the 19th of March, the Executive Committee held a meeting for the purpose of taking this matter into consideration; and then appointed a sub-committee, consisting of the President and Treasurer of the Society and Mr. David Schram, to investigate the subject of the enlargement or removal of the grounds.

On the 9th of April the sub-committee made their report to an adjourned meeting of the Executive Committee, setting forth what might be done in the way of enlarging the old fair grounds, and the estimated cost of the same. They also reported that they had received a proposition from Mr. Wm. Hamilton to sell to the Society a tract of land, about twenty-five acres more or less, sixty rods nearer the court-house than the old grounds, for the sum of $10,000, payable at the option of the Society at any time within ten years, with ten per cent interest, payable annually. The report recommended the purchase of this land, as well adapted to the wants of the Society, and that the old grounds should be sold, it being expected that they would bring about $6,000.

After visiting together and examining the land offered by Mr. Hamilton, the Executive Committee adopted the report and recommendations of the sub-committee, and appointed the same sub-committee to complete the arrangements for the purchase and removal. On being surveyed, the tract in question was found to contain 26½ acres, and the amount of the consideration, $10,600.

To bring the ground into an improved shape for the uses of the Society, a small portion of the land at one side was exchanged with Mr. L. Wesson, for an equivalent in land at another point.

At a subsequent meeting of the Executive Committee, held on the 23d of April, Mr. Artemas Thayer attended and offered the following written proposition:

FLINT, April 19, 1870.

To the Executive Committee of the Genesee County Agricultural Society:

We, the undersigned, being desirous of aiding your Society in getting up and perfecting, on your projected new fair grounds, a first-class half-mile race-track and riding-park for the joint interest of your Society and for the undersigned, who may be more particularly interested in a riding-park and race-track, and the improvement of horses; and having been informed that your Society would be willing to make any reasonable arrangement, not inconsistent with your interests, for the furtherance of such an object, we would therefore most respectfully make to your committee the following proposition (as being acceptable to us) for your consideration:

1st. We will lease of your Society the use of the track and grounds for the special purpose of general horse show and exhibition, and the trial of the speed of horses, for the term of ten years, under the following considerations, conditions, mutual agreements, and restrictions:

2d In consideration of said lease, we will proceed at once in the work of grading and putting in first-class condition the projected half-mile track, and have the same completed and finished by the first of July next, at our own cost and expense, the grade and survey to be made and established by you,—the work to be done to your entire satisfaction.

3d. In consideration aforesaid, we also propose to be to one-half the expense of building a first-class grand stand and judges' stand, and such stabling capacity on the ground as may be deemed for the joint use and necessity of your interest and ours, and also pay one-half the expense of fencing and draining the grounds. The whole of the above works and improvements to be done under the direction of a joint committee of the two interests.

4th. Proposition for a general riding park,—how to be got up, and how supported:

A gate-keeper or overseer of the grounds to be appointed by you, who is to have entire charge of the grounds, and to keep the same open for ingress or egress under such rules and regulations, both as to tariff of rates for season tickets, and gate fee for common entrance, as you may deem best calculated to encourage the use of said grounds, and make the same a popular place of resort as a riding park.

No season park tickets to give any right to entry upon the grounds on any show or exhibition day, or time when the gates shall have been directed to be closed by your Society or ours.

The entire receipts thereof to be sacredly dedicated to paying the gate-keeper, keeping in repair the track and the stands, sheds, and fences, and the building such further buildings as may be deemed for the joint interest of the two Societies. If said fund should not be sufficient, the expense to be borne equally between said parties.

Your Society to put the grounds generally in good condition, and keep the same in such condition. The undersigned, by themselves or their agents, giving notice to the gate-keeper that they want the grounds for public exhibition or show, or for the trial or the test of speed of horses, to have the right to have the full and entire control of the entire grounds, and to establish such rates of gate fee, and entrance upon the grand stand, and the leasing the grounds or buildings on such days, as they may deem fit. The entire avails of such days to belong to the undersigned, or their successors or assigns.

Your Society reserving to itself twenty days preceding and fifteen days succeeding your annual fair, in which no exhibition or horse show is to be had by us on said ground.

Your Society reserving also to itself the right to hold a sheep-shearing, and show exhibition of stock and stock horses, any time in the month of May or June (for one week); first giving twenty days' notice to the undersigned of the time of holding said show.

The old fence on the old fair grounds to be used, so far as may be deemed suitable, on new grounds, for the joint interest of the two interests.

Respectfully yours,

ARTEMAS THAYER,
CHARLES SMITH,
LEVI WALKER,
GEO. M. WALKER,
W. H. TOUSEY,
WM. HAMILTON.

After due consideration and discussion, the foregoing proposition was accepted, and the sub-committee already appointed were authorized to make a contract on the basis of said proposition. At the same time, Messrs. Oren Stone and Charles Smith were appointed agents for the disposal of the old fair-grounds by contract or otherwise, on the ground being re-platted, for which they were to receive ten per cent on the amount of sales.

A lease, embodying the provisions set forth in the foregoing proposition, was executed to the Riding Park Association, and the two Societies proceeded to fit up and prepare the new grounds for use, according to its terms. The expense has unavoidably been large, but your committee think the advantages gained amply compensate for the outlay. This Society can now congratulate itself upon possessing—we think we are safe in asserting—the best county fair grounds in the State, all things considered; large enough for all the future wants of the Society, handsomely fitted up, and most conveniently located within the city limits.

As this Society has never been incorporated, and therefore

has no existence in the eye of the law, we think the time has arrived when its incorporation should no longer be delayed; and we therefore recommend that immediate steps be taken to effect that object.

In preparing the list of premiums to be offered for competition at the fair of 1870, several new classes were added and others enlarged and extended, and the premiums in all classes of stock, and in domestic manufactures, were thrown open to the competition of all the counties bordering on Genesee. Arrangements have also been made to provide diplomas in addition to the cash premiums, which will soon be ready for delivery, and will be found highly ornamental and permanent memorials of success in competition for our premiums.

A vacancy having occurred in the Executive Committee, owing to the removal of D. H. Stone, to reside in Oakland county, Mr. Phineas Thompson of Grand Blanc was appointed to fill his place.

At the desire of many of the owners and agents of reaping and mowing machines, arrangements were made for holding a mowing match, under the auspices of this Society, in the month of June. It took place on the farm of Mr. Dewey, one mile north of the city, and elicited considerable interest.

The annual fair was fixed for four days, commencing on Tuesday, the 14th of October. That day, however, proving very wet and inclement, the fair was extended to the fifth day, and the entry books were held open throughout Wednesday, for the accommodation of those kept away by the storm of Tuesday. The weather cleared off fine, and the fair was remarkably successful. The entries for competition exceeded those of the previous year by about 300, and the increaae in the sale of membership tickets was 162. The exhibition of horses, sheep, fruit, vegetables, and the display in Floral Hall never was larger or better; while the show of swine, poultry, carriages, and miscellaneous articles was far ahead of those at any previous fair held by this Society. The superiority of many of the animals in the cattle classes was noted by visitors from abroad. Among the most admired were several Shorthorns and a yearling Hereford bull from the Crapo herds, which were not in competition for premiums, but exhibited by Mr. Crapo to help to show what our county is maintaining in the line of blooded stock.

As the sale of the old fair grounds in separate lots was found to proceed but slowly, and it seemed very desirable to realize from them in order to appropriate the proceeds to relieve the Society from a part of its debt and accruing interest for the new grounds, it was deemed advisable to advertise for propositions for the purchase of the whole in a body, which was accordingly done. At the time appointed for receiving the bids, the only proposal made was from Mr. Oren Stone, who offered $6,000, payable in annual installments; with interest at ten per cent, which offer was accepted. The practical effect of this sale is to reduce the indebtedness of the Society for its real estate in the said sum of $6,000, leaving the amount of its liabilities for land, to be provided for out of its other resources, about $5,000.

For additional information in regard to the finances of the Society, we beg to refer to the report of the Treasurer.

All of which is respectfully submitted.

For the Executive Committee.

F. H. RANKIN, *Secretary.*

FLINT, December 31, 1870.

HURON COUNTY.

The Huron County Agricultural, Horticultural, and Arts Society was organized in 1869. It was a matter of serious doubt to the projectors of the enterprise, whether the state of agriculture in the county, and public sentiment, would sustain a society in such a condition as to confer that benefit and awaken that spirit of generous rivalry which were the chief objects sought in its organization. As the capital in the county was chiefly invested in lumbering operations from its first settlement, to the total neglect of agriculture, till about twelve years ago, it was feared that neither farmers nor capitalists would feel themselves sufficiently interested to contribute of their means, time, and labor to get up a respectable exhibition. By the exercise of some energy and perseverance, a portion of the county was canvassed and sufficient stock subscribed to warrant the holding of a fair in 1869. One hundred and seventy-six dollars were offered in premiums at that time, only eighty-one dollars of which were awarded. In 1870 about two hundred dollars were offered, and one hundred and twenty eight dollars awarded. At these exhibitions, the quality of the agricultural and mechanical products was good, and manifested a progressive interest on the part of patrons of the Society. A very good number of entries was made, and in 1870 quite an advance on the preceding year. The Board of Supervisors have this year voted us one-tenth of a mill assessed on all taxable property in the county, which, added to stock subscribed, will soon enable us to make a show that will compare favorably with that of some older counties. The officers are elected at the annual meetings, held the third Friday in January. The Board consists of a President, Secretary, Treasurer, and five Directors.

OFFICERS FOR 1871.—President, Watson Robinson; Treasurer, Jeremiah Ludington; Secretary, J. Ross, M. D.; Directors, Robert Scott, Waterhouse Whitman, Alex. Harrison, Donald Carrie, and Rev. J. B. Wilson. The accounts show a balance on hand of $41 86.

HILLSDALE.

In compliance with statutory provisions, we herewith transmit the twentieth annual report of the Hillsdale County Agricultural Society, accompanied by such notes as seem to be proper under the circumstances. We do this the more cheerfully from the fact that the year has been one of prosperity to all our people engaged in agriculture or mechanic arts. Favored as we have been above many sections of the country, with rains and sunshine to mature our crops, with health and strength to secure the same, and a reasonable demand for all our products, just encouragement is given to the producers of our county to press forward in their noble calling.

With pleasure we record the fact that a majority of our farmers are seeking a higher standard of agriculture. The inquiry comes from every quarter, How may I make my farm more productive with the means at my command? To what stock, or of what breeds, may I look for the best returns? How may I so change this or that field, this or that course of cropping, this or that surrounding of dwelling or barns as to give profit to labor and add comfort and pleasure to the household? How may I most profitably return to the soil the few hundred dollars heretofore extracted from it, and which has been wrapt up in mortgages or bonds stored away in some old family chest or bank vault? We look upon this unrest as an omen for good to the future of our county. The inquisitive minds of our Yankee population will not stop short of legitimate results. Progress is being indelibly stamped on all our actions and doings, on none more than those of the farmer and mechanic. Else why the call for so much agricultural and scientific knowledge as is being distributed weekly by the press in almost every household in our land? Why the avidity with which every new species or variety of seed, plant, animal, or idea, is sought after by the agriculturist and mechanic, if not in hope of bettering his condition? Why the annual voting of taxes for the dissemination of books and premiums for the advancement of agricultural and mechanical science, if not already convinced that it pays, and pays largely, for the outlay?

The peculiar and diversified character of soil, the numerous lakes, rivers, and rivulets, with the great altitude of the country, pre-eminently adapt it to mixed husbandry, and especially so to fruit culture. We are satisfied, from a residence and experience of more than thirty years, of the peculiar adaptability of our county to fruit culture. More reliable from year to year than our wheat crop, and in quality not excelled even by the lake counties, which, with the great facility to markets, offers inducements to our people enjoyed by few other localities. And we are pleased to report the fact that our orchards and vineyards are multiplying, and from the greater care and attention bestowed upon them the results are very gratifying. Our crops, as previously stated, were very good. Grass got an early start, turning a yield of hay of superior quality, averaging one and a half tons per acre.

Wheat, owing to the early setting in of winter, and at a time when making its most vigorous growth, was badly winter-killed on old lands where indifferently put in or sown broadcast, but the very favorable spring told for good to it, and the result was a very excellent quality of grain, with an average yield of eighteen bushels per acre.

Oats are being raised in the county more than formerly. Farmers have learned the exhausting quality of the crop, and they very systematically prepare for it by using an abundance of fertilizers. The result is very favorable, giving an average yield of forty bushels.

Rye is but very little raised, not sufficiently so to be noted.

Peas do well on our clayey and firm soils. They are raised with profit in about half the county. The average yield is about twenty-five bushels per acre.

Buckwheat is raised by many farmers in small quantities. Owing to drouth and excessive hot weather when filling, the crop was but little above a half one. Yield, about fifteen bushels per acre.

Corn was never better. A large average. Cultivation generally good. Quality superior, and the yield about forty-five bushels, shelled, per acre.

Potatoes were a failure for the first time since the county was settled. All late varieties were seriously affected by hot weather and drouth just as the tubers were forming. And in addition to this, many localities were visited by the Colorado beetle, destroying in many instances whole fields. They appeared in small numbers as early as the middle of June, multiplied very fast, until potato vines were dead from the scorching sun. As a preventative, many picked them, others used Paris green. Both failed of preserving the crop, the first being too expensive, the other failing to kill except where they partook of a hearty meal.

Of hops there were no yards cultivated, the reaction in the article being too great for those having them planted.

Barley is but an indifferent crop of the county; it as often falls under ten bushels per acre as it exceeds it. This year it will average fifteen bushels per acre. *Query.*—Is it because of the high altitude of our county, giving extremes of heat and dry air, that this crop makes so small growth; or is it in the composition of our soil, or both?

Stock—Our cattle are largely represented by the Shorthorns and their grades. Few, if any, counties in the State can show better types of the thorough-bred than this, and they are telling for good with a large number of our best farmers.

Our horses are a very excellent class for all work, being descendants from, and retaining, the Morgan, Black Hawk, Messenger, and Duroc bloods. Our people seem content with the useful, without risking their chances for the *noted.*

Sheep.—In this class of stock there has been a steady decline since 1866. We must report a falling off of 20 per cent over last year. Some have changed from fine to long-wooled sheep, but it is problematical, as yet, whether it will pay, as the difference in keep of the one over the other will fully use up the difference in price of wool. And to feed for mutton so far from the mutton market, is still more problematical with our farmers.

Swine are in goodly number and represented by several distinct breeds, with their grades and crosses. Each has its advocates. The prevailing breed is a mongrel, made up of the Byfield and old English Leicestershire, and goes by the name of Chester White. These are a large, rangy hog, given to legs and ears, and if properly fed will ripen in two years. A better type is found, originating in Chester county, Pennsylvania, composed of part English Suffolk and part Yorkshire, which, from its Suffolk blood, matures early and ripens profitably for the market. Another class, much sought after, for easy keeping, early maturity, and excellence for family use, because of its fineness in bone and on the spit, is the Essex. They are being disseminated through the county, and telling for good in crossing on the larger breeds. Another of the big-boned class is being introduced into the county, known as the Magie hog, from Butler county, Ohio, made up of the big-boned China, Poland, Irish Grazier, and Berkshire breeds These are represented to mature early and attain great weight, being easy feeders.*

Poultry is found in almost every yard, whether in city or country. Each lass or lad will exercise their own judgment as to the kind, number, and quality that shall hover around them. Still, the surplus from this medley, shipped to other markets, will more than buy the best farm of 160 acres in the county, every year.

---

*Note.—Is not this questionable? Is it in accordance with the historic facts of either or any of the breeds from which they inherit? And is it not questionable as to the profit on all the large breeds, where the first cost of corn is over a penny a pound?

A very important interest to the county, and of late date, is her cheese manufactories, of which, at this writing, there are eight, the oldest of which is the Allen factory. It has been in operation four years, Reading three years, and Pittsford two years; Cambria, Hillsdale, Fayette, Osseo, and Adams, each one year. The three first, for the past season, used the milk of a thousand cows; the other five, about five hundred all told, it being their first year, and most of them starting late. All are sanguine, from first experience, that it will be a paying business in the future.

Mills.—From the numerous rivers finding their source in our county, we have every facility for manufacturing the large surplus of our grain crop for the Eastern market. There are sixteen flouring mills in different parts of the county, running forty burrs,—two of these on a large scale, valued at $50,000 each.

Mechanics.—Among this class of our producers, several new and important industries have come into notice, the most important of which has been inaugurated by Messrs. Phelps & Pettis, of the city of Hillsdale, for the manufacture of farm machinery, portable steam saw-mills, engines, boilers, threshers, cutters, and all implements for cultivating the soil, even to shovels and spades, giving employment to a large number of ironmongers. These, with the many other manufactories mentioned in former reports, furnish our people with facilities surpassed by few communities.

Another important interest is the opening up of the Detroit, Hillsdale & Indiana Railroad, connecting Hillsdale with Detroit, by way of Adams, Moscow, Somerset, Brooklyn, Manchester, Saline, Ypsilanti, and Michigan Central, to Detroit, making nearly an air line, and which will tell largely in the future in the matter of freights and market to the county at large, and especially to the northeast towns of our county, heretofore deprived of these advantages.

In speaking of the doings of our Society for the year so

soon to close, we would rather that our works speak for us, lest we fail to be understood; but they being silent monitors, unintelligible and unappreciable except by sight and touch, we will attempt the work, depending upon our records for assistance. From long experience and observation in the management of fairs, especially our own, we became fully satisfied of the necessity of systematizing the work more fully if we attained the ends contemplated by the act of bringing them into being. To this end we prepared new rules and regulations, a new system of classification and entry, a new principle of judging, affixing a responsibility upon the office of Judge that enlisted his or her good judgment, and made it a pleasure to them to exercise it in so conspicuous a place. This plan our Board of Directors readily adopted, and the experience of one year gives us full confidence in the system.

Our fair opened the fourth of October, and closed the seventh. The first two days were very discouraging, a slow, drizzling rain most of the time retarding, in a measure, our entries, and preventing many from bringing forward their offerings at all. The third and fourth days were all we could ask or desire. The old and young, rich and poor, came and mingled together, all putting on their best smiles, and living intensely for the day.

The number of entries in the several divisions, with the number of awards made to each, will best convey an idea as to the extent and merit of the exhibition, for be it understood that no premium is awarded except on articles or animals in the opinion of the Judges strictly meritorious.

There was 452 exhibitors, making 1,348 entries, as follows: Cattle, 62 entries, 49 awards. Horses, 146 entries, 61 awards. Sheep, 56 pens entries, 38 pens awards. Swine, 28 pens entries, 17 pens awards. Poultry, 29 coops entries, 19 coops awards. Manufactures of wood and iron, 152 entries, 60 awards. Manufactures of leather, 18 entries 11 awards.

Textile fabrics, 164 entries, 79 awards. Fine arts, 50 entries, 23 awards. Floriculture, 45 entries, 21 awards. Fruit, 356 entries, 75 awards. Conserve products, 107 entries, 39 awards. Grain and vegetables, 135 entries, 42 awards. Total—1,348 entries, 534 awards.

From the above, it will be seen that the exhibition, in all its parts, was well represented, the department of fruit taking the lead, and making of itself a magnificent show.

Financially, our Society has been a success since its location in 1858, and reorganization under the general law in 1864. It has been the aim and purpose of the managers to build up a local habitation for the Society second to none in the State, around which it could rally from year to year, meeting the comforts, as they come together, so oft experienced by well-regulated families whose members have been long absent and now return.

The location of our fair grounds was a very fortunate proceeding. The land, aside from improvements, has enhanced in value more than two hundred per cent. This, with the improvements, brings to the Society a realty, the cash value of which may be estimated at $21,000; our indebtedness toward the same, $4,500. The receipts from our last fair were $3,553 25; expenses of same, and premiums, $1,891 46; leaving a balance of $1,661 79 as the net profits of the fair. This has been applied, together with the county appropriation, $242 25, to the payment of past indebtedness, and the improvement fund.

The officers of the Society for the year 1871 are as follows: Col. F. Fowler, President; Hon. R. Worden, Vice President; F. M. Holloway, Secretary; Daniel Beebe, Treasurer; and a Board of 20 Directors, one from each town in the county, and two from the city of Hillsdale.

INGHAM COUNTY.

Geo. W. Bristol, Esq., of Mason, Secretary, sends the following report of the Ingham County Agricultural Society, and of the county at large:

The year 1870 has been a year of general good crops in this county, with the exception of the potato crop, which, owing to the ravages of the potato-bug, has been about one-half an average. The potato-bug made its appearance in this section as soon as the potatoes were out of the ground, and remained until about the middle of September. The only remedy found to be effectual here, was to gather the bugs with their eggs, by shaking them into a pan, and burning them. Eternal vigilance has been the price of potatoes in this section.

The wheat crop was more than an average yield, but, owing to the wet season, was somewhat damaged.

Indian corn was a very large crop, and of excellent quality. Oats were very heavy; never better.

The fruit crop was very good, especially apples. The trade in apples is fast increasing in this county. Large quantities were shipped this season to Western markets, at remunerative prices.

The sixteenth exhibition of the Society was held on their grounds at Mason on the 6th and 7th of October, and was more numerously attended than any previous exhibition, notwithstanding the close proximity of the Central Michigan Fair at Lansing and the Michigan State Fair at Jackson. Every department of the exhibition was well represented, especially that of stock; and a decided improvement in the quality of horses, cattle, and swine was a subject of general eral remark, and a gratifying evidence of the good our animal fairs have already accomplished. There was a larger show than usual of swine, and some very fine specimens of the Essex and Chester White breeds. The show of sheep was quite large. There were several exhibitors of Cotswolds, and a number of very fine animals on the ground. Grain and vegetables were shown in abundance.

The display of fruits and flowers was very large, and attracted great attention. Floral Hall was made particularly attractive by the good taste displayed by Rev. Geo. W. Barlow, the superintendent, in the arrangement of the articles on exhibition. There was also a large show of farm implements, for a county fair. The annual address was delivered by Prof. W. H. Perrine of Lansing, on the subject of "Farm-houses and Grounds," and contained much matter that is particularly deserving of the attention of the farmers of Ingham county.

The whole number of entries was 810; and the receipts from all sources amounted to $1,013 58.

Our fair grounds are situated within the corporate limits of the village of Mason, and embrace eighteen acres, worth about $2,500. The Society has erected on the grounds, the past year, a very fine building, for an office and committee-room; also, a building suitable for a judges' stand. The successful result of the last exhibition has enabled the Society to entirely pay for the new buildings, defray all its expenses, pay all the premiums awarded at the fair, and discharge the last payment and interest upon the grounds. This places the Society, for the first time in its existence, entirely out of debt, with good show-grounds, and we trust that the time will soon come when the Ingham County Agricultural Society will rank second to none in the State.

The receipts and expenditures of the Society for the year ending January 18th, 1871, are as follows:

RECEIPTS.

Receipts from annual exhibition	$893 58
Received from the sale of hay	20 00
Received from the county	100 00
	$1,013 58

EXPENDITURES.

Paid for new buildings	$247 00
Paid for printing	90 00
Paid for premiums	430 00
Paid last payment and interest on grounds	90 00
Paid expenses of exhibition	156 58
	$1,013 58

IONIA COUNTY.

W. D. Arnold, Secretary, of Ionia, writes that the annual exhibition of the Ionia County Agricultural Society was a decided success, being largely attended, with a good show of stock, grains, vegetables, fruits, and agricultural implements.

The Treasurer's report shows the receipts to have been $1,930 89, and the expenditures $1,781 85, leaving a balance of $149 04.

The officers for 1871 are: President, A. F. Kelsey; Secretary, W. D. Arnold; Treasurer, J. W. Loomis; Directors, Erie LeValley, Southwick Merritt, Jerry Spaulding, Henry Hitchcock, Wm. S. Bates.

JACKSON COUNTY.

The officers of the Jackson County Agricultural Society for 1871 are as follows: President, Ephraim Van Horn; Secretary, R. D. Knowles; Treasurer, Chester Warriner.

There are twenty-three Vice-Presidents, one for each township and each ward in the city of Jackson.

The Executive Committee is composed of six members.

KENT COUNTY.

The twenty-second annual fair of the Kent County Agricultural Society was held, in connection with the State Pomological Society, at Grand Rapids, September 27th, 28th, 29th, and 30th, 1870.

The weather was delightful, the attendance large, and the exhibition in all respects good, in some superior.

The display of cattle was small, but quite good. Among the horses were some excellent ones for all work. Also among the mares and colts, and the saddle and carriage horses.

All classes were represented in about the same way. While in each class there were not a very large number of entries, some of the specimens were very fine.

The millers of the county offered a special premium for the best crop of white winter wheat, five acres or more. The piece that took the first premium averaged 42 bushels and 26 lbs. per acre, and the second best 31 bushels.

The indebtedness at the beginning of the year, and the expenses for the year, amount to about $4,147 84, which is covered by the receipts for the year.

The Society owns its grounds and buildings, free from indebtedness; and owing to the complete success of the last exhibition, and the liberality of the present Board of Supervisors, it is now entirely free from any incumbrance.

OFFICERS FOR 1871.—President, Geo. W. Griggs; Secretary, Edwin A. Burlingame; Treasurer, Omar H. Simons, with an Executive Committee of five members.

LAPEER COUNTY.

The Lapeer County Agricultural Society held its thirteenth annual exhibition at the city of Lapeer on September 28th, 29th, and 30th, 1870. The exhibition was good, but a heavy rain the third day prevented any attendance, and the total receipts were light.

The total number of entries was 879, as follows: Cattle, 164; horses, 153; sheep, 54; swine, 23; poultry, 25; farm implements, 55; wagons, carriages, etc., 14; domestic manufactures, 79; fine arts, needle-work, etc., 81; fruit, 27; vegetables, 92; flowers, 30; grain, 39; miscellaneous, 43.

The exhibition in most departments showed a marked improvement, especially in cattle, swine, poultry, and farm implements manufactured in the county.

The receipts and expenditures of the Society for 1870 are as follows:

RECEIPTS.

From admittance to exhibition	$684 24
From rents, etc.	60 00
From county	150 00
	$894 24

EXPENDITURES.

Rent of show grounds	$100 00
Printing	97 00
Other expenses of exhibition	172 55
Paid premium checks, 1869	95 00
Premium checks, 1870	612 50
	$1,077 05
	894 24
	$182 81
The Society owes	125 00
Making the indebtedness	$307 81

OFFICERS FOR 1871.—President, Gardner Dexter; Vice-President, B. F. Moore; Secretary, Ira H. Butterfield, Jr.; Treasurer, Oliver Nichols.

The season for this county has with most crops been very productive. The early spring was dry and cool, the summer wet and very warm. I have no measurement of the rainfall, but think that more fell in July than any other month. Wheat was, however, secured in good order. Hay generally overripe. Corn is better than for some years, being well ripened and sound. Oats and barley were about an average crop. Clover seed is fair; not as well filled as in some seasons. Apples a bountiful crop. Winter apples are not keeping well. The cultivation of hops is nearly abandoned.

The potato-bug made its appearance early in the season, and

did some damage in gardens, but did not find its way to the fields till late, and did not affect the yield to any extent. I am not aware that any remedies were used except to pick them off by hand. They seemed to work only when it was dry, wet weather almost stopping them.

The autumn was unusually mild and free from frost, there being none to injure delicate plants and vines until November 1st, a month later than usual. A number of drains have been made in the county during the year, under the direction of the Drain Commissioner, which will assist many farmers in draining their low lands.

The manufacture of lumber and shingles is a prominent business in this county, but I have no statistics.

IRA H. BUTTERFIELD, Jr., *Secretary.*

LIVINGSTON COUNTY.

Albert Tooley, Esq., of Howell, Secretary of the Livingston County Agricultural Society, writes that their annual fair was held the 4th, 5th, 6th, and 7th of October, 1870, and proved very successful. Each department was well represented by those interested in the welfare of the Society. A debt of $3,700 was reduced by more than one-half.

Pure-bred and grade Short-horns were well represented; also, cows, working oxen, and fat cattle. There were some excellent sheep, and the horses made a fine show.

Officers of 1871.—President, Ira D. Crouse; Secretary, Albert Tooley; Treasurer, Asa Van Kleicle; Executive Board —E. J. Hardy, Linus Reed, William White, Lewis Meyer, H. G. W. Fry.

MACOMB COUNTY.

The twenty-first annual fair of the Macomb County Agricultural Society was held at Romeo October 5th, 6th, and 7th, 1870. The weather was propitious, the attendance large, and the number of entries 732,—nearly equal to last year's number. Nearly all the departments were well represented. The display of agricultural implements was larger than usual. The exhibit of flowers not equal to that of the previous year. The fair came too late in the season to secure many offerings of the numerous varieties of flowers which flourish in the open air of the summer in this vicinity. Our people are not wanting in the love of flowers, or in those domestic sentiments which their cultivation implies. We rejoice to report, however, that at no preceding fair of this Society, within our knowledge, have there been exhibited better evidences of a growing taste in works of art, and we hail this fact as one of many signs of an improvement in a right direction.

R. F. Johnstone, Esq., Secretary of the State Agricultural Society, increased the pleasures of the occasion by his kind and social intercourse, as well as by his able and instructive address. The financial condition of the Society, as shown by the Treasurer's report, is as follows:

RECEIPTS.

Money on hand January 1st, 1870	$3 00
Money received at annual fair	1,268 51
Money collected by Secretary	154 15
	$1,425 66

EXPENDITURES.

Amount of premium and other orders 1870 (cashed)	$1,273 62
Amount of premium and other orders 1869, "	17 50
Amount on hand December 8th, 1870	124 54
	$1,425 66

The amount in the treasury, when all outstanding orders are cashed, will not exceed $30.

Officers for 1871.—Willard A. Wales, President; Silas A. Colby, Vice-President; Daniel B. Briggs, Secretary; William W. Andrews, Treasurer; with a Board of Directors fourteen in number.

At the winter meeting of the Society the constitution was so amended as to increase the number of the board from eight to fourteen members, seven to be elected annually, to seven for two years.

The potato crop was not to any considerable extent injured by the Colorado potato-beetle. It first made its appearance in this county the present season. Hand-picking the old beetles from the vines, and crushing or burning them, has been the prevailing mode of exterminating the "ugly creatures."

OTTAWA COUNTY.

The Ottawa County Agricultural Society is in a flourishing condition. The fifteenth annual fair was held October 4th, 5th, and 6th, on their new fair grounds, near the village of Berlin. The Society occupy a beautiful piece of land of 27 acres, for which they paid $2,000, all substantially fenced, with a good floral hall building that cost $700. Much credit is due the officers for accomplishing so much in so short a time. They have upwards of five hundred members. The receipts of the fair were $813 50.

The officers for the year 1871 are: S. A. Vansleycke, President; Henry Pennoyer, Vice President; Robt. B. McCulloch, Secretary; Martin Philleo, Treasurer, and an Executive Committee of seven members.

SHIAWASSEE COUNTY.

C. Y. Osburn, Esq., of Owosso, Secretary, writes that the Shiawassee County Agricultural Society held its eleventh annual fair at Owosso, October 5th, 6th and 7th. The greatly increased interest shown, and the large number of entries in all departments, were good evidence of the increasing prosperity of the county. Improved breeds of stock are receiving attention from leading farmers, and thorough-breds, both of horses and cattle, were well represented. Good success is being attained in this direction. Total entries for the year, 930; the membership list shows 518; receipts for the year, $1,011 71; expenditures, including premiums, $1,005 43.

OFFICERS FOR 1871.—President, S. A. Yerkes; Treasurer, A. B. Chipman; Secretary, C. Y. Osburn; with an Executive Board of five members.

The Secretary also furnishes considerable information concerning the potato-bug. He says it first appeared in that section in 1869, but only to a slight extent, and did but little injury. In 1870 it appeared in myriads in some places, mostly, however, in gardens of cities and villages. The main crop does not appear to have been much injured. Of all the remedies tried the only sure one is Paris green mixed with plaster or flour. Picking and brushing the bugs off the vines was practiced some, but was pronounced too tedious. Recently a garden was plowed near here, and thousands of the bugs discovered burrowed in the ground. Perhaps if this plan could be adopted generally, late in the season when the cold would make them dormant, it might result in great benefit by freezing them during the winter.

ST. JOSEPH COUNTY.

The twentieth annual fair was holden on the grounds of the Society at Centreville, on Sept. 28th, 29th, and 30th, 1870. So far as concerned the number of competitors for premiums, and the general excellence of the stock, farm products, and the evidence of skill and good taste by our mechanics and artists, the fair was the most successful ever held in the county.

The entries were nearly 900. There were a large number of fine horses,—grade and thorough-bred,—from fancy, light steppers for the carriage, to the square, solid, serviceable animals for farm work and heavy draught. There was an excellent show of cattle; a very large number, also, of sheep; and the show of swine, both in number and quality, we think exceeded that at the State Fair. The amount of farm machinery on exhibition was large enough in number and variety to make a fine and useful display. Fruit was well represented; rather better than the previous year. The department of domestic manufactures and industries, was full to overflowing. A fine display of flowers and house plants added great beauty to the Floral Hall.

The attendance of people the first and second days was quite equal to expectation, and would have been immense the third day, but for a drizzling rain that set in early and continued all day. As a result, the fair was not financially as successful as usual, although otherwise a great success.

The total receipts were	$2,161 13
Expenditures and premiums	2,193 18
Leaving amount due treasurer	$32 05

OFFICERS FOR 1871.—President, D. D. Antes; Secretary, L. A. Clapp; Treasurer, Edmond Stears, and a Board of five Directors.

TUSCOLA COUNTY.

In compliance with the provisions of Article 11 of the By-laws of the Tuscola County Agricultural Society, the Executive Committee submit the following remarks in reference to the general condition of agriculture, horticulture, and manufactures, and a synopsis of their proceedings for the past year. Although the season has been in some respects, perhaps, less favorable than others, yet, upon the whole, there is little cause for complaint. The last winter set in some six weeks earlier than usual, and was followed by open weather in mid-winter, causing, by alternate freezing and thawing, considerable damage to winter crops. The spring opened early, and continued without late or untimely frosts.

The summer was excessively warm, and after about the first of July was quite rainy. September afforded an excellent opportunity for seeding, which was generally well improved. The later months presented the usual changes of weather that ordinarily herald the approach of rigorous winter. Considering the impossibility of any one season occurring in which all the various kinds of grain and grass would be likely to do well, it is a matter of congratulation when we are favored with one like the past, when most sorts cultivated in the county can be successfully raised.

Wheat.—The crop upon which our farmers rely very largely for profit, as well as for home consumption, was hardly equal to that of most former years, in quantity at least, and probably no better than the average in quality. The early appearance of cold weather in the fall of 1869 prevented the crop from getting sufficient top to protect the roots, in any material degree, and during the winter, weather was much of the time very unfavorable; sometimes the ground would be partially thawed, again frozen, then covered with snow, then again left without any protection whatever, and a portion of the time the wheat was coated with ice. The favorable weather in spring and early summer brought the crop rapidly forward, and an average of about fifteen bushels per acre was secured. Rainy weather during harvest time occasioned considerable damage. In some instances the kernel was slightly grown, and in others, where proper attention was not given after threshing, it became a little musty in the bin.

Corn succeeded remarkably well, better, probably, than for many previous years. No frosts came to injure it during its growth, and the continued warm weather matured the crop rapidly, and an unusually good crop as regards both quantity and quality was secured. We estimate the average of shelled corn at fifty bushels per acre.

Oats were fully up to the average, and very good quality. Some new varieties are being introduced, but have not been sufficiently tested to permanently establish their value in our locality.

Buckwheat has done very well; considerable was sown, and nearly or quite an average crop was raised. The very hot weather during the latter part of summer prevented its filling in some measure, but on acount of the uncommonly rank growth of straw, there was probably about the usual amount per acre harvested.

Hay.—Of this crop, about an average per acre was secured. The quality is good except that it was in some instances injured by rain while being harvested. That cut early was secured in better condition than that which was cut later in the season. There was not that rankness in its growth that in the estimation of many farmers so injures its spending qualities.

Potatoes suffered considerably on low land on account of wet weather, rotting badly in many localities. The crop was usually good as regards quantity, except in cases where the potato-bug partially or entirely destroyed them. The appearance of this pest is very properly regarded as a serious matter. Taking into account the importance of the potato crop, and the known difficulty of checking the bug in its destructive course, it becomes us to examine carefully its habits, manner of propagation, and time of making its appearance in its destructive state, with the view of saving in some measure so important a part of the farmer's resources. It is believed by some that particular sorts of potatoes are less liable to be attacked by it than others.

Fruit.—Apples were not perhaps as abundant as in some former seasons; still there was a fair crop, both in quantity and quality. Peaches were very scarce, owing probably to the excessive crop of the previous year. Small fruits of most sorts were plenty and of good quality. We notice a very commendable discretion used very generally throughout the county, in selection of choice varieties of fruits. Llittle or no poor fruit is raised, and trees bearing inferior sorts are very properly rejected.

Cattle.—The disposition noticed in former reports to improve our stock of cattle is still noticeable. Although wheat in many sections of the county is regarded as of prime importance, there is a very large proportion of it where the raising of stock and dairying would be attended with greater profit to the producer than exclusive raising of cereal crops.

Horses are rapidly increasing in numbers and the quality probably keeps pace with the increase. There was a very creditable exhibition of young horses and colts at the late fair, evincing that farmers have an eye not only to the increasing of numbers, but also to the improving upon the blood. Horses for farm and other heavy work will always be in demand in the county, and raising such should engage the careful attention of the farmer.

Sheep.—In sheep there seems at present to be no very special interest. Prudent farmers see in the present a very suitable time for improving their flocks, and are carefully weeding out inferior animals and introducing valuable ones, thus preparing at no very distant day for a reaction in the wool business, which will be pretty sure to take place. There still seems to be a disposition to raise long-wool sheep, although some are systematically working up their flocks of Merinoes to a better and more profitable standard.

Swine.—In quality are manifestly improving. Some fine specimens were exhibited at the late fair. Improved breeds have been introduced, and are being annually diffused throughout the county.

Manufactures.—In lumber, shingles, etc., there is a large amount of business done. Flouring mills abound, while foundries and shops of various sorts continue to produce at home the implements and other necessaries for the convenience of our population. The woolen mills at Vassar are being improved yearly, and are working up a large share of the wool clip of the county, thus furnishing another item for the convenience of the people at their homes.

TRANSACTIONS.

The fourth annual meeting of the Society was held at Watrousville, January 10, 1870; the President, A. B. Weaver, in the chair. At said meeting the following persons were elected to office for the year 1870: President, A. B. Weaver, of Juniata; Secretary, C. C. Stoddard, Fair Grove; Treasurer, R. C. Burtis, Juniata; Auditors, B. A. Wood and Wm. Buddington, Juniata. Executive Committee—Wm. King, Juniata; Wm. Johnson, Vassar; John McPherson, Tuscola; R. R. Atwood, Millington; M. D. Orr, Aylmer; also a Vice-President from each town in the county.

The fair was held on the grounds of the Society, near the village of Watrousville, on Wednesday, Thursday, and Friday, September 28th, 29th, and 30th. Owing to rainy weather the last day of the fair, and the threatening appearance the second day, there was not as large an attendance as might otherwise have been expected. The receipts of the fair were about the same as last year, while the entries were slightly in excess. An address was prepared for the occasion by Hon. B. W. Huston, but owing to the rain it was not delivered. For a full account of the finances of the Society, we respectfully refer you to the report of the treasurer. The time for which the present site for the ground was leased will expire next May. In anticipation of that fact, the Executive Committee at its meeting in March last, appointed a sub-committee to select a new site, with leave to report at the next meeting. At a subsequent meeting that committee reported that a site could be obtained one mile north of the village, which report was adopted, and a site of twelve acres has been secured at the expense of seven hundred and twenty dollars. The distance from the village was considered by some to be an objection. The clayey nature of the soil, making the old ground almost unendurable in extremely dry weather, and equally objectionable in wet weather, reminded the committee that a more gravelly soil would be preferable. This has been obtained, and with the further advantage of living water, we consider that in these particulars the new location will be more desirable than the old one. The committee also came to the conclusion that a site conveniently near the village sufficiently large to admit the construction of a trotting course, could not be obtained without involving the Society hopelessly in debt, while the new site has been purchased sufficiently extensive for all purposes, upon terms which we may reasonably hope to be able to meet within two or three years at most.

The Treasurer's report shows the receipts to have been	$1,064 60
And the expenditures	1,011 14
Leaving a balance on hand	$53 46

Officers for 1871.—President, Townsend North, Vassar; Secretary, C. C. Stoddard, Fair Grove; Treasurer, R. C. Burtis, Juniata; with an Executive Committee of five members, a Vice-President for each township, and two Auditors.

VAN BUREN COUNTY.

O. H. P. Sheldon, Secretary of the Van Buren County Agricultural Society, states that the fair was held on their new grounds just west of Paw Paw, which have been fitted up in a tasteful and agreeable manner.

A good degree of interest was manifested in all departments, there being about 1,100 entries.

The receipts of the fair were about $1,200. Most of those who drew premiums generously donated the same to the Society, which was quite a help, as it was considerably in debt for the new grounds and fixtures.

The buildings are fitted up and painted in first-class style, and are pronounced equal to any of the kind in the State.

OFFICERS FOR 1871.—President, F. M. Manning, Paw Paw; Vice President, David Woodman, Paw Paw; Treasurer, Joseph Kilburn, Paw Paw; Secretary, O. H. P. Sheldon, Paw Paw; with a Board of twelve Directors.

The Secretary also furnishes some facts concerning the county at large. The agricultural interests of the county are principally grain-growing, except in the western or lake-shore region, which is devoted to fruit. Stock-raising has not become very general, but is increasing. There are two dairying establishments in the county. All kinds of fruit except plums are raised in great abundance.

WESTERN MICHIGAN AGRICULTURAL AND HORTICULTURAL SOCIETY.

The sixth exhibition of this Society was held at Horticultural Hall, Spring Lake, September 29th, 1870, and the following day.

The first day was principally devoted to making entries and arranging articles for exhibition. Over 100 entries were made, consisting of grapes, apples, pears, canned peaches, strawberries, and other fruits. Also, a good display of vegetables, tomatoes, potatoes, onions, egg-plants, squash, pumpkins, sweet corn, seed corn, pop-corn, etc.

House plants, consisting of salvia, coleus, fuchsias, geraniums, single and double, and numerous other choice plants, the foliage plants taking the lead for novelty. A Russian sunflower was exhibited measuring 16 inches in diameter, four feet circumference, thirteen feet stalk, the stalk being three inches thick. This was grown on the sandy soil of Grand Haven.

Mr. John H. Newcomb, the President, exhibited a fine assortment of apples, including the "Spring Lake Pippin," a new and promising variety.

Mr. Thomas Petty was on hand with some very fine grapes and vegetables.

Mr. Wm. Penn exhibited a number of choice pears, consisting of six varieties, also some fine apples and Isabella and Catawba grapes, which were quite ripe.

Mr. H. G. Smith exhibited some fine apples and canned fruit.

Mr. Bilz, some beautiful canned pears and strawberries.

Mr. Chas. E. Soule and Mr. Timothy Hall exhibited squash, watermelons, etc., of gigantic proportions.

The house plants were exhibited by H. S. Clubb, also the large sunflower.

The second day brought a fine display of grapes and some peaches, apples, pears, etc.

Mr. Hunter Savidge had eight boxes of Catawba, Isabella, Concord, and Delaware grapes. These were all well ripened and suitable for table grapes. We never saw finer Catawbas. The Delaware were, as usual in this region, larger than represented by the pictures of Eastern nurseries.

Mr. H. G. Smith exhibited a very fine box of Rebeccas, Ionas, Catawbas, Dianas, and Clintons. The Rebecca were small bunches, but peculiarly rich looking. The Iona showed quite an improvement on last year's. The bunches, although not yet perfect according to Grant's picture, are compact, and some slightly shouldered. We shall expect perfect Iona bunches from Mr. Smith's vines next year. The Dianas are large and very compact bunches, and in this respect superior to the Catawba of the same vineyard.

Mr. Thos. Dennison exhibited eleven varieties of grapes: Diana, Concord, Delaware, Adirondac, Hartford Prolific, Catawba, Isabella, Clinton, Martha, Progress No. 22, and a very large black grape labeled, "Native California," similar to Isabella grapes. This show of varieties is valuable, as showing that these varieties can be grown in this climate and ripened. The Diana, Concord, Delaware, and Adirondac were fine large bunches, and appeared nearly perfect; the others more or less imperfect, which may be owing more to the youth of the vines than any climate causes.

Messrs. J. K. Kneeland & Co., exhibited Concords, Catawbas, and Delawares. The two former were perfect bunches; the latter the bunches were small, but the grape fine.

Mr. George Seagrove exhibited a large assortment of grapes,

including Delawares, Concords, Rogers' No. 3 and 33, (the latter very large perfect bunches and berries), and Clintons. These were all handsome looking grapes.

Mr. H. D. Scott of Crockery showed some seedling wine grapes which promise superior to the Clinton, being two weeks earlier and larger in size.

Mr. H. S. Young of Crockery exhibited some very large Sicilian Crabs; they are a very dark red; excellent bearers. Also, a fine box of honey.

Mr. John H. Newcomb exhibited some very large Stump-the-World peaches.

Mr. Seagrove showed samples of several varieties of peaches.

Mrs. J. W. Barns, Grand Haven, had a fine bouquet of verbenas.

Mrs. Hungerford exhibited several handsome bouquets, in which were dahlias, zinias, gladiolis, asters, etc., beautifully arranged.

Mr. Bloodgood exhibited two fine coxcombs, one red and the other yellow.

In the vegetable department were very large cabbages, turnips, beets, rhubarb, sweet potatoes, egg-plants, parsnips, carrots, etc.

DISCUSSION.

After the various objects of interest had been fully examined, and the large company present seemed disposed to rest from their promenading, John H. Newcomb, Esq., President of the Society, ascended the platform, called the assembly to order, and announced that no preparation had been made for an address, but for the want of anything better, Capt. H. S. Clubb and Col. Ferry had just been invited to say a few words, being the best that could be done under the circumstances. Henry S. Clubb said:

Ladies and Gentlemen :—The subject of horticulture is so extensive and diversified, that it is best only to attempt one or two branches at a time, especially as in the present instance only a few minutes can be occupied with the discussion. Peach-growing was at first regarded as the only subject worthy of attention on this Lake Shore region, owing, perhaps, to the peculiarly sandy character of the soil, but the experience of the past six years has shown that the capabilities of the Lake Shore region are much more varied. The peach for the past two years has not been a heavy crop, and by some was regarded as too uncertain to be profitable. I do not pretend to give scientific reasons for this circumstance. I came from a country where seven years are considered necessary to learn any trade or profession, and this Society is only yet in its sixth year, and is not yet out of its apprenticeship. In fact, all fruit-growers here, with one or two prominent exceptions, must acknowledge to be at present only learning the business. The locality and soil differing from those of other fruit regions, books alone cannot teach practical peach culture. It must come by actual experience. The varieties adapted to this region must be taken into consideration. It is remarkable that while the Barnard, Early York, Smock, and all seedlings, bore this year, Crawfords are almost an entire failure. Seedlings are almost always a sure crop. To raise good seedlings must become the great study of peach-growers who desire an annual return for their labors. It has been suggested by a practical peach-grower, that in planting orchards every other peach tree should be a seedling. All who have good seedling trees this year will appreciate this suggestion. My seedling peaches are heavily laden with a very fair quality of fruit, while my Crawfords, that bore well last year, are now taking a year's rest. I am satisfied progress is being made in peach-growing.

Grapes next claim our attention. If nothing else has been demonstrated in the past six years, the fact that this region of country is clearly shown by practical results to be adapted to the growth of the best American grapes, is enough to satisfy me for the past years of labor. It is now known to be a fact, beyond all question, that the Delaware and the Concord, two of best grapes grown in this country, can not only be grown here, but can be grown to a good profit. This fact alone is enough to show that this is a country worth living in; a country where a good income can be secured by that ancient but pleasant occupation, vine culture. Orchards and vineyards improve the whole face of the country, and add pleasantness and beauty to profitable and healthful occupation. But the present exhibition shows not only that the Delaware and Concord can be grown, but also that twelve other varieties arrive at perfection here!

Of the light-colored grapes I place the Delaware first, because its success here is fully established. The Diana is next, and I have rejoiced to see some very fine bunches of this variety on exhibition. The Catawba has arrived at its high perfection here this year, as is clearly demonstrated before us on that table. The Rebecca, Iona, Martha, and Rogers 22, are all produced here ripe and rich, and probably fully equal to the best in flavor, although, owing to the vines being young, the bunches are not yet of full size. I have watched the progress of these choice varieties, and especially the Iona, and I am glad to see here more perfect bunches every year; and I have no doubt next year the Iona will be produced at our fair in all its beauty of bunch as well as rich lusciousness of flavor.

Of the dark varieties, I place the Concord at the head, because its success here is thoroughly established. The Hartford Prolific is next in importance, although an earlier grape. Its success here is well established, but the superior size of the Concord will always cause the latter to be preferred as a market grape. The Isabella has done nobly this year, and ranks high. The Clinton has also shown itself well worthy of cultivation. The three other grapes I noticed here are the California, a very large, black grape, Rogers No. 3, and 33. These have very well perfected berries and good-sized bunches, but the vines are evidently too young to produce perfect bunches, and although it is a great point gained to demonstrate, as is done here to-day, that these varieties can be grown here and perfected, that they can be grown to a profit remains yet to be shown by experience. It is enough to know that the grape crop has become as certain here as any crop can be, and its general culture may be regarded as a safe and profitable investment.

A few words on flowers, and I have done.

I confess, I am surprised there is no competition here in flowers. Nothing adds so much to the interest of these exhibitions as flowers. Fruits appeal to the appetite, and in some degree to the sense of sight. There is beauty in the ripe peach and the half-transparent grape, but these are produced for their flavor rather than for their beauty; but flowers appeal directly to the love of the beautiful, which is a passion that can never be destroyed. It is true, as Mrs. Howitt says,

> "God might have made the earth bring forth
> Enough for great and small:
> The peach tree and the grape vine,
> And not a flower at all,"

but he has not done it. He has strewn the earth with flowers, and they are calculated to cultivate taste and to refine the senses. Why should we not secure for our next exhibition a good show of flowers? Let us have competition in the different varieties. Let us have a premium for the best dahlia, the best zenia, the best coleus, the best verbena, and the best fuchsia. Let us emulate the example of older societies and keep up a lively interest in these matters, making our homes and gardens attractive and our exhibitions a blaze of glory in the display of both fruits and flowers. [Cheers].

Col. Wm. M. Ferry said:

I am reminded, by the introduction given to Capt. Clubb and myself, by our President, of the lady who wanted to make excuses for her calling, and who was met by the response: "No excuses, my dear madam, are necessary;

we are situated so that we are glad to see anybody." [Laughter.] We are very glad to know that you are so situated that you are glad to hear from almost anybody; therefore can say what I please to you. I have been gratified in coming here, to see (what has been omitted, to be noticed by the preceding speaker), the grand display of vegetables that we have along that table and under it. It is not long ago that an Englishman, who came to New York, on looking at some pumpkins, remarked, "These will do passably well for pumpkins," when a Yankee replied, "Hold on, Johnny, they are wortleberries." Look at those sugar beets and those large watermelons. St. Paul said to Timothy, "Take a little wine for your stomach's sake," but if you want to get corned, go to the end of that table and you will see corn that would surprise a stranger to the capabilities of our soil. I care not where you go throughout the length and breadth of the land, even the corn that grows on this sand on the eastern shore of Lake Michigan is of a superior quality to anything you can find elsewhere, and I feel it is my privilege to call attention to it, lest in the multitude of other things you lose sight of the staff of life. I do not feel like giving up peach culture. For eighteen years I have had peaches every year on my place at Ferrysburg. Not every year a large crop, but always sufficient to satisfy the wants of my family, and some years we have had a very large crop. When we know this to be a fact, and knowing as we do that there are localities in this and adjoining States where peaches cannot be grown, it is evident we have atmospheric influences here which make peach-growing successful. We cannot expect to get peaches as big as your fist without labor and effort. Emigrants have been told that in this country they would find roasters running about the streets ready roasted, with knives and forks in their backs. You cannot expect fruit to come tumbling right into your baskets without experience and cultivation. Before a garden can yield profitable results, there must be labor, toil, and careful attention. It is as true in this locality as in any other, a man must reap these benefits by the sweat of his brow; and those who anticipated so largely to be making fortunes without labor, will of course be disappointed. There will be recurring years of profitable crops. I take no discouragement whatever. I do not view the matter in that light. It takes energy and continued application, and the time will come when these embryo orchards along the lake shore here will become very valuable investments. What may bear a profitable crop one year may rest the year after, but taking one season with another, and one crop with another, there is no doubt you will receive a rich reward for your investment.

The Colonel continued for some time to point out the features of encouragement around us, and referred to the publication of the report of the Spring Lake Convention of last winter, in the State Agricultural Report, as furnishing the means of laying the advantages of this portion of the State before the minds of others, showing that the results of our labors here are appreciated and considered worthy of extensive publicity. He concluded by saying he had no feeling of discouragement, but regarded the prospect ahead bright and glorious.

REV. L. M. S. SMITH.—I suppose that after having an address from a Captain and a Colonel, you will not expect much from a plain citizen. Some nine years ago we had a society at Grand Haven, which, however, suspended during the war. The first meeting to organize this Horticultural Association was held at my little sanctum in Grand Haven. It was called a meeting of citizens of Grand Haven to organize a Horticultural Association. There were ten from Spring Lake and two from Grand Haven, but as the Spring Lake people had the majority, they had a right to remove its meetings to Spring Lake, I have had too much to do with Spring Lake to feel rivalry or jealousy at its progress and prosperity. If a peach crop is secured once in three years, such as has been raised here, it will make the growers rich. With regard to apples, a gentleman told me he did not think the trees could grow on this poor soil. My reply was, that experience showed that the trees grow too rapidly. The soil is too rich. I agree with the first speaker in reference to flowers. We love flowers. I have often asked the question why God did not produce fruit without flowers? Flowers are to produce refinement of taste, and the cultivation of flowers should be encouraged. But there are little human flowers which we also take great delight in. I cannot better conclude than in the language of the poet:

A seedling sprung from Adam's side,
A most celestial shoot,
It came of Paradise the pride,
And bore a world of fruit.

[Cheers.]

HON. TOWNSEND E. GIDLEY.—I would dislike, Mr. President, very much, to have our friends leave this hall under the impression that any remarks made in relation to peaches should be regarded in the light of a disparagement or discouragement of the cultivation of that important fruit, for in reality the remarks have not been of that character, and were not so intended; nor, rightly understood, would they have that tendency. I think this season, because we have had no peaches to amount to anything, to tax the growth of my trees, is worth at least $500 to me, on account of the superior growth my trees have made to what they would have made had they borne fruit. We cannot always grow fruit. Ask any agriculturist here if he is always successful with his wheat and corn, and he will tell you that he is only eminently successful in these crops once in three or four years. I speak advisedly, having been an agriculturist the greater part of my life. With regard to the peach, it is not a crop to be relied on as a crop every year. It is of a very delicate nature. At St. Joe, it is my observation that the peach men admit that if they get a heavy paying crop once in three years, they are getting wealthy. Land paying $1,000 to $1,200 an acre in a season, cannot be expected to yield such a crop every year. I shall plant out my 2,000 peach trees the next season. I believe there is a much larger profit, acre for acre, on peaches, than on any other fruit crop you can raise. But we have abundance of small fruits as well, which have been proved successful here,—strawberries, raspberries, and grapes, if you please. They are all successful here. It is all right. We evidently have a favored locality here, and it will be the means of enriching those who embark in these beautiful enterprises. But a man can give attention to twenty acres of peach trees, while, if he can take care of half an acre of grapes, he is a smarter man than I am. I want you to go from here with confidence in this locality, for it is best for the trees now and then to get a season of rest, and then next year you get abundantly paid for it. [Cheers.]

REV. H. BECKWITH.—In speaking for Spring Lake, we are thankful for the display that has been made to-day. It is beyond our own expectation. It is as good an exhibition as we usually find in other parts of the country. Even in vegetables, I say, you beat them all. Some are discouraged. We have not expected to have returns in so short a time. This has been a very fair display of fruit. I hope we shall return home satisfied with the bounties of Providence. I think in a very few years this region will be one vast succession of peach orchards and vineyards. [Cheers.]

CAPT. FASSETT.—This has been a season of test for varieties. The Crawford has been shown to be unreliable, while the Barnards and Smock have proved themselves good, hardy varieties, and almost equal to seedlings for regular bearing.

The Secretary, Mr. W. G. Sinclair, then announced the awards.

After a few words from the President, the large audience, who had been attentive throughout, gradually dispersed.

The Society has a large Horticultural Hall at Spring Lake, where fruit conventions and discussions are held.

At the annual meeting of the Society held December 12th, 1870, the following officers were elected for the ensuing year: President, Capt. Chas. S. Fassett, Spring Lake; Secretary, Capt. Henry S. Clubb, Grand Haven; Treasurer, Thomas D. Denison, Spring Lake. Directors—Hunter Savidge, George Seagrove, Alonz Bilz, Timothy Hall, Spring Lake; and Henry S. Clubb, Grand Haven.

At the adjourned annual meeting held on Monday, Dec. 19, 1870, President Newcomb in the chair, Henry S. Clubb tendered his resignation as Secretary, which was accepted, and Chas. E. Soule of Spring Lake was elected to fill the vacancy.

OBJECTS OF THE SOCIETY.

Capt. Chas. S. Fassett, President elect, said:

This Society has proved a success, in this: that it has attained many of the objects for which it was organized. Most of the fairs we have held have done credit to us. This building may be termed a success. Again, this being about the first effective Horticultural Society organized in Michigan, it easily exerted that influence which has called into existence scores of other similar organizations, some of which, to-day, are not merely our co-workers, but our rivals. Our members having so great faith in fruit culture, every statement that has emanated from this Society has been bold, confident, and persuasive. We have talked success,—we have thought of nothing short.

Its headquarters being located at Spring Lake, it has directed public attention to this place, and from this cause it has been of immediate advantage to us. The growth and prosperity of our village in the past four years, are sufficient evidences of the worth of this organization. Spring Lake of to-day, with its elegant residences, its four lofty church-spires, its convenient school-house, and its numerous trading and manufacturing establishments, would be the uninviting Mill Point of four years ago, had it not received the impetus which the fruit-growing interest excited. Within the last five years, the value of most of the real estate in your village has more than quadrupled. Your stripped and unimproved pine lands, then deemed almost valueless, are now selling at from $10 to $150 per acre.

At the time our honorable President, Mr. Newcombe, first set foot on this soil, there were here existing the same natural adaptation and facilities for growing and marketing the tender varieties of fruit as we find here to-day. Indeed, we have no reason to suppose but that our natural advantages have existed here as long as at St. Joseph, and I believe the principal, if not the only reason why we have not orchards and vineyards as extensive, old, and profitable as those in the vicinity of St. Joseph, is that the particular advantages here offered were not known to men of means wishing to engage in the growing of fruit. Certainly, our facilities for marketing are to-day, and have been for years, superior to those of St. Joseph. But then this was of no advantage to us so long as the outside world were ignorant of the fact. Indeed, we never had any means of advertising until this Society was organized. Since its organization, its efforts have been such as have secured for Spring Lake, and especially for a large extent of this northern portion of the eastern shore of Lake Michigan, not merely a State but a national recognition of the peculiar advantages here offered for profitable fruit culture.

Our intelligent landowners and business men, foreseeing this particular advantage the Society would have, were first in its organization and chief in its support. Indeed, we fear this Society might long since have ceased to exist, had it not been for the very liberal support of these men.

Our business men have in a great measure attained the object they had in view, of bringing this place into notice. Have our practical fruit-growers gained as much? If they have not, it is certainly not the Society's fault. One of the grandest objects for which this Society was organized and sustained was that it might become a school for fruit-growers. Now, my brother fruit-growers, we have most shiftlessly neglected this object, which is all-important to our success. If we ever succeed in profitably growing the peach, and other tender varieties of fruit, we will find something beside muscle, perseverance, and patience are requisite, and that is intelligence. Certainly, during the long evenings of the winter we will have sufficient leisure, and this Society, with its comfortable hall, offers every inducement for us to meet here once a week, and devote an hour to discussion and practical talk on subjects pertaining to our particular pursuit.

Although we have read much and talked much in the past four and five years, still we are probably in want of light. The horticultural books and papers to which we have access furnish us many valuable hints, yet if we are not careful they will certainly lead us astray; for systems of culture which have been adopted and found successful in other places prove failures here; for certain systems of pruning and cultivation, adapted to some soils, are not to ours; to some climates, but not to ours; to some seasons, but not to others. So in regard to varieties of fruit. Those that are found profitable in some locations have proved valueless in others.

Again, it is remarkable that the majority of our agricultural and horticultural writers and preachers are noted rather more for their scholastic attainments than for the callous on their hands. They can wield the pen more gracefully than the hoe. It is much more pleasant for them to gain a livelihood by theorizing, than by the sweat of the brow in reducing theories to practice. If there is anything wicked, it is in a writer or speaker publishing to the world, with an air of authority, directions for setting and caring for fruit plantations, when he knows not whereof he speaks. Hundreds and thousands may adopt his plans and follow his teachings: losses, failures, and discouragement are the result.

No one of us can speak with much authority as to the growing of fruit in this section, for we are pioneers in the business here; but each of us has gained some little personal experience in the past four years, and if we would meet here and talk over our failures and successes, the plans we have tried, with their results, we would find that one each, in the aggregate, would amount to considerable. Each failure we meet with might furnish us a wonderful amount of material for earnest thought and speculative calculation. So each success we meet with should lead us to study the cause of that success, and see if the result might not have been tenfold better.

Then if we would meet here once in a week, not to talk "buncombe," but to frankly speak of our failures as well as of our successes, and try to divine the reason of each, I believe it would be of untold advantage to us, and through the medium of the Grand Haven *Herald*, it might be of benefit to others.

That our meetings may be made entertaining and pleasing to all, we are in favor of having our exercises interspersed with music, and we would not insist that all the subjects spoken upon be agricultural, or horticultural, or pomological, but that we may have a variety of subjects, literary, as well as practical; this will enlist more in sustaining our meetings, and make our exercises far more entertaining. Although our society is called a Horticultural Society, yet every man and woman in our midst is interested in its prosperity, and it is no more than right; and all who feel disposed may become members and active workers for its good. I have much I wish to say under the head of the subject the committee gave me. I have crowded as much of it into the time allowed as possible. I regret very much I cannot enlarge upon the idea of this Society being a means of advertising this place, and to have noticed the fact, that at our first or second year, Capt. Clubb's speech being afterwards published in pamphlet form and distributed as circulars throughout the country, Capt.

Craw took advantage of that to advertise his land. Indeed, those circulars were the first that directed public attention to this place, and the officers of this Society have kept this object steadily in view. The result has been favorable to the merchant, to the manufacturer, and every land-holder. And yet there is still more to be accomplished in this direction. We want more fruit-growers here; we want more acres planted. Here let us all, for all of our citizens have an interest here, unite to make this society effective and valuable, so as to aid to extend its influence. [Applause.]

LOCATION OF ORCHARDS.

Mr. Chas. E. Soule said:

I have failed to find anything new on this subject, but it may be a benefit to bring up the points hitherto discussed: First, soil; second, height of land; third, shelter. Before I came here I heard it said the soil was fit for peach trees and for nothing else. My experience has been that our best soils have grown the peach most successfully. The trees growing on the richest land have made the finest trees and the most and best fruit. Trees near the garden, where the land is rich, have been from three to five to one better bearers than those in the center of the orchard. But I should not be satisfied with land enriched with manure. A good strong natural soil would be the one to pick upon. Mr. Dowling says a rich, mellow soil for a peach. When the trees are young they will do well on thin, sandy soil, but the trees will not last as long. We should get a soil enriched with alkalies, and keep it in good condition. A peach tree seven years old, in good healthy condition, and in a good season for fruit, ought to bear from five to six bushels of peaches. Those peaches would weigh, say 300 pounds. To grow any vegetable and raise a crop equal to that, you would have to cover the soil a foot thick with manure. It is equal to a thousand bushels to the acre.

All agree that trees must not be planted on soil that has water near the surface; hence, I would say get high land; so high that there is no danger of the roots reaching into the water. I have in my mind an orchard where on the high land the trees show a respectable growth, while those on the low land are in an almost dying condition. Whether it is the frost settling there, or the thawing of the snow in winter, or the ice about the bark in thawing and freezing in spring, I do not know. Rolling land, as a matter of principle, would be the best upon which to plant an orchard; as a rule rolling land is thoroughly drained. Avoid sink-holes. Rolling land is better than level land unless the level land is well drained.

The peach is a native of the oases of Persia, and is a tender tree. I find that when exposed to our heavy southwest winds, those winds will whip the trees to such an extent that the leaves fall off too soon and the trees will not thrive. Shelter is imperatively demanded for an orchard from southwest and west winds. What this shelter should be would be another question, In this locality of cheap lumber an orchard would be well protected by a high board fence.

THE PEACH.

MR. JACOB GANZHORN—I will take up the subject of the peach crop of the present season. Many have had the idea that, the crop being light, it was a failure on the part of the soil or climate. I admit the crop was light, but this was not the reason. The principal varieties in bearing here are the Crawfords, and being a tender kind, it was affected by a severe storm last spring, followed by freezing and ice on the trees. It was supposed to have caused this failure, but the Barnard, Smock, Early York, and Coolidge's Favorite bore well, and had we more of these varieties our crop would not have been so limited. In Southern Illinois they choose Stump-the-World, Smock, Hale's Early, and Early York sometimes, but the Crawfords they discard. The Smock is more universally planted in Illinois than any other. In the Eastern States the Crawfords are discarded, and other varieties planted in their stead. One variety of peach does not make a locality a successful peach region, but varieties are selected such as will stand the more severe weather in the winter. We have had experience with sufficient varieties from which to select.

It has been said that trees in light, sandy soil will not continue long lived. This is not my experience. In Holland, I have seen trees planted eleven years ago in a coarse sandy loam; the trees annually shortened in each year's growth from one-third to one-half, keeping the tree compact so that the limbs were prevented from growing sprawly, so that they would not branch down. These trees now look vigorous and healthy, and will bear many years to come. The umbrella form was the shape selected at first, but it was found to tend to split the trunk at the point where the limbs struck out. This was abandoned and the pyramid form adopted, which is nature's form of training a tree. The tree in this way grows in width as it grows in height. Such trees need but little pruning, merely to preserve the pyramid shape of the tree.

When grapes were first introduced, the European varieties were tried, failed, and abandoned. Our native grapes succeeded better, and it is now claimed that the wine made here is equal to the best made in Europe. On this principle, I would raise seedling peaches here. By raising seedlings we improve our varieties and at the same time secure hardiness and adaptedness to the soil and climate where we experiment in. I believe if every peach-grower would plant peach pits every year, we would secure many varieties of great value in this climate.

THE PEAR.

MR. H. BECKWITH—I have experimented on pears in Wisconsin, but I failed and got discouraged, but in my native hills of New Hampshire, without any particular soil, they raised pears with great abundance. In Massachusetts, the poorest soil produces the best of pears. It has been supposed that we could not raise pears here because the soil is light, but it is much better soil than that of New England, where pears are successful. The dwarf pear has been considered a failure, but now this is considered a mistake, and nurserymen are raising them in large quantities. For three years past I have had six or eight pear trees that have borne every year, the Bartlett and a Russet pear. To my horticultural friends I wish to say that I have concluded to set out pears. The dwarf pear is grafted in the quince stock. They throw out the roots near the surface, and the roots are apt to be frozen. The trees are lifeless during the time they are frozen, and frequently the life of the dwarf pear is destroyed. The remedy is to protect the roots by heavy mulching. The reason for preferring the dwarf is that it comes earlier into bearing. It is with fruit-growers as with the ladies, we follow foreign fashions and import foreign varieties. The Flemish Beauty will withstand even the Wisconsin climate. I think, by a proper protection to the pear, you can grow it here as well as elsewhere. If our fruit-growers will select a few hardy varieties they will find a pear orchard a profitable investment. We have our marl beds here to supply the nourishment, and we can adapt our soil to pears, apples, or any other crop. The pear does best with the orchard seeded down. The grass does protect the pear trees. At a farm 100 years old, East, the pears stood 40 years and bore well. The horticulturists of Western New York are making pears a profitable crop.

GRAPES.

MR. GEORGE SEAGROVE—I would much rather take my shears and go into the vineyard and prune, than to come here and explain how it is done. It is thought grape-growing is too much trouble. It was remarked by a prominent fruit man that half an acre was all he could attend to. Now I have to attend to five or six acres, besides other work. I let them grow without pruning the first year; the second year, if any arms are long enough for bearing, I allow them to remain, but those that are not, I cut back. I find by taking a medium course it does better than over-pruning. Pruning the Concord to one foot is all I dare to. The system I pursue is to have a new arm every year, and I find I get a much better crop by doing so. Every man in this place should have an acre of grapes. I know of $200 being realized this year on less than an acre, and the owner employed in other business. The cost of starting a vineyard is much less than a few years ago. If he cannot buy wire or posts he can train to stakes until he can. The idea that a man can take care of 20 acres of peaches to one of grapes, I cannot agree to. Taking the hunting of borers and shortening in, there is as much work as in growing grapes. I go in for grape-growing.

I esteem the Concord the highest of all, even above the Delaware, for profit.

MR. A. BILZ—Is not the Isabella a good grape here?

MR. SEAGROVE—The Concord is the best for profit, as it ripens more evenly than the Isabella, and will fetch a better price in the market than the old Isabella. My Concord grapes fetched two cents a pound more than any other in the market, thanks to our facilities for shipping. As soon as grapes are planted we ought to have our eye to the manure heap. No man can take crop after crop from the soil without replenishing it with something. If we fertilize we shall succeed; if not, we shall not. I would not manure in large quantities and then leave it alone for a few years, but a little every year is the plan. I have seen orchards at other places with large crops but no market near enough to make it pay. Now we have no such trouble in Spring Lake. I always go for the insects. I provide boxes on the tops of the posts for the birds, and they, the birds, take the worms. [Cheers.]

SMALL FRUITS.

The subject of small fruits was discussed by Henry S. Clubb, as follows:

Of late years several small fruits have become cultivated for market purposes that were previously only regarded as wild, and unworthy of cultivation. The blackberry, the cranberry, and the blackcap raspberry have lately become important market berries, and are by their intrinsic merits, both as wholesome and delicious food and as dietetic elements of health, forcing their way into the position of permanent garden fruits. I predict that it will not be long before the dewberry, the whortleberry, and the wintergreen berry will also be made objects of improvement and cultivation. Already these fruits are admitted to a place in our city markets, although no pains have yet been taken with their improvement and cultivation. They abound in the woods of Michigan, and on account of their commonness are not much thought of in the country, but place them on the stalls of a city fruit-dealer and they command a good price. Bestow upon them the same attention that has been devoted to the cultivated blackberry and raspberry, and we shall soon find them in similar esteen.

THE DEWBERRY.

The dewberry ripens a little earlier than the blackberry, and will fill a niche of time between the raspberry and the blackberry when the market needs a good berry.

THE WHORTLEBERRY.

The whortleberry comes next and occupies an important position as a market berry. Its improvement in size and flavor by suitable soil and culture, would make the berry season one continual period of marketing, from the strawberry to the blackberry. It has great advantage on account of transportation.

THE WINTERGREEN.

The wintergreen berry, in places where the soil is rich, is as large as the cranberry, and would undoubtedly, by proper cultivation, become a popular berry. If it cannot rival other berries in its lusciousness, its wholesome and agreeable character, its beauty of form and color, its hardiness and lateness, coming when the other berries are over, render it exceedingly desirable. It would not only be a valuable late berry as a winter fruit, but its keeping good under snow until the spring, makes it desirable as a fruit coming earlier than any fruit at present cultivated. Those who have only tasted wintergreen berries grown and stunted on poor soil can form little idea of the character of the fruit when grown on rich soil. As a preserve, the wintergreen berry is also excellent, and will undoubtedly command a ready sale if introduced in sufficient quantities to attract attention.

These berries require no "test" in Michigan; they already grow profusely in our woods and marshes. The most dreary wilderness in Michigan at this most dreary season of the year is made cheerful and interesting by the beautiful little scarlet berry nestled in the folliage, so appropriately named "wintergreen." Its beauty at this season is only equaled by the holly in Europe and in the Southern States; but the holly is beautiful only, while the wintergreen is valuable as a fruit, and its leaves produce a favorite extract, which both for flavor and medicinal property is unrivaled.

But while I refer to these prospective additions to our cultivated market berrries as desirable, I do not wish to be understood as recommending an attempt on the part of small-fruit growers generally to introduce them. I maintain that experiments should only be made by those who have abundant means and time to devote to them. The practical fruit-grower, whose living and family depend on the amount of fruit he raises for market, has no business with mere experiments, either with untried varieties of established fruits or in the cultivation of fruits not hitherto cultivated. The amateur who follows fruit-growing for pleasure rather than for profit, should make these experiments and give to the world the benefit of his success or failure.

The small-fruit grower for profit should select only varieties that have been proved suitable for this climate, and grow enough of each kind and variety to make it worth while to ship his crop to the nearest or best market. In selecting these varieties regard to the time of ripening is also important, so as to keep up a supply during as many months as possible.

STRAWBERRIES.

If I could commence with two acres of strawberries I would plant half an acre Hovey's Seedling, one acre of Wilson's Albany, and half an acre of Jucunda. This would be a safe investment, and would, in ordinary seasons, maintain a good supply of berries from the 15th of June till about the 6th of July.

RASPBERRIES.

There are many varieties of the raspberry, but those which appear to be the safest here at present for general market cultivation are the Doolittle Blackcap, Miami, and the Philadelphia. I would plant these in about equal quantities, and on as rich soil as I could find.

THE GOOSEBERRY.

Although one of the richest of berries, especially cooked or preserved, the gooseberry is not in this country popular with growers. Our most successful varieties, the Houghton and American Seedling, are good bearers, and so far as I have ever seen, free from mildew, the only objection to them being their size. They are too small. In Lancashire, England, gooseberries are made objects of special culture, and gooseberry shows are held at the season of ripening, at which berries are exhibited nearly as large as common hen's eggs. They grow in a peculiarly humid atmosphere, and in a rich, black soil; the smoke from the factory chimneys loads the humid air with carbon, and probably destroys any tendency which the gooseberry might otherwise have to mildew. The operatives nurse their gooseberry bushes with great care, and thin out the berries in order in order to produce a few of large size. I find that even the Houghton Seedling of this country will attain quite a respectable size, if thinned out. Its natural tendency is to over-bear every other year. Poultry will thin out the fruit most faithfully, but should not be allowed to remain too long at the job. When the berries get a moderate size, if still too thick, a portion should be gathered when yet quite green; they are good even then, cooked, and those which remain will be much larger berries when matured. Standard bushes are best for gooseberries.

CURRANTS.

Everybody having a garden will grow more or less of the "dwarf grape," as the currant is sometimes called. The strong winds of this region are apt to break down the tender wood of the currant bushes; they cannot therefore be trained as standards, but must be cultivated in canes, five or six to the bush. Whether currants can be made profitable as a market fruit here when grown on an extensive scale, is a question yet to be decided by experience. Currants have been produced in great abundance in many private gardens, and therefore there is not much risk in planting a few acres. For domestic or medicinal wine, for jelly, and for canning, the currant will always command a ready sale and we see no reason why its cultivation should not be greatly extended.

BLACKBERRIES.

The general opinion is that the Lawton blackberry is not hardy enough for this latitude, but the Kittatinny, and perhaps the Wilson, it is hoped may succeed here. I would not advise a very extensive plantation of blackberries until one of the cultivated varieties has been sufficiently tested. The immense supply of wild blackberries in Michigan, of such excellent quality, will for some years be the principal reliance of dealers.

CRANBERRIES.

The cultivation of cranberries has not made very rapid progress in this region, but the success of drainage in some parts of the county is leading to the preparation of cranberry lands, and if drainage should be allowed to progress, we may reasonably expect that the cranberry will become a staple product of our county.

THE GRAPE.

The grape is undoubtedly the most important of the small fruits, but as it is made the special subject of one of our most experienced fruit-growers, I shall merely say that I regard its culture as second only in importance to that of peaches in this region.

ADVANTAGES OF SMALL-FRUIT CULTURE.

The great advantage of small-fruit culture is that land which would otherwise be unproductive can be profitably employed in this business. The grape will grow on the steep hill-side, the cranberry on overflowed marshes, the strawberry between rows of standard fruit trees, and currants under peaches and apples, with advantage to both, proper care being taken not to impoverish the soil.

Another and a still more important advantage of small-fruit culture is the fact that it produces a much more immediate return for the first outlay than the larger fruits. This is particularly the case with the strawberry; I have gathered a good crop of strawberries from plants within ten months of the time of planting. In quickness of return the Philadelphia Raspberry is perhaps next to the strawberry.

Fruit-growers commencing with very little capital will do well to pay special attention to these two crops, as they will bring in money in less time than any other fruit, and as quick as almost any ordinary farm crop. I know fruit-growers who have had to sell off portions of their farms to obtain the means of cultivating peaches and apples, who, if they had started with strawberries and raspberries between their rows of peaches and apples, or on separate patches, would have been in funds the second year after planting, and fully able to wait the growing of their peach and apple trees.

Instead of being discouraged at hard times and slow returns for large investments, the fruit-grower should go intelligently to work with these quick-bearing fruits, and he will find that fruit-growing will yield as quick a return for his labor and capital as almost any other productive business, and with as little risk.

Another advantage of small-fruit growing is the employment of women and children. After plowing, nearly all the labor of strawberry and small-fruit culture can be performed by these "light hands" while the men are engaged in the heavy work of the farm.

It would detract nothing from the dignity of woman to be engaged in the cultivation, gathering, and packing of small fruits, and in training children to the same occupation, while it would promote health and vivacity, and add a new charm to existence. If the thousands of women who are now wasting their lives in cities at poorly paid occupations, in ill-ventilated rooms, were to come to Michigan and engage earnestly in small-fruit culture, they would soon become objects of admiration rather than of pity; would not only realize independence for themselves, but would promote the prosperity and increase the wealth of the State. There is a wide field open for this branch of industry; it is only just in its infancy. Ten thousand such hands distributed through the Michigan fruit region would find homes of comfort, profitable occupation, and bright prospects for the future. Why do they not come?

If our Society does its duty these facts will be made known, and the thousands of men and women who are now unprofitably employed will come here, settle on our lands, make homes of comfort and gardens of profit, building up on this eastern shore of Lake Michigan a country of unsurpassed fruitfulness, prosperity, and wealth.

OTTAWA COUNTY FARMERS' CLUB.

In January, 1870, pursuant to a call of the Executive Committee of the Ottawa County Agricultural Society, a club was organized, consisting of many of the most active, intelligent agriculturists of Ottawa county, and monthly meetings were held during the year. The officers elected were: President G. W. Danforth; Secretary, Simeon Hazelton; Treasurer, Hon. W. F. Storrs; all of Coopersville.

From their proceedings we select the following particulars:

DRAINAGE.

Mr. Pennoyer thought that the subject of drainage should be discussed. A man could be induced to come here with his machinery for manufacturing drain-tiles. The Club could arrange this. There was no land so high as not to be capable of being benefited by drainage. Our wet lands would be immensely improved by drainage.

THE CHAIRMAN—Yes, and dry lands too.

Henry S. Clubb said, when at the State Fair last fall he met the inventor of the "Pioneer Drain-Tile Machine," who offered liberal inducements for men who had suitable clay, to purchase the Pioneer machine, offering to set it up and manufacture, with the aid of a man and a boy, 5,000 two-inch drain-tiles in a day, which, at $10 a thousand, could be made to pay for the machine in a very short time.

Mr. Pennoyer said that was only half the price paid for drain-tiles in Grand Rapids. The drain-tile factory there was doubling its business every year.

Mr. C. S. Randall said he had tried drains made of scantling and slabs and wood poles. He found the drains kept open for four years.

THE CHAIRMAN—Take hemlock boards in the form of a V.

Mr. MCNAUGHTON—Wood will not rot if kept wet and from the air, and will last twenty years.

Mr. Pennoyer said that clay tiles were the cheapest. There were some men who drained every two rods through their driest land, and who had been greatly benefited.

Mr. P. D. MCNAUGHTON—That's where the benefit of Farmers' Clubs is: bringing out experiments made by farmers, and letting their neighbors know of them.

Mr. Gifford Sherman said that he had seen drains made with cobble-stones, in a black-ash swamp, and the result was magical in the production of carrots and other vegetables.

Mr. C. S. Randall knew of an apple orchard that had been drained, and since then it had become very productive. Had sold 100 barrels of apples last season from it.

CHEESE FACTORY—MILK.

Mr. Pennoyer said that if the Farmers' Club is organized, it would not be long before a cheese factory was established here. Coopersville here could supply Grand Haven with milk, and it would be cheaper to the people there than keeping cows could be. It would then be kept from churning by filling the cans full. All agreed that this would be a good undertaking.

A BAD PRACTICE.

MR. PENNOYER—One great evil was growing, which would be of very serious result. Farmers are selling off all their hay and driving off their young stock. It was a bad sign to see farms stripped of their produce. He did not like to see through the cracks of a barn with nothing in it. If farmers knew their own interest they would keep as much hay and grain as possible, and feed it to stock on the farms, so as to fertilize instead of impoverishing their farms. There were many subjects of this character which could be discussed with advantage to farmers. There would have been a much larger attendance to-day, but farmers were trying to make the best use of the sleighing.

STOCK-RAISING.

The subject appointed for discussion was embodied in the following resolution :

Resolved, That stock-raising should be the general policy of the farmers of Ottawa county.

HON. W. F. STORRS—I am not perhaps a farmer sufficiently to enter upon the subject so thoroughly as it should be discussed. My view is, that our soil is particularly adapted to grass-growing. The first thing that comes up in a clearing is white clover, whereas wheat is not a certain crop. The tendency now is to seek more ease; with stock there is but little to do all the summer season besides seeing to the watering of the stock. During the summer a man can be a gentleman farmer. The stock is growing and the land improving all the time, with but little labor. Wherever you go in a stock-raising country, you will see it is a richer country; farms are not worn out as they are where tilling is extensively carried on. If our soil is equally calculated for stock as it is for grain, we have the advantage settled at once. I can go to places in Ohio, where they do not plow more than two acres to a farm, and the farmers are wealthy and in easy circumstances. A farmer now raising an acre or two of wheat, or a patch of potatoes, for his own use, is not properly a mixed farmer as I understood.

MR. HAZLETON—I do not believe it is more beneficial for the farmers to go into stock-raising altogether. The affirmative of the question, however, has good ground, as this county is almost destitute of stock. Some time ago the stock was sold off, and the sheep business is now played out. My estimate for raising wheat has been $10 per acre. The cost of everything constituting expenditures in keeping house and carrying on a farm is increased. I am not prepared to say what are the profits of stock-raising, but I am prepared to put against the affirmative of this question, dairying. The average of cheese from a cow is, say 460 lbs. It is pretty well understood with farmers, that 200 lbs. of butter from each cow is a fair average. The cow taking, we will say, two acres for pasture and one for hay, I figured up and made a dairy of sixty acres and ten in timber. I reckon the interest on the cows and of the land at $73 per acre, and I made the net income $1,000 a year.

MR. B. F. LILLIE—It has been my theory to have rotation in farming, but I believe, with Mr. Storrs, that the soil of this county being adapted to the grasses, stock-raising will be of the best advantage to farming. I tell you how we have to do it. We are not going to buy stock, but we must raise it. We cannot go into it on a big scale. According to the rules of stock-raising East, I am aware that we can raise more stock to the acre than they can there. It takes about four acres East to keep a cow. I am aware that we can come in under that one-quarter; three acres will keep a cow all the year here. I have fed 20 cows off of 22 acres of meadow. On this meadow the grasses are growing better as the land grows older. I see plain enough by land I have cropped 26 years. I cut last year four tons to the acre. The last crop of wheat I have grown, where I have raised 40 bushels to the acre, has come down to 13 bushels, although it grows as much straw as ever. Besides, the quality of the wheat is inferior. The quality of our wheat used to be good and plump, but now it is becoming inferior. A two-year-old steer requires about as much feed as a cow. In summer the objection to the dairy business is, the period of drought would dry up the cows, but this objection should be overcome by growing corn for green feed, or even by growing clover. Either would give feed and keep the cows in milk. The Russian mammoth sunflower is reported to produce double the amount of green feed to corn.

Not only as a green crop, gathering the leaves, but the seed will be excellent for winter feed. A good spring pig, well kept on milk during summer, would average 200 lbs. This is an additional source of profit in dairy farming.

Another advantage of stock-raising is, it continually adds to the value and improves the condition of the farm. In Chautauqua county, N. Y., the farmers were all in debt, but when they commenced stock-raising and dairying, they were soon enabled to pay off their debts, and soon had money to loan. Mr. L. instanced several farmers, who while grain-raising were in debt, but who had seeded down their farms, and had to raise their stock gradually. When about 16 cows were raised they began to support a family in comfort, and when they had 30 they were saving money. His own experience was also instanced. He never knew until the money was in his hands how much he realized by growing wheat. The crop here is too precarious to think of getting money by.

Mr. M. Cole—When we have to fodder six to eight months to keep our cows the other four or six, it changes the subject. I think we have a man in this room who has been successful in raising grain, and would like to hear from him. The West can beat us in raising stock, where cattle can take care of themselves ten or eleven months in the year.

Mr. H. S. Clubb inquired for information, whether it was not necessary to raise something besides hay to keep stock profitably all winter? Also, whether the extra labor in winter preparing food, feeding stock in-doors, and taking care of the animals during winter months, did not off-set the easy time enjoyed during the summer, when cattle could shift for themselves. Machinery, no doubt, for cutting roots and fodder, could be obtained by large farmers, and labor materially lessened, but all this involves expense and must be charged against stock-raising. The prairies west and southwest possess advantages in cheap pasturage and long summers, which make it almost impossible for Michigan, with pastures worth $75 to

$100 an acre, to compete with them in the large markets. Although stock-raising, to a certain extent, would be beneficial in improving the condition of the land when made universal, as proposed, it might be found unprofitable, owing to the competition of those boundless prairies, where keeping stock within a certain range and hunting animals when required, was about all the expense attending them.

Gifford Sherman said the feeding of stock in winter all on hay, would be expensive. In the East, the way they make money is by raising roots, and using machinery for cutting. One acre of ground in roots will keep double the stock it will with grass. I think we will all do better to go into stock, corn, and roots. One acre of corn for fodder is worth more than an acre of hay, besides the corn. The corn can be raised more conveniently than roots, with our facilities. My experience is that I have made double on stock that I have on wheat. I bought four sheep and they did more in a few years towards getting me out of debt than wheat ever did.

Mr. W. F. Storrs—I do not know that it is the design to confine ourselves to the resolution. The question as now debated is between stock-raising and dairying. Mr. Clubb's views correspond with those of Mr. Cole, as to the long winters rendering an offset to the little labor of summer. I think in settling this matter we should take into account the price we pay for labor. I think in the East labor is cheaper, and they can better afford to pay for raising roots than we can. They may deem it in the East an advantage to raise roots, while here it would not be so profitable. With regard to dairying, we take it as it is here, without a factory, and it would require the labor of all our girls and wives. Is it for the interest of the community to adopt a plan which will exhaust our families? The question is, What will produce the greatest amount of health or wealth and pleasure as we go along? I admit there is money made by dairying, but it is at the expense of the health and comfort of our families. If you can show that in making of cheese and butter, you make most money, and yet it is to the injury of our families, I shall object that wealth is no consideration for such cost. I do not believe that it is even necessary to plow up a stock farm after it has once become seeded down. Top dressing will always keep grass in good condition, without renewing it.

MR. RICHARDS—You spend the same extra labor on dairying that you do on grain-raising, and it will pay better. It is not necessary for our women to do it. Men can do the work in a dairy and it will pay. I believe in sheltering cattle in winter, and I do not believe in selling hay. We should feed every ton of hay on our farm. George Luther used to ask me why I did not raise wheat? I told him it did not pay. I have never altered my mind.

MR. HAZELTON—I agree with my friend Richards. We are not going back to grain-raising. I hope to see the time when we shall have a cheese factory established in every school district. The affirmative of this question have produced no figures to show the advantage of stock-raising. The farming community, in order to go into dairying, have got to provide themselves with cows. It takes about three acres, two in summer and one in winter, per cow. Allowing three acres, the interest on the land makes $22.50 a year, reckoning land worth $75 an acre, just for the interest, without other expenses. How much would a three-year-old creature at that age be worth? It can hardly average above the interest of the land on which it fed. I believe dairying can be made more profitable. Mr. Sherman understands horses, and he can purchase a colt for $25 or $30, and in one or two years sell it for $200 or $300, but could I do that? Can farmers generally do that? Very few are such good judges of horses as Mr. Sherman.

MR. STORRS.—It is well known that Mr. Sherman is a horse-jockey anyway. [Laughter.]

Mr. Sherman hoped he would have an opportunity to reply. [Laughter.]

Mr. Hazelton—The County Farm, when purchased, was all grown to June grass, and was worthless. That part has been plowed up over and over again, and it has been completely renovated by sowing plaster, clover, and raising hay. It is well known that farmers can get some two tons of clover the first cutting; and the second cutting, undoubtedly, it is best to leave standing until it blossoms, and then turn in stock. By use of roots, one acre will winter two cows.

Mr. Storrs—A great deal has been said about drawing from the soil in raising grain. Is it not equally a fact that in dairying you are drawing from the soil, principally in the phosphate of lime? We have got to supply this or we will lose as much in deterioration as we make with the produce.

Mr. Hazleton said the chemists told us that of the elements obtained from the air, from one-third to one-half goes to the nourishment of plants. Here in raising clover, two-thirds is obtained from the atmosphere, and what proportion of that obtained from the soil is taken away in the milk?

Mr. Sherman thought it was necessary to get up early in the morning to raise grain, as well as to do dairying, and as some cows must be kept anyway, the women folks, or whoever has to attend to them, must get up early. If wheat was raised on a farm continually for ten years, he believed it would result in a mortgage on his farm.

Mr. I. Hunting read extracts from the California correspondent of the Utica *Herald:*

"In 1867, California produced 6,000,000 pounds of butter and 3,000,000 of cheese. This year the product is estimated at 9,000,000 pounds of butter, and 4,000,000 pounds of cheese. The cheese cures very rapidly. The rennets used by the cheese-makers are imported from Germany.

"The dairy farm of Laird & Kellogg is situated in Santa Cruz county, and is stocked with 400 cows. During the past year the milk of 200 was used for butter, and that of the other 200 for cheese. The cows milked for butter turned out

20,000 pounds, which was sold at an average price of 45 cents, amounting to $9,000. The cows milked for cheese produced 90,000 pounds, which was sold at 16 cents, amounting to $14,400, making the gross returns of the 400 cows $23,400, besides the calves raised and the pork produced from the whey. The cows are worth on an average $40 each."

The correspondent adds that "there are millions and millions of acres of unoccupied land in this State, with every possible facility for the dairy business." So it appears quite certain that California will produce her own butter and cheese, and may become a rival of the Eastern States in the market of the world.

Mr. Lillie referred to the fact that long feeding in winter would decrease the profit in keeping stock, and said: I think the objection of long feeding is not a very strong objection. I think we get along in winter in a good deal better shape than they do East, and I claim we can produce more grass to the acre than they do in the State of New York. Lands fitted up in Michigan to as high a state as in New York, and we can beat them, as our winters are lighter.

Mr. Richards considered the winters of Michigan more favorable to stock than those of Ohio.

Mr. Hazelton said our winters are shorter than in St. Lawrence county, N. Y., and in that county the farmers had become wealthy by dairying.

Mr. Storrs—My father used to say, if after making a living you raise but one calf, you are increasing in wealth.

Mr. Lillie heard of a man who used to raise a dozen or so of calves a year. One year he neglected to buy up any calves, and found himself minus $300 or $400.

Mr. Hazelton advocated the keeping of cows as more profitable than raising calves or keeping stock.

SHOULD FARMERS GROW FRUIT?

Mr. Henry S. Clubb said he had selected this subject with a view to draw out the facts and experience of farmers, rather than with a view to any instruction which he expected to give on the subject. By farmers it was understood, men who devoted their attention to the cultivation of large crops of grain, hay, and the raising of stock, production of butter and other dairy articles. Now, would it be worth while for persons so engaged to raise fruit? In reply to this question, the location must be considered. A man ten miles from railroad or navigation, and from a market, should not indulge in raising fruit beyond the requirements of his own family or neighborhood, but those farmers who were so fortunate as to live near a railroad, or near our magnificent Grand river, or its tributaries, or whose land bordered any of our beautiful inland lakes, or on our Lake Michigan, could undoubtedly raise fruit to a good profit.

A gentlemen residing at Lamont, well situated for knowing the facts, told me that the apple crop alone saved the credit of the farmers last fall. The price of wheat being low did not meet the expense of raising, and had it not been for the apples the farmers would have been unable to pay their grocery and dry goods bills.

He would not recommend the growing of a great variety of fruit; a good selection of the best varieties of apples should be all that a farmer busily engaged in agriculture need attempt, unless he had plenty of help, in which case a few acres of small fruit, especially strawberries, and but one or two varieties of them, would be found profitable.

The advantage of small fruit over apples and other tree fruit was, the short time required for bringing it into bearing. It produced an income so quickly that it became a source of ready cash, of great convenience at a season of the year several weeks before the farmer can realize from his grain or hay, and many years before apples come into bearing.

What would be the condition of the farmers of Ottawa county to-day, if ten and twenty years ago, when starting, instead of planting 25 to 100 apple trees, they had planted 500 to 1,000 apple trees each? An apple tree bearing from ten to thirty bushels of apples, each bushel worth a dollar, it needed no great arithmetical figuring to arrive at the result. All the farmers present could figure this out themselves. What farm crop was equal to it for little labor and large produce?

A great advantage a farmer had over those who only grow fruit was, the supply of straw and barnyard manure for mulching, an important item for fruit growing.

Mr. Hazelton—In many respects I agree with my friend who has spoken; in other respects I differ. A farmer should have a large apple orchard,—it is a source of income of great importance. The fact that it takes a long time for apples to grow, makes them unlikely ever to glut the market. I would not, however, recommend a general cultivation of small fruits. Persons engaged in small-fruit culture must give it their whole attention. I have found this to be the case, and for that reason I am offering portions of my farm for sale, as I prefer to grow small fruit.

Hon. W. F. Storrs—I have had these speculations in my head on farming. My plan is to keep a dairy. We have settled that something besides grain must be resorted to. To keep a dairy a man must have a large amount of help, and it seems to me that the raising of strawberries or small fruit can be made more profitable. It will pay better than raising corn and potatoes. He can make his labor pay to very great advantage. He can more than make it pay in yielding strawberries. They have always paid. I think a mixture of dairying with the smaller fruits will pay better than anything else. It pays better than an apple crop, even when apples come into bearing. It is a fact that you can hardly overdo with a thing of this kind, whether dairying or fruit-growing. I believe the only safe way is to connect it with some other kind of farming.

Mr. Cady—I believe I can corroborate the statement of Mr. Clubb. The farmers have for some years past depended on wool, which failed, and the apple crop came last year and saved the farmers. The apple is the main fruit crop for farmers. If they cannot sell them at once, they can keep them and sell for good price. I also think, as stated by Mr. Storrs, that the dairying business along with the fruit crop is just the thing. Mr. Luther gathered 50 or 60 bushels from two apple trees.

Mr. W. C. Phileo of Big Rapids—I agree with Mr. Storrs, that the time occupied by dairying not being all day, the later part of the day could be profitably employed with the small-fruit culture. Since the organization of cheese factories, the price of milk has increased. Cheese is exported to Europe, and money has already been brought into this country in return therefor. I know that this county is good for fruit-growing, and even in the Northern Peninsula the apple is a sure crop. The apple will live a great many years, while the peach will last only four or five years. The failure of peaches has generally been after the production of a heavy crop, and not on account of frost.

Mr. Storrs said, in the Western Reserve, Ohio, the peach used to bear abundantly, but now it is seldom that a peach tree can be seen there.

Mr. Hazelton—You have often heard about the old saying, "Jack of all trades does not succeed in any." My friend, Mr. Storrs, thinks the dairying business will leave a good deal of time. I think if he had tried the experiment he would find it different. I think it would be necessary to raise root crops for his cows in time of drought or in winter. Now, in regard to over-stocking the market with small fruit, it is well known that fruit is running high, that the whole of Western Michigan is going into it. Suppose all the farmers from Grand Traverse to St. Jo. go into it, how long will it be before the strawberries, for instance, will be a glut upon the market? We shipped here one day 145 bushels of berries, and the price run down from 21 to 8 cents a quart. A few years ago the hop crop was worth $1,000 to the acre. A great many went into it. But it failed. It would be so with strawberries. If one make a specialty of the business he can do better than anybody else.

Mr. Storrs said that ten cows for one man to milk is a great quantity. It will take three men for thirty cows. I am considering all the time of dairying with cheese factories. It is certain there would be considerable time to spend in fruit-raising. Now with regard to markets, the country is filling up; and now the facilities for keeping small fruit fresh for a month or so is being invented, we may conclude it will make a great difference. With regard to the drop in prices referred to, it was partly owing to the wet and to the condition of the berries when shipped, and the fact that they had to be carried over Crockery creek in a manner that injured them. Taken as a whole, I believe the prospect for profits on small fruit is as good as anything I can conceive of.

Mr. Clubb responded, stating that the comparison of fruit and hops was hardly correct. The number of persons consuming hops was small in comparison with the number that could be taught to consume fruit. Strawberries in the market in abundance would create a demand which could not be said of hops. The way to secure a large demand for small fruit was to produce the best kinds in sufficient quantity to make it an object for business men to engage in the shipping and forwarding business. Not a family in the county but would consume two quarts of strawberries a day if placed before them. The lines of railroad and steamboats in every direction would convey them to remote places if sufficient quantity were produced. There was no business without drawbacks, and exceptions only proved the rule. An occasional fall of rain or a railroad accident, causing loss, could not be set down specially as against that business.

CHEESE-MAKING AND DAIRYING.

Mr. Benj. F. Lillie opened the subject of cheese-making. He said it seemed to him that there was only one thing to hinder Ottawa county farmers from going into dairying, and that was the high price of cows. There was now some hope, however, that crops having failed in some Eastern States, cows would be cheaper there. Even for making butter, he was satisfied a cow would pay $50 a head a year. According to Eastern statements, cheese-making is nearly as profitable again as butter-making; why should we not go into that business? He could not see why farmers should not join together and get up a cheese factory. Cheese would make cows worth at least a third more, making a cow worth at least $75 a year.

Mr. Hazelton was not much acquainted with dairying, although he came from a country where dairying made the farmers rich. It was a cold country. It was said the average was between 350 and 450 lbs. of cheese to the cow, and in addition to the cheese, there was considerable butter, which, together with calves and pork, made the produce per cow $100, or on 20 cows $2,000, to be made on 60 acres of dairy land. A team and that number of cows could be kept on that number of acres. Every farmer knows that raising grain diminishes the value of the farm, and wheat not being profitable at present prices, it is the duty of the farmer to make his farm into a dairy farm as soon as possible. He believed if established at Coopersville, 100 cows could be furnished first year, and 200 the second, but on inquiry among the farmers he had found a larger number could be had to commence with.

President McNaughton said the average was 150 lbs. of milk to produce 20 lbs. of cheese.

Mr. Lillie, in reply to a question of Mr. Easton, stated his estimate of $50 was based on present prices of butter,—not less than 25 cents per pound, and reckoning the value of calves; also a litter of pigs, three months old in August, fed entirely on milk. These together with the calves added, it is plain the cows will average $50 each. The pigs are now worth $10 each. If I were going to put it to the highest point, I am satisfied it would be $60.

Mr. Richards said he should lay it at $75 a head.

Mr. Lillie was satisfied it would not be less than $60.

Mr. Hazelton reckoned this way: 400 lbs. of cheese $60, and then 30 or 40 lbs. of butter during the winter, and then the pork and the calves, adding altogether $40, making $100 in all for each cow.

Mr. Joel Walters said in making butter there was no detriment to calves, but in making cheese you cannot count much on the value of the calves. Cannot make cheese without a factory.

Rev. H. Beckwith—When a boy I was initiated to the business of cheese and butter making. I became dissatisfied with the business from the fact of having to be the cow-boy, and from that time to the present I have never kept but one cow, but as a business I am satisfied it is a lucrative one. Chester, Mass., was celebrated for its large cheese presented to Thomas Jefferson by Elder Kneeland. Some of the wealthiest men made their wealth by that business. Ira Richardson made his wealth that way. In Wisconsin they had raised wheat year after year for 20 years, and their farms had deteriorated. The boys not only objected to cheese-making and butter-making, but the women objected to it also. They will tell you the dread of marrying farmers grew out of the fact of having so much churning to do; but modern methods did away with this objection, and they would as soon marry a farmer as a lawyer or doctor, and a little rather, as they would be sure to have plenty to eat. You establish a factory in the center of a town and the women have nothing to do with it, and that would be a great reason for going into this business. It not only exempts them from this burden, but is a great saving in point of labor.

Mr. Pennoyer said a great deal had been written and said upon the subject of cheese-making. We should consider not only how we should enrich ourselves,—how we may live the easiest and improve our farms,—but, in a national point of view, Is it best?—would it be right? I contend not, because it would tend to effeminacy. We are retrogressing. When we had to struggle hard we were more hardy. Let the whole nation go into it as a body, and we would become so effeminate that the 400,000 Prussians could drive us to the Pacific. That occupation is the best that engages man's whole strength. Women who are wholly engaged in business are always more intelligent than those who are dallying in the parlors all the time. I always enjoyed being cow-boy till I was twenty-five years of age. I never considered it a hardship. I would now rather sit down to milk a cow than do anything else. Mr. Lillie is right. I know I have realized more than $80 from the products of a cow, each year. At forty cents a pound, the price I have realized a greater part of the time, 200 pounds would amount to $80. Last February a cow came in. She fed on dry feed. I milked into a patent pail, and the milk half filled the pail. We kept it by itself, and Mrs. Pennoyer weighed the butter. For the first two weeks the milk was given to the calf. After that, some weeks she made nine pounds, and some weeks,—as soon as the cow got a little green feed,—it increased to ten pounds. By the 15th of May the amount of milk increased to even a pailful. From five cows I obtained eleven pounds of butter a week, each, for five months. Those cows wére milked nine months in the year. Find your location and start a factory. It can be done. I don't know but Mr. Walter has a good place. I know I have a place where cold water can be run, in a two-inch pipe, into the second story. This is not the location, however, but Coopersville is central enough. Farmers can add to their wealth more by going into the dairy business than anything else; but, in a national point of view, I condemn it. Don't let us hand down effeminacy to our children.

Mr. Richards did not agree with Mr. Pennoyer. He believed dairying was a hardening business. Michigan, Illinois, and Indiana had as good hardy sons in the war as Vermont. He wanted nothing said against Vermont. He was raised where he had to look twice to see the tops of the hills. As to his sons and daughters, he was ready to risk their hardiness.

Mr. Walters said he was not afraid of living easy. He came west with the idea of getting an easy living, but had not seen it yet. It made too much hard work to make butter by hand, but a cheese factory would undoubtedly reduce the labor greatly. There is no trouble in making a cow pay at least $60.

Mr. Pennoyer asked the cost of wintering a cow.

Mr. Walters said, there was a great deal of coarse stuff raised on a farm; did not think it would cost to exceed $25 on a farm. Cornstalks will winter a cow, and are better than hay while she is giving milk. If you fed hay entirely, it would cost $40 at present prices of hay. A cow would eat a ton and a half of hay during winter. That has been $12 or $14 a ton.

Mr. Richards said that two tons would be required.

Mr. Pennoyer—I always use two and a half tons.

President McNaughton confirmed Mr. Walters' estimate of the cost.

Mr. Hazelton said, in his estimate of three acres to keep a cow there would be one acre to mow. Reckon land worth $100; cost of cutting, $5; that would, with interest on value of land, make $30 to a cow. The great question with farmers is, What shall we do? Taxes are away up, and bid fair to take our farms. It is necessary to make a change. We must save the interest on the farm, and if we can do it by dairying, that is what we have to do. From 400 to 500 pounds of cheese can be realized per cow, if a factory is established. A cow would average about 200 pounds of butter. Some factories are now making both butter and cheese.

Mr. Golden thought a cow would make from six to ten pounds of butter a week.

Mr. Easton had a relative in Tioga county, N. Y., who has been in the dairying business at home for some years. Last year he used a cheese factory four miles off; had seventeen cows; sent his milk five months, and it netted $65 a cow, after paying expenses of hauling it four miles. His own experience from two cows: Butter realized $90; sold $20 worth of pork, besides keeping enough for family. Fed bran and mill-feed as well as hay. Thought there was no business would pay as well as a cheese factory. Mr. Pennoyer said it could be done by speaking the word. He would speak the word. [Laughter.]

REGISTER OF
METEOROLOGICAL OBSERVATIONS
FOR THE YEAR 1870,

TAKEN AT THE

State Agricultural College of Michigan,

BY R. C. KEDZIE,

PROFESSOR OF CHEMISTRY.

LATITUDE 42° 42′ 24″; LONGITUDE 7° 33′ 19″ WEST OF WASHINGTON.
Height above the Sea, 895 Feet.

METEOROLOGICAL OBSERVATIONS FOR THE MONTH OF JANUARY, 1870.

Day of Month.	Thermometer in the open air. 7 A. M.	2 P. M.	9 P. M.	Mean.	Rain and Snow. Time of beginning of rain or snow.	Time of ending of rain or snow.	Amount of rain or melted snow in gauge, in inches.	Depth of snow, in inches.	Clouds. 7 A. M. Amount of cloudin's.	Kind of clouds.	2 P. M. Amount of cloudin's.	Kind of clouds.	9 P. M. Amount of cloudin's.	Kind of clouds.
1	26	40	32	32⅔	7 P. M.	-------	-------	-------	100	Cu. St.	100	Cu. St.	100	Nim.
2	29	32	30	30⅓	-------	-------	-------	-------	100	Nim.	100	Nim.	100	Nim.
3	18	19	20	19	-------	5 A. M.	.330	3½	100	Cu. St.	100	Cu. St.	100	Cu.
4	21	24	19	21⅓	-------	-------	-------	-------	100	Cu. St.	100	Cu. St.	100	Cu. St.
5	19	32	26	25⅔	-------	-------	-------	-------	100	Cu. St.	100	Cu. St.	100	Cu. St.
6	28	22	14	21⅓	-------	-------	-------	-------	100	Cu. St.	90	Cu.	100	Cu.
7	14	19	12	15	1 P. M.	-------	-------	-------	100	Cu. St.	100	Nim.	100	Nim.
8	14	19	4	12⅓	-------	-------	-------	-------	100	Nim.	90	Cu.	100	Nim.
9	2	10	11	7⅔	-------	-------	-------	-------	100	Cu. St.	100	Cu. St.	10	Cir.Cu.
10	22	41	36	33	-------	9 A. M.	.190	4½	100	Nim.	50	Cu.	100	Cu.
11	22	41	36	33	3 P. M.	-------	-------	-------	30	Cir. St.	100	Nim.	100	Nim.
12	41	32	27	33⅓	-------	-------	-------	-------	100	Nim.	100	Cu.	100	Cu.
13	13	14	6	11	-------	8 A. M.	.800	1	100	Nim.	50	Cu.	00	-------
14	10	41	19	23⅓	In night.	-------	.250	-------	100	Cu. St.	100	Cu. St.	100	Nim.
15	42	42	25	36⅓	-------	-------	-------	-------	100	Cu.	100	Cu.	100	Cu.
16	27	36	42	35	8 P. M.	-------	-------	-------	90	Cu. St.	100	Cu.	100	Nim.
17	46	40	12	32⅔	-------	11 P. M.	.170	-------	100	Nim.	100	Cu.	50	Cu.
18	6	14	10	10	-------	-------	-------	-------	30	Cu. St.	100	Cu.	50	Cu. St.
19	10	32	24	22	-------	-------	-------	-------	10	St.	00	-------	00	-------
20	26	40	26	30⅔	-------	-------	-------	-------	90	Cir. St.	30	Cir. St.	100	Cu.
21	19	35	20	24⅔	-------	-------	-------	-------	90	Cu. St.	100	Cu.	100	Cu.
22	37	40	39	38⅔	7 P. M.	-------	-------	-------	100	Cu. St.	100	Cu. St.	100	Nim.
23	28	25	19	24	-------	11 A. M.	.130	-------	100	Nim.	100	Cu. St.	00	-------
24	25	30	23	26	-------	-------	-------	-------	50	Cu. St.	90	Cu.	20	Cu.
25	18	38	35	30⅓	-------	-------	-------	-------	100	Cu.	30	Cu. St.	80	Cu.
26	29	30	23	27⅓	-------	-------	-------	-------	100	Cu. St.	100	Cu.	00	-------
27	12	30	22	21⅓	-------	-------	-------	-------	30	Cu. St.	20	Cu. St.	50	Cu. St.
28	13	34	29	25⅓	-------	-------	-------	-------	50	Cir. St.	50	Cir. St.	100	Cu.
29	32	34	23	29⅔	-------	-------	-------	-------	100	Cu. St.	100	Cu.	100	Cu.
30	26	30	18	24⅔	-------	-------	-------	-------	100	Cu. St.	100	Cu. St.	00	-------
31	25	34	28	29	Squalls	-------	.060	½	100	Cu. St.	100	Nim.	100	Nim.
Sums	----	----	----	----	-------	-------	1.930	9½	----	-------	----	-------	----	-------
Me'ns	----	----	----	25°.37	-------	-------	-------	-------	86	-------	84	-------	70	-------
Average	-------				-------				80					

Day of Month.	Winds. 7 A. M. Direction.	Force.	2 P. M. Direction.	Force.	9 P. M. Direction.	Force.	Register Thermom'tr. Max.	Min.	Barometer Height Reduced to Freezing Point. 7 A. M.	2 P. M.	9 P. M.	Mean.	Force or Pressure of Vapor, in Inches. 7 A. M.	2 P. M.	9 P. M.	Relative Humid'y, or Fraction of Saturation. 7 A. M.	2 P. M.	9 P. M.
1	S W	1	S W	2	E	2	40	16	28.735	28.731	28.731	28.732	.105	.182	.162	75	73	89
2	N E	3	N W	3	W	5	32	12	28.550	28.053	28.047	28.216	.160	.181	.167	100	100	100
3	W	4	S W	3	S W	2	20	10	28.372	28.574	28.544	28.496	.098	.087	.108	100	84	100
4	S W	2	W	3	S W	2	24	10	28.615	28.626	28.801	28.681	.096	.111	.103	85	86	100
5	W	1	------	0	S W	3	32	12	28.956	28.860	28.802	28.872	.103	.181	.105	100	100	75
6	W	2	W	2	W	3	28	3	28.851	28.926	29.072	28.949	.117	.118	.082	76	100	100
7	W	2	S W	2	W	3	19	5	29.068	28.944	28.871	28.961	.082	.103	.075	100	100	100
8	W	2	S W	3	W	2	19	-17	28.867	28.974	29.191	29.010	.082	.071	.052	100	69	100
9	S W	2	S W	3	S W	3	20	-5	29.182	28.946	28.841	28.989	.048	.068	.071	100	100	100
10	S W	3	S W	2	------	0	41	13	28.703	28.679	28.896	28.759	.118	.126	.191	100	49	90
11	S E	1	S E	1	S W	1	41	18	29.006	28.946	28.846	28.932	.084	.257	.212	71	100	100
12	S W	1	N W	2	S W	3	42	8	28.735	28.838	28.961	28.844	.257	.181	.147	100	100	100
13	N E	3	N E	2	E	1	15	1	29.021	29.247	29.311	29.193	.078	.082	.057	100	100	100
14	S E	2	S W	1	S W	2	41	5	29.271	28.919	28.655	28.948	.068	.212	.103	100	82	100
15	S W	3	S W	3	W	3	42	18	28.495	28.523	28.761	28.593	.244	.222	.100	91	83	74
16	S E	2	S E	2	S E	3	45	22	28.951	28.746	28.507	28.734	.111	.149	.248	75	71	100
17	S E	3	S W	3	S W	4	46	00	28.319	28.744	28.888	28.650	.311	.160	.075	100	64	100
18	S W	3	S W	3	S W	2	14	00	29.292	29.248	29.378	29.306	.057	.082	.068	100	100	100
19	------	0	S W	3	S W	2	33	5	29.342	29.204	29.101	29.215	.068	.181	.129	100	100	100
20	S W	2	S W	2	N W	3	40	14	28.968	28.861	28.946	28.925	.088	.139	.088	62	56	62
21	W	2	S W	1	W	2	35	7	29.144	28.934	28.940	29.006	.071	.108	.108	69	53	100
22	S W	1	S W	1	N E	1	40	14	28.894	28.896	28.888	28.892	.178	.182	.157	81	73	71
23	N E	1	N E	1	E	1	29	24	28.906	29.107	29.198	29.070	.135	.135	.103	88	100	100
24	N E	2	E	3	N E	1	30	32	29.106	28.930	28.830	28.955	.100	.148	.123	74	89	100
25	E	2	S W	2	S W	3	38	13	28.671	28.611	28.547	28.609	.082	.186	.162	84	81	80
26	S W	2	W	3	S W	2	30	5	28.771	28.869	28.961	28.867	.123	.130	.106	77	78	86
27	S W	1	S W	3	W	2	30	6	29.097	29.118	29.137	29.117	.045	.093	.084	62	56	71
28	S	2	S W	2	S W	1	34	3	29.134	29.078	28.880	29.030	.048	.155	.142	62	79	88
29	S W	2	W	2	W	3	34	20	28.578	28.687	28.838	28.701	.162	.155	.106	89	79	86
30	N E	1	E	1	------	0	30	7	28.866	28.858	28.867	28.863	.123	.130	.098	87	78	100
31	S W	1	------	0	W	2	34	-1	28.820	28.727	28.795	28.780	.117	.155	.158	87	79	100
Sums	------	----	------	----	------	----	----	----	-------	-------	-------	-------	----	----	----	----	----	----
Me'ns	------	----	------	----	------	----	32°.19	9°.03	-------	-------	-------	28.867	.114	.144	.119	83	82	92
Average	-------						-------		-------				125			85		

METEOROLOGICAL TABLES FOR THE MONTH OF FEBRUARY, 1870.

Day of Month.	Thermometer in the open air. 7 A. M.	2 P. M.	9 P. M.	Mean.	Rain and Snow. Time of beginning of rain or snow.	Time of ending of rain or snow.	Amount of rain or melted snow in gauge, in inches.	Depth of snow in inches.	Clouds. 7 A. M. Amount of cloudin's.	Kind of clouds.	2 P. M. Amount of cloudin's.	Kind of clouds.	9 P. M. Amount of cloudin's.	Kind of clouds.
1	4	30	28	19					30	Cir. St.	50	Cu.	100	Cu.
2	28	30	26	28					100	Nim.	60	Cu. St.	100	Cu.
3	16	20	14	16⅔					100	Cu.	50	Cu. St.	100	Cu.
4	14	30	19	21					100	Cu.	100	Cu.	30	Cu. St.
5	19	29	26	24⅔					100	Cu. St.	100	Cu. St.	100	Cu.
6	23	36	20	26⅓					100	Cu. St.	50	Cir. St.	00	
7	18	44	24	32					50	Cir. St.	80	Cir. Cu.	10	Cir. St.
8	22	30	30	27⅓					10	St.	100	Cu. St.	100	Cu.
9	28	30	30	29⅓					50	Cu.	100	Cu. St.	100	Cu.
10	25	41	32	32⅔					50	Cu.	30	Cir. St.	100	Cu.
11	36	38	38	37⅓					100	Cu.	50	Cir. St.	10	Cir. St.
12	18	35	14	22⅓					30	Cu.	50	Cu.	90	Cu.
13	12	26	27	21⅔					90	Cir. St.	100	Cu. St.	100	Cu.
14	31	41	34	35⅓	3 A. M.	11 A. M.	.150	1	100	Nim.	90	Cu.	100	Cu.
15	17	36	26	26⅓					10	Cu. St.	10	Cu. St.	00	
16	14	41	32	29					10	Cir. St.	90	Cu.	50	Cu.
17	35	40	28	34⅓	4 P. M.	10 P. M.	.128	1	100	Cu.	100	Cu.	100	Nim.
18	9	29	22	20	11 A. M.				50	Cu.	10	Cu.	100	Cu.
19	18	22	20	30					100	Cu.	100	Nim.	100	Nim.
20	−2	15	1	4⅔		2 A. M.	.075	3	30	Cir. St.	50	Cu.	00	
21	−13	14	−2	−⅓					30	Cu.	30	Cu.	00	
22	0	26	2	9⅓					100	Cu. St.	100	Cu. St.	00	
23	−6	35	22	17					10	St.	00		00	
24	18	32	17	22⅓					00	Cu. St.	100	Cu.	00	
25	8	40	18	22					100		100	Cu.	100	Cu. St.
26	30	35	30	31⅔	4 A. M.				100	Nim.	100	Nim.	100	Nim.
27	32	34	32	32⅔					100	Nim.	100	Nim.	100	Nim.
28	26	30	22	26			.850	3	100	Nim.	100	Nim.	10	Cu.
29														
30														
31														
Sums							1.203	8						
Me'ns				24°.25					66		71		60	
Average											65			

Day of Month.	Winds. 7 A. M. Direction.	Force.	2 P. M. Direction.	Force.	9 P. M. Direction.	Force.	Register Thermom'te. Max.	Min.	Barometer Height Reduced to Freezing Point. 7 A. M.	2 P. M.	9 P. M.	Mean.	Force or Pressure of Vapor, in Inches. 7 A. M.	2 P. M.	9 P. M.	Relative Humid'y, or Fraction of Saturation. 7 A. M.	2 P. M.	9 P. M.
1		0		0	S	2	30	−1	28.958	28.930	28.800	28.896	.052	.148	.084	100	89	71
2	W	2		0	E	2	30	10	28.688	28.919	28.965	28.857	.153	.130	.141	100	78	100
3	N E	1	E	2	E	1	20	5	29.130	28.769	29.281	29.060	.090	.108	.082	100	100	100
4	E	1	E	1	S E	1	30	6	29.307	29.160	29.163	29.210	.082	.167	.103	100	100	100
5	S W	1	S W	1	S W	1	29	18	29.147	29.091	29.078	29.105	.103	.160	.105	100	100	75
6		0	S W	1		0	36	10	29.091	29.070	29.042	29.067	.123	.115	.108	100	54	100
7		0	W	1		0	44	14	29.064	29.040	28.954	29.019	.098	.108	.129	100	37	100
8	N	1	N E	1		0	30	12	28.834	28.821	28.811	28.822	.101	.167	.148	86	100	89
9	S W	3	N W	2	W	3	30	10	28.645	28.547	28.669	28.620	.117	.155	.155	76	89	89
10	W	1	W	1	S	2	41	20	28.900	28.929	28.854	28.894	.100	.064	.143	74	25	79
11	S E	2	S W	2	S W	2	45	13	28.525	28.391	28.382	28.432	.115	.165	.123	54	72	54
12	W	5	W	4	W	3	35	−2	28.605	28.781	28.824	28.736	.098	.108	.082	100	58	100
13	S E	2	S W	2	S	3	28	7	29.058	28.824	28.711	28.864	.075	.088	.076	100	62	52
14	S E	1	S	1	W	2	41	13	28.416	28.338	28.541	28.331	.174	.212	.155	100	82	79
15	N W	1	E	1		0	36	9	28.800	28.930	28.951	28.893	.094	.191	.088	100	90	62
16		0	S E	2	S W	3	41	10	29.068	29.097	29.006	29.057	.082	.147	.162	100	57	89
17	S W	2	W	2	W	1	40	3	28.749	28.760	28.801	28.770	.188	.103	.099	90	45	64
18	N W	1	W	2	W	1	29	3	28.857	28.820	28.801	28.826	.065	.105	.118	100	66	100
19	S E	2	N E	1	N E	3	25	−10	28.648	28.616	28.729	28.664	.072	.118	.075	75	100	70
20	W	1	N W	2	N W	1	15	−28	29.050	29.019	29.127	29.065	.040	.086	.046	100	100	100
21		0	W	3		0	14	−22	29.156	29.114	29.028	29.099	.025	.082	.040	100	100	100
22		0	S W	1		0	26	−11	28.995	29.078	28.801	28.958	.044	.071	.048	100	51	100
23		0	W	2	W	1	36	−10	28.739	28.686	28.655	28.693	.033	.127	.052	100	62	44
24	N W	1	W	2		0	32	−3	28.731	28.719	28.727	28.725	.072	.181	.063	75	100	67
25		0	W	2	S W	3	40	5	28.824	28.719	28.708	28.750	.062	.097	.098	100	39	100
26		0	N W	3	W	2	35	25	28.657	28.426	28.302	28.461	.167	.162	.167	100	80	100
27		0	W	3		0	34	20	28.092	28.181	28.134	28.135	.181	.196	.181	100	100	100
28	W	1	N W	2		0	35	18	28.478	28.644	28.717	28.613	.141	.167	.118	100	100	100
29																		
30																		
31																		
Sums																		
Me'ns							32°.39	4°.79				28.807	.097	.133	.106	93	76	85
Average														.112			84	

METEOROLOGICAL OBSERVATIONS FOR THE MONTH OF MARCH, 1870.

Day of Month.	Thermometer in the open air.				Rain and Snow.				Clouds.					
	7 A. M.	2 P. M.	9 P. M.	Mean.	Time of beginning of rain or snow.	Time of ending of rain or snow.	Amount of rain or melted snow in gauge, in inches.	Depth of snow, in inches.	7 A. M. Amount of cloudin's.	7 A. M. Kind of clouds.	2 P. M. Amount of cloudin's.	2 P. M. Kind of clouds.	9 P. M. Amount of cloudin's.	9 P. M. Kind of clouds.
1	24	33	23	26⅔		11 P. M.		1	100	Cu. St.	100	Cu.	100	Nim.
2	18	30	25	24⅓					100	Cu. St.	50	Cir.Cu.	100	Cu.
3	18	31	26	25					50	Cu. St.	100	Cu.	50	Cu. St.
4	19	35	23	25⅔					60	Cu.	90	Cu.	100	Cu.
5	31	31	24	28⅔					90	Cu.	100	Cu. St.	100	Cu.
6	24	31	24	26⅓	10 A. M.				100	Cu. St.	100	Nim.	90	Nim.
7	21	29	23	24⅓					100	Cu.	100	Cu.	100	Nim.
8	19	31	20	23⅓		2 P. M.	.150	5	90	Cu. St.	50	Cu.	50	Cu. St.
9	19	36	32	29					30	Cir. St.	90	Cu. St.	90	Cu.
10	32	26	20	26					90	Cu. St.	100	Cu.	50	Cu.
11	13	36	18	22⅓					50	Cir. St.	100	Cu.	100	Cu.
12	18	20	18	18⅔	4 A. M.				100	Nim.	100	Cu.	100	Nim.
13	18	25	24	22⅓		6 P. M.	.815	5	100	Cu.	100	Nim.	90	Cu.
14	20	41	28	29⅔					50	Cu.	30	Cir. St.	80	Cir.Cu.
15	24	32	20	25⅓	1 P. M.	7 P. M.	.360	6	100	Cir. St.	100	Nim.	50	Cu.
16	-4	32	16	14⅔	Squalls			1	10	Cir. St.	50	Cir.Cu.	100	Nim.
17	32	50	35	39					90	Cu. St.	00		00	
18	10	50	28	29⅓					00		00		10	St.
19	20	44	42	35⅓					90	Cir. St.	50	Cu.	100	Cu.
20	38	45	41	41⅓	Squalls		.180		100	Nim.	100	Cu.	100	Cu.
21	34	48	32	38					100	Cu. St.	100	Cu. St.	100	Cu.
22	30	44	34	36					100	Cu. St.	100	Cir.Cu.	00	
23	22	39	24	28⅓					00		00		00	
24	18	48	34	33⅓					10	St.	90	Cir. St.	00	
25	32	46	35	37⅔					10	Cir. St.	50	Cir. St.	100	Cu.
26	34	45	39	39⅓	1 A. M.				90	Cir.Cu.	100	Nim.	100	Nim.
27	37	37	32	35½					100	Nim.	100	Nim.	100	Nim.
28	32	47	34	37⅔		3 P. M.	.802		100	Nim.	100	Nim.	100	Cu.
29	34	44	38	38⅔					90	Cu.	100	Cu.	100	Cu.
30	34	42	38	38	8 P. M.				100	Cu. St.	100	Cu. St.	100	Nim.
31	34	42	41	39		12 M.	.700		100	Nim.	100	Cu.	100	Nim.
Sums							3.007	18						
Me'ns				30°.28					70		79		76	
Average									75					

Day of Month.	Winds.						Register Thermom'tr		Barometer Height reduced to freezing point.				Force or Pressure of Vapor, in Inches.			Relative Humid'y, or Fraction of Saturation.		
	7 A. M. Direction.	7 A. M. Force.	2 P. M. Direction.	2 P. M. Force.	9 P. M. Direction.	9 P. M. Force.	Max.	Min.	7 A. M.	2 P. M.	9 P. M.	Mean.	7 A. M.	2 P. M.	9 P. M.	7 A. M.	2 P. M.	9 P. M.
1	W	1		0	W	3	33	12	28.728	28.769	28.838	28.778	.094	.150	.106	73	80	86
2	N W	2	W	2		0	30	10	28.956	29.027	29.053	29.005	.098	.167	.135	100	100	100
3	W	1	S W	2	W	2	31	10	29.070	29.091	29.106	29.089	.098	.155	.141	100	89	100
4	W	2		0	W	1	35	15	29.073	29.109	29.140	29.107	.103	.142	.106	100	70	86
5		0	N E	2	E	2	31	12	29.094	29.097	29.046	29.079	.155	.155	.094	89	89	73
6	S E	1	S E	3	E	1	31	10	28.951	28.803	28.729	28.827	.094	.113	.196	73	100	100
7	N W	1		0	W	1	31	13	28.688	28.757	28.862	28.769	.113	.142	.106	100	88	86
8	W	1	N W	1		0	38	8	28.927	29.063	29.030	29.006	.087	.136	.108	84	78	100
9	W	2	S W	2	S W	2	36	13	29.066	28.994	29.017	29.025	.087	.149	.162	84	71	69
10		0	N W	2		0	32	7	28.786	28.902	28.977	28.888	.143	.105	.075	79	75	70
11	S E	1	E	2	E	3	36	8	28.831	28.747	28.743	28.773	.063	.170	.098	81	80	100
12	E	3	E	3	E	3	20	10	28.597	28.451	28.238	28.428	.098	.108	.098	100	100	100
13	E	3	E	2	N E	!	25	13	28.289	28.581	28.721	28.513	.098	.100	.094	100	74	73
14		0	E	2	E	2	41	14	28.916	28.929	28.902	28,915	.091	.190	.082	85	74	58
15	E	3		0	N E	2	32	-15	28.719	28.684	28.539	28.647	.111	.181	.108	86	100	100
16	N E	1	S W	1	W	3	32	-11	28.570	28.529	28.552	28.550	.036	.106	.090	100	58	100
17	N E	1	N E	2	N E	1	52	4	28.6.1	28.669	28.883	28.721	.143	.210	.142	79	58	70
18		0		0		0	64	5	29.001	29.061	29.051	29.037	.089	.258	.117	57	71	76
19		0	W	2	S W	2	44	14	29.064	28.979	28.911	28.984	.075	.151	.222	70	52	88
20	S W	2	S W	2	S W	3	45	17	28.695	28.613	28.499	28.602	.229	.138	.235	100	46	91
21	S W	2	W	2	W	1	48	25	28.672	28.817	28.902	28.797	.155	.285	.181	79	85	100
22	N	1	W	1		0	49	11	28.970	29.089	29.137	29.048	.181	.151	.155	100	52	79
23	S W	1	N E	1		0	52	11	29.326	29.387	29.390	29.367	.084	.216	.094	71	91	78
24		0	S E	2	S E	2	49	13	29.399	29.370	29.307	29.358	.098	.260	.155	100	78	79
25	S E	2	N E	2	S E	1	48	25	29.262	29.164	29.044	29.156	.181	.125	.181	100	40	100
26	S E	2		0	E	2	45	29	28.952	28.838	28.638	28.797	.155	.278	.238	79	100	100
27	N W	1	N E	2	N W	2	39	24	28.439	28.441	28.467	28.449	.238	.221	.181	100	100	100
28	W	2	W	1		0	47	25	28.636	28.705	28.793	28.714	.181	.156	.175	100	48	89
29		0		0	N E	1	44	19	28.937	28.941	28.964	28.947	.155	.218	.186	79	76	81
30	E	2	S E	1	N E	2	42	30	28.964	28.927	28.903	28.931	.196	.222	.208	100	88	91
31	E	1	E	2	S E	1	42	29	28.887	28.899	28.893	28.893	.196	.222	.257	100	88	100
Sums																		
Me'ns							39°.48	13°.22				28.845	.125	.173	.146	88	77	87
Average													.148			84		

METEOROLOGICAL OBSERVATIONS FOR THE MONTH OF APRIL, 1870.

Day of Month.	Thermometer in the open air.				Rain and Snow.				Clouds.						Winds.						Register Thermom'tr		Barometer Height Reduced to Freezing Point.				Force or Pressure of Vapor, in Inches.			Relative Humid'y, or Fraction of Saturation.		
									7 A. M.		2 P. M.		9 P. M.		7 A. M.		2 P. M.		9 P. M.													
	7 A. M.	2 P. M.	9 P. M.	Mean.	Time of beginning of rain or snow.	Time of ending of rain or snow.	Amount of rain or melted snow in gauge, in inches.	Depth of snow, in inches.	Amount of cloudiness.	Kind of clouds.	Amount of cloudiness.	Kind of clouds.	Amount of cloudiness.	Kind of clouds.	Direction.	Force.	Direction.	Force.	Direction.	Force.	Max.	Min.	7 A. M.	2 P. M.	9 P. M.	Mean.	7 A. M.	2 P. M.	9 P. M.	7 A. M.	2 P. M.	9 P. M.
1	37	47	40	41⅓					90	Cu. St.	100	Nim.	100	Cu.		0	N E	2	N E	1	47	26	28.964	28.959	28.959	28.960	.221	.249	.203	100	77	82
2	39	54	42	45					90	Cir. St.	50	Cu.	80	Cu. St.	E	1	S E	2	N E	2	54	27	29.005	29.202	28.960	28.989	.181	.206	.222	43	49	83
3	38	48	32	39⅓					10	St.	00		00		N E	3	N W	3	S E	3	49	18	29.084	29.109	29.133	29.108	.144	.120	.143	63	36	79
4	30	56	30	38⅔					00		00		00		N E	1		0		0	56	27	29.138	29.090	29.086	29.104	.167	.179	.167	100	40	100
5	40	58	30	42⅔					00		10	Cir. St.	00		W	2	N	2	S W	1	58	27	29.019	28.979	28.981	28.993	.263	.129	.167	82	27	100
	52	60	40	50⅔					10	St.	00		00			0		0		0	60	20	28.997	28.994	28.989	28.993	.111	.203	.160	67	39	64
7	52	63	50	55					00		00		00			0	N	1		0	64	23	28.983	29.001	28.990	28.991	.181	.189	.258	100	33	71
8	34	67	50	50⅓					10	St.	00		00			0		0	S E	1	67	22	29.023	29.008	28.985	29.005	.175	.393	.210	89	59	58
9	33	67	54	51⅓					10	St.	10	Cir. St.	20	Cir. St.	N W	1	N E	2	E	2	67	50	29.003	28.983	28.929	28.971	.168	.190	.203	89	29	32
10	52	66	50	56	8 A. M.	11 P. M.	.320		100	Cu. St.	100	Nim.	100	Nim.	S E	1		0		0	66	29	28.926	28.910	28.905	28.913	.232	.376	.361	60	59	100
11	50	63	54	55⅔					100	Cu.	50	Cir. Cu.	00			0	N E	3	N W	2	64	29	28.879	28.882	28.908	28.889	.334	.189	.256	86	33	61
12	48	61	54	54⅓					00		40	Cir. St.	30	Cir.	W	2		0	E	1	71	41	29.008	29.007	28.984	28.999	.260	.219	.362	78	31	87
13	54	65	58	59					100	Cu. St.	90	Cu. St.	90	Cir.	S E	2	S W	2		0	72	41	28.916	28.850	28.808	28.858	.256	.451	.365	61	73	76
14	54	76	60	63⅓	Squalls				100	Nim.	100	Cu.	90	Cir.	W	2	S W	4	S W	2	77	52	28.810	28.797	28.777	28.794	.308	.358	.456	74	46	88
15	58	62	35	51⅔					100	Cu. St.	50	Cu.	00		S W	3	S W	3	W	1	62	30	28.822	28.779	28.760	28.787	.423	.399	.142	88	72	70
16	34	38	31	34⅓					100	Cu.	90	Cu. St	100	Cu. St.	S W	1	W	1	W	1	38	23	28.728	28.751	28.800	28.759	.196	.123	.174	100	54	100
17	33	36	34	34⅓					100	Cu.	100	Cu.	50	Cu.	W	1	N E	2	W	1	38	25	28.782	28.777	28.762	28.773	.168	.191	.196	89	90	100
18	34	42	42	39⅓	10 A. M.				100	Cu. St.	100	Nim.	100	Nim.	N W	1	N W	2	N W	2	44	29	28.692	28.671	28.687	28.683	.175	.244	.267	89	91	100
19	44	48	48	46⅔					100	Nim.	100	Nim.	100	Nim.	W	1	S W	1	W	1	48	39	28.648	28.606	28.595	28.616	.289	.335	.335	100	100	100
20	46	52	46	48					100	Nim.	100	Nim.	100	Cu.	N E	1		0	N W	2	52	39	28.590	28.565	28.532	28.562	.311	.384	.311	100	86	100
21	45	52	48	48⅓		8 P. M.	.800		100	Cu.	100	Cu.	100	Cu.		0	S W	1		0	52	38	28.717	28.765	28.775	28.752	.275	.334	.335	52	86	100
22	50	72	58	60					30	Cu.	00		00		S E	1	S W	3	S W	2	72	39	28.797	28.746	28.731	28.758	.361	.489	.423	100	62	88
23	55	83	66	64⅔					10	Cu.	00		00		S W	1	W	2		0	83	48	28.810	28.891	28.960	28.887	.349	.677	.391	81	60	87
24	69	80	45	64⅔	6 P. M.				90	Cir. Cu.	100	Cu.	100	Nim.	S W	2	N	3	N E	3	82	35	28.942	28.875	28.970	28.929	.430	.417	.300	61	41	100
25	42	56	38	45⅓		2 A. M.	.540		40	Cu.	90	Cu.	10	Cu.	N E	3	N	2		0	56	25	29.220	29.183	29.213	29.205	.199	.230	.229	74	51	100
26	42	60	52	51⅓					00		50	Cu. St.	10	Cu. St.	S E	1	S W	2	S W	1	60	30	29.219	29.110	29.002	29.110	.222	.176	.334	83	29	86
27	62	83	72	72⅓	11 P. M.				90	Cir. St.	100	Cu.	100	Cu.	S W	1	S W	3	S W	3	83	35	28.890	28.782	28.599	28.757	.399	.597	.422	72	53	54
28	44	56	45	48⅓		2 A. M.	.360		10	Cir. St.	50	Cir. St.	10	Cu.	S W	3	N E	3	S W	2	57	25	28.917	28.927	28.995	28.945	.196	.179	.182	68	40	61
29	43	59	40	47⅓					00		10	St.	00		S E	2	S W	2	W	1	59	30	29.096	29.040	28.905	29.013	.209	.242	.139	75	48	56
30	40	60	57	52⅓					80	Cir. Cu.	50	Cu.	00		N W	1	W	3	S W	2	60	29	29.010	28.971	28.919	28.966	.203	.310	.322	82	60	69
31																																
Sums							2.020																									
Me'ns				50°.39					55		54		43								60°.60	31°.70				28.902	.244	.290	.267	80	55	81
Average									50																		.267			76		

METEOROLOGICAL OBSERVATIONS FOR THE MONTH OF MAY, 1870.

Day of Month.	Thermometer in the open air. 7 A. M.	2 P. M.	9 P. M.	Mean.	Rain and Snow. Time of beginning of rain or snow.	Time of ending of rain or snow.	Amount of rain or melted snow in gauge, in inches.	Depth of snow in inches.	Clouds. 7 A. M. Amount of cloudin's.	Kind of clouds.	2 P. M. Amount of cloudin's.	Kind of clouds.	9 P. M. Amount of cloudin's.	Kind of clouds.
1	68	80	62	70					10	Cu.	10	Cir. St.	00	
2	70	80	62	70⅔					00		10	Cir.	00	
3	70	86	60	72					100	Cu.	50	Cir. Cu.	00	
4	76	78	56	70	3 P. M.				90	Cu.	50	Cir. Cu.	80	Cu. St.
5	55	84	64	61					100	Cu.	100	Nim.	90	Cir. St.
6	62	70	54	62		4 P. M.	.360		90	Cu.	100	Nim.	10	Cir. St.
7	58	60	48	58⅔					100	Cu. St.	10	Cu.	50	Cu. St.
8	56	66	58	60					50	Cir. Cu.	50	Cir. St.	90	Cir. Cu.
9	60	67	48	55					30	Cir. St.	40	Cu.	20	Cir. St.
10	48	60	48	52	3 P. M.				10	Cir. St.	90	Cir. Cu.	90	Cir. Cu.
11	48	62	52	54		11 P. M.	.160		100	Nim.	90	Cu.	100	Cu.
12	48	65	48	53⅔	Squalls		.040		90	Cu.	100	Cu.	30	Cir. St.
13	52	86	52	63⅓					10	Cir.	00		00	
14	64	88	54	68⅔					00		00		00	
15	70	84	62	72					00		00		10	Cir. St.
16	60	86	64	70					00		30	Cu.	00	
17	54	84	67	63⅓					100	Cu.	00		00	
18	64	86	58	69⅓					00		50	Cir. Cu.	00	
19	72	78	64	71⅓					00		90	Cir. Cu.	00	
20	62	85	66	71					90	Cir. St.	50	Cu.	100	Cu.
21	62	83	65	70	11 P. M.				90	Cu.	100	Cu. St.	100	Cu.
22	62	74	66	67⅓					100	Cu. St.	100	Cir. St.	100	Nim.
23	63	75	68	68⅔		3 A. M.	.600		100	Cu.	00		00	
24	60	60	44	54⅔					90	Cu. St.	100	Cu.	00	
25	60	73	64	65⅔					10	Cir. St.	00		30	Cu. St.
26	47	66	52	55					90	Cir. St.	50	Cir. St.	10	Cir. St.
27	48	80	46	58					20	Cir.	00		00	
28	57	80	52	63					00		00		00	
29	60	82	63	68⅓					00		00		00	
30	62	83	58	67⅔					00		00		00	
31	60	83	60	67⅔					00		30	Cir. Cu.	00	
Sums							1.160							
Me'ns				64°.32					47		42		29	
Average											39			

Day of Month.	Winds. 7 A. M. Direction.	Force.	2 P. M. Direction.	Force.	9 P. M. Direction.	Force.	Register Thermom'tr Max.	Min.	Barometer Height Reduced to Freezing Point. 7 A. M.	2 P. M.	9 P. M.	Mean.	Force or Pressure of Vapor, in Inches. 7 A. M.	2 P. M.	9 P. M.	Relative Humid'y, or Fraction of Saturation. 7 A. M.	2 P. M.	9 P. M.
1	W	1	S W	2		0	82	35	29.019	29.036	29.004	29.019	.476	.382	.399	69	37	72
2	S W	2	W	3	W	2	80	34	29.069	28.956	28.910	28.978	.353	.561	.340	48	55	61
3	N W	2	S W	3		0	86	51	28.908	28.785	28.744	28.812	482	.518	.487	66	42	94
4	W	3	S W	2	S W	1	88	49	28.723	28.701	28.748	28.724	.577	.704	.391	64	73	87
5	E	2	N E	1	N E	2	84	48	28.790	28.691	28.645	28.708	.433	.584	.464	100	50	77
6	E	3	N E	2		0	70	45	28.433	28.490	28.393	28.438	.491	.353	.348	88	48	93
7	N E	1	E	2	E	1	60	40	28.364	28.396	28.461	28.410	.375	.518	.335	93	100	100
8	N	1	S W	1	S W	1	68	36	28.597	28.604	28.653	28.618	.308	.376	.423	69	59	88
9	S W	1		0		0	67	29	28.699	28.694	28.710	28.701	.283	.489	.189	54	75	56
10	S W	1	S W	1		0	65	41	28.729	28.700	28.725	28.718	.260	.288	.285	78	54	85
11		0		0	N	2	64	40	28.716	28.715	28.755	28.728	.335	.370	.282	100	66	73
12	N W	2	N W	2		0	67	33	28.878	28.892	28.889	28.886	.310	.216	.335	92	35	100
13		0		0		0	86	35	28.950	28.934	28.934	28.939	.282	.480	.334	73	39	86
14	S W	2		0		0	88	45	29.014	28.991	28.997	29.000	.464	.453	.362	77	34	87
15	S W	2	S W	3		0	85	40	29.001	28.904	28.907	28.937	.449	.470	.429	61	40	77
16		0	S W	2		0	88	47	28.894	28.804	28.836	28.844	.518	.850	.529	100	68	89
17	S E	1	S W	1	S E	3	85	48	28.935	28.906	28.907	28.916	.418	.789	.596	100	68	100
18	S W	2	S W	2		0	86	40	28.921	28.904	28.962	28.929	.464	.677	.483	77	54	100
19	S W	1	N E	2		0	80	40	29.019	28.977	28.976	28.990	.668	.443	.596	86	46	100
20		0		0		0	85	55	29.065	29.013	29.000	29.026	.556	.531	.570	100	44	89
21		0	W	1		0	88	44	28.921	28.980	29.061	28.987	.556	.558	.483	100	50	78
22		0	E	2		0	74	40	29.091	29.041	28.891	29.007	.491	.463	.502	88	56	78
23		0	W	1	S W	1	75	44	28.894	28.865	28.871	28.876	.478	.318	.685	83	37	100
24	S W	2	S W	3		0	60	32	28.749	28.759	28.819	28.775	.367	.456	.239	71	88	100
25		0	W	1	E	2	73	47	28.909	28.881	28.871	28.887	.338	.345	.314	65	42	53
26	N E	2	N E	2	N E	2	66	33	28.915	28.891	28.897	28.901	.328	.316	.334	100	49	86
27	S W	2	W	2		0	80	33	28.854	28.991	28.891	28.912	.626	.315	.311	65	31	100
28	E	1	E	1		0	80	39	28.940	28.891	28.884	28.905	.378	.417	.388	81	41	100
29	E	1	S E	1		0	82	36	28.891	28.945	28.951	28.929	.518	.497	.543	100	45	94
30		0		0		0	83	46	29.015	29.013	29.047	29.025	.429	.520	.423	77	46	88
31		0	E	2		0	83	45	29.079	29.054	29.035	29.056	.456	.483	.456	88	43	88
Sums																		
Me'ns							77°.52	40°.97				28.857	.484	.475	.414	81	52	86
Average														.441			73	

METEOROLOGICAL OBSERVATIONS FOR THE MONTH OF JUNE, 1870.

Day of Month.	Thermometer in the open air.				Rain and Snow.				Clouds.						Winds.						Register Thermom'tr		Barometer Height Reduced to Freezing Point.				Force or Pressure of Vapor, in Inches.			Relative Humid'y, or Fraction of Saturation.		
									7 A. M.		2 P. M.		9 P. M.		7 A. M.		2 P. M.		9 P. M.													
	7 A. M.	2 P. M.	9 P. M.	Mean.	Time of beginning of rain or snow.	Time of ending of rain or snow.	Amount of rain or melted snow in gauge, in inches.	Depth of snow, in inches.	Amount of cloudin's.	Kind of clouds.	Amount of cloudin's.	Kind of clouds.	Amount of cloudin's.	Kind of clouds.	Direction.	Force.	Direction.	Force.	Direction.	Force.	Max.	Min.	7 A. M.	2 P. M.	9 P. M.	Mean.	7 A. M.	2 P. M.	9 P. M.	7 A. M.	2 P. M.	9 P. M.
1	66	83	67	72					50	Cir.	00		10	St.		0	N E	2		0	84	45	29.027	29.009	28.932	28.989	.536	.534	.489	84	49	75
2	68	86	64	72⅔					00		00		00		S E	1	N E	1		0	86	45	28.951	28.924	28.884	28.903	.509	.518	.596	75	42	100
3	63	88	66	72⅓	Shower		.003		00		50	Cir.Cu.	00			0	S W	1		0	85	55	28.838	28.775	28.799	28.804	.576	.650	.570	100	49	89
4	73	87	67	75⅔					00		80	Cir.	80	Cir.Cu.	S W	1	S W	2		0	85	58	28.823	28.795	28.796	28.804	.476	.623	.626	59	48	95
5	78	80	66	74⅔	Shower		.003		90	Cir.Cu.	90	Cu.	90	Cu. St.	S W	1	W	2	N E	2	80	48	28.756	28.747	28.762	28.755	.717	.349	.536	77	34	84
6	67	80	54	67					90	Cir.Cu.	80	Cir.Cu.	00		S W	2	S W	3		0	81	50	28.861	28.857	28.849	28.855	.522	.677	.418	79	66	100
7	64	77	49	63⅓					70	Cir.Cu.	50	Cu. St.	10	Cir.		0	S W	2		0	77	52	28.929	28.923	28.946	28.932	.464	.601	.348	77	65	100
8	64	56	50	56⅔	1 P. M.				90	Cu. St.	100	Nim.	100	Cu.	S W	1	W	1		0	64	42	28.931	28.886	28.891	28.902	.464	.282	.309	77	63	85
9	60	66	56	60⅔		5 P. M.	1.120		100	Nim.	90	Cu.	90	Cir.Cu.		0	N W	1	S W	2	66	45	28.838	28.777	28.777	28.797	.487	.502	.396	94	78	75
10	54	68	54	58⅔					100	Cu.	90	Cu.	90	Cu. St.	S W	1	S W	2		0	70	45	28.782	28.773	28.734	28.766	.418	.476	.418	100	69	100
11	52	66	50	56					90	Cu.	90	Cir.Cu.	50	Cu.	S E	2	N E	2		0	70	45	28.768	28.749	28.741	28.752	.388	.470	.309	100	73	85
12	70	69	58	65⅔	8 P. M.				50	Cu.	100	Cu.	100	Nim.	S W	1	S W	2		0	72	52	28.722	28.725	28.661	28.702	.621	.529	.483	85	75	100
13	59	67	56	60⅔					100	Cu.	100	Nim.	50	Cu.	S W	3		0	S E	3	67	50	28.514	28.647	28.493	28.551	.469	.393	.449	94	59	100
14	56	63	59	59⅓					100	Nim.	100	Nim.	90	Cu.	S W	3	S W	2	W	1	67	57	28.540	28.547	28.561	28.549	.449	.497	.469	100	83	94
15	60	72	60	64		3 P. M.	2.144		100	Cu.	90	Cu.	10	Cu.	S W	1	S W	2	W	1	72	52	28.677	28.698	28.791	28.722	.487	.668	.367	94	86	71
16	59	75	58	64					100	Cu.	70	Cu. St.	00		W	2		0		0	75	49	28.869	28.883	28.951	28.901	.469	.666	.483	94	77	100
17	58	80	64	67⅓					00		00		00			0	E	1		0	84	54	28.991	29.005	29.002	28.999	.483	.677	.596	100	66	100
18	74	94	64	77⅓					00		00		00			0		0		0	95	52	29.066	28.990	28.996	29.014	.568	.653	.596	67	41	100
19	72	98	66	78⅔					00		50	Cu.	10	Cu. St.	W	2	S W	1		0	98	50	29.017	28.943	28.914	28.958	.681	.983	.604	81	70	94
20	70	82	68	73⅓					00		10	Cir.Cu.	10	Cir.		0	N W	3	N	2	90	56	28.883	28.864	28.883	28.876	.658	.610	.543	90	56	79
21	61	77	55	64⅓					20	Cir. St.	00		10	St.	N E	2		0		0	77	57	28.858	28.741	28.894	28.829	.505	.816	.612	94	75	90
22	63	83	60	68⅔					10	Cir. St.	10	St.	00			0	S W	3		0	84	44	29.091	29.064	29.083	29.079	.505	.520	.405	94	46	94
23	72	95	68	78⅓					00		00		00		S W	1		0		0	95	60	29.167	29.135	29.140	29.147	.446	.759	.543	77	67	79
24	74	96	69	79⅔					00		00		00		S W	1		0		0	96	54	29.144	29.078	29.071	29.097	.559	1.001	.685	72	59	100
25	84	99	74	85⅔					10	St.	50	Cu. St.	00		N E	1	S W	1		0	99	62	29.077	29.055	29.064	29.065	.680	.719	.680	81	39	81
26	86	95	76	85⅔					00		50	Cu.	00		S W	2	W	2	S W	2	96	63	29.064	29.016	29.010	29.030	.789	.820	.718	68	50	86
27	78	98	72	82⅔					00		80	Cir.Cu.	10	Cu. St.	S W	2	S W	2		0	98	62	29.005	28.956	28.968	28.976	.805	1.119	.731	65	68	81
28	75	96	74	81⅔	2 P. M.				50	Cir.Cu.	100	Nim.	100	Nim.		0	S W	2	N E	3	96	62	28.972	28.943	28.950	28.955	.870	.779	.706	91	43	90
29	70	84	77	77					50	Cu. St.	10	Cu.	100	Nim.	S W	2	S W	2	N E	3	91	60	28.949	28.916	28.861	28.908	.745	1.016	.799	86	87	86
30	77	94	77	82⅔			4.000		10	Cir. St.	50	Cu.	20	Cu. St.	E	1	S E	1	N E	1	92	63	28.826	28.841	28.764	28.810	.799	1.355	.799	86	85	86
31																																
Sums							7.270																									
Me'ns				70°.87					42		51		34								83°.06	52°.96				28.880	.569	.653	.542	84	62	86
Average									42																		.588			77		

METEOROLOGICAL OBSERVATIONS FOR THE MONTH OF JULY, 1870.

Day of Month.	Thermometer in the open air: 7 A. M.	2 P. M.	9 P. M.	Mean.	Rain and Snow: Time of beginning of rain or snow.	Time of ending of rain or snow.	Amount of rain or melted snow in gauge, in inches.	Depth of snow, in inches.	Clouds 7 A. M.: Amount of cloudin's.	Kind of clouds.	Clouds 2 P. M.: Amount of cloudin's.	Kind of clouds.	Clouds 9 P. M.: Amount of cloudin's.	Kind of clouds.
1	63	65	78	68⅔		11 A. M.	.264		100	Nim.	100	Cu. St.	00	
2	59	80	67	68⅔					10	Cu. St.	50	Cu.	00	
3	60	90	64	71⅓					00		50	Cir.	00	
4	72	88	60	73⅓					00		80	Cir.Cu.	00	
5	71	80	60	70⅓					00		50	Cir. St.	00	
6	68	82	67	72⅓	3 P. M.				100	Nim.	100	Nim.	100	Nim.
7	65	74	60	66⅓		6 P. M.	2.300		100	Cu.	70	Cir.Cu.	40	Cu.
8	61	74	58	64⅓					10	St.	50	Cu.	30	Cir. St.
9	68	80	66	71⅓					10	St.	50	Cu.	00	
10	75	84	68	75⅔					80	Cu.	70	Cu.	80	Cir.Cu.
11	70	69	68	69	9 A. M.				100	Nim.	90	Cu.	30	Cir. St.
12	71	82	68	73⅔		3 A. M.	1.890		100	Nim.	10	Cu.	00	
13	71	84	66	73⅔					90	Cir.	90	Cir. St.	100	Cu.
14	63	76	70	69⅔	Shower		.060		100	Cu.	90	Cir. St.	100	Cu.
15	67	79	70	72	Shower		.220		100	Cu.	90	Cir. Cu.	90	Cir.Cu.
16	78	93	71	80⅔					100	Cu.	50	Cu.	00	
17	75	92	72	79⅔	1 P. M.				100	Cu.	90	Cu. St.	100	Nim.
18	73	88	74	78⅓		6 A. M.	.760		90	Cu.	50	Cu.	10	Cir. Cu.
19	74	90	70	78					00		50	Cir. Cu.	00	
20	73	92	74	79⅔					00		30	Cu.	50	Cu.
21	71	86	65	71⅔					20	Cu.	00		00	
22	65	89	80	78					30	Cir. St.	50	Cu.	00	
23	71	92	76	79⅔					00		10	Cu.	00	
24	73	84	70	75⅔	8½ A.M				100	Cu. St.	90	Cu St.	100	Nim.
25	71	92	74	79		4 A. M.	.960		100	Cu.	50	Cu.	00	
26	80	92	80	84					80	Cir. St.	50	Cu. St.	50	Cu.
27	72	90	76	79⅓	9 A. M.				90	Cu.	100	Nim.	100	Nim.
28	69	84	67	73⅓					100	Cu.	50	Cu. St.	10	Cu.
29	74	74	73	73⅔		8 A. M.	1.570		100	Nim.	100	Cu.	00	
30	72	83	60	71⅔					00		50	Cu.	10	Cu.
31	73	85	64	74					20	Cir.Cu.	00		10	Cu.
Sums							8.024							
Me'ns				74°.40					56		58		32	
Average									48					

Day of Month.	Winds 7 A. M.: Direction.	Force.	Winds 2 P. M.: Direction.	Force.	Winds 9 P. M.: Direction.	Force.	Register Thermom'tr: Max.	Min.	Barometer height reduced to freezing point: 7 A. M.	2 P. M.	9 P. M.	Mean.	Force or pressure of vapor, in inches: 7 A. M.	2 P. M.	9 P. M.	Relative humid'y, or fraction of saturation: 7 A. M.	2 P. M.	9 P. M.
1	S W	2	S W	2	S W	1	78	59	28.746	28.724	28.753	28.741	.543	.549	.785	94	89	82
2	N E	1	N E	3		0	80	55	28.773	28.742	28.836	28.788	.352	.561	.489	70	55	75
3		0	W	2	S W	1	90	59	28.877	28.850	28.865	28.864	.518	.841	.563	100	60	94
4		0	S W	3		0	88	50	28.866	28.723	28.747	28.778	.706	.609	.518	90	46	100
5		0	N W	2	W	1	86	59	28.801	28.844	28.837	28.827	.572	.530	.456	76	40	88
6	S W	1	W	1		0	82	57	28.810	28.766	28.792	28.789	.612	.950	.662	90	87	100
7	S W	2	S W	2		0	74	44	28.591	28.567	28.647	28.601	.549	.604	.518	89	73	100
8	N E	1	S W	1		0	75	55	28.782	28.767	28.881	28.810	.442	.604	.483	83	73	100
9	S W	2	S W	1	S W	1	82	56	28.915	28.897	28.907	28.906	.443	.599	.570	65	59	89
10	S W	1	S W	2		0	84	56	28.946	28.934	28.927	28.935	.666	.545	.685	77	47	100
11	W	2	N W	2		0	70	53	28.836	28.770	28.707	28.771	.551	.599	.577	75	85	85
12		0	S W	2	S W	2	83	59	28.743	28.767	28.780	28.763	.503	.516	.543	66	75	79
13	S W	2	S W	3	S W	1	84	57	28.768	28.735	28.735	28.746	.572	.623	.639	76	53	100
14	S W	1	S E	1		0	78	55	28.783	28.757	28.746	28.762	.543	.577	.621	94	64	85
15	S E	1	S W	2	S W	2	79	55	28.750	28.736	28.732	28.739	.626	.651	.658	95	66	90
16	S W	2	S W	2		0	93	60	28.802	28.780	28.814	28.798	.704	.942	.644	73	61	86
17		0	N	1	S W	2	92	64	28.863	28.850	28.869	28.860	.554	.695	.785	64	95	100
18		0	S W	1		0	88	62	28.856	28.842	28.863	28.853	.771	.915	.839	95	69	100
19	W	2	N W	1		0	90	60	28.893	28.873	28.811	28.859	.680	.887	.658	81	63	90
20		0	S W	3		0	92	63	28.764	28.706	28.831	28.767	.617	1.005	.718	77	67	86
21		0	S W	1		0	86	57	28.986	29.004	28.988	28.992	.537	.518	.618	71	42	100
22	S W	2	W	3	S W	1	89	63	28.900	28.955	28.978	28.944	.483	.516	.758	78	38	74
23		0	S W	3		0	92	64	28.929	28.844	28.815	28.829	.682	.956	.812	90	64	91
24	S W	2	S W	2	N W	3	90	63	28.844	28.835	28.866	28.848	.617	.877	.733	77	75	100
25		0	W	1		0	92	63	28.934	28.913	28.955	28.934	.759	.904	.758	100	83	90
26	S W	2	S W	3	S W	2	92	65	28.949	28.699	28.895	28.847	.848	.908	.843	88	60	83
27	S W	2	W	2	N W	2	90	62	28.894	28.846	28.830	28.856	.745	.887	.812	95	63	91
28		0	S W	2	N W	3	85	63	28.825	28.756	28.805	28.795	.671	.789	.522	95	68	79
29	S W	2	W	2		0	80	62	28.851	28.884	28.983	28.906	.680	.641	.732	81	77	90
30		0	S W	2		0	85	60	29.006	29.057	29.034	29.032	.631	.597	.518	81	53	100
31	S W	2		0	W	2	85	55	29.051	28.955	28.972	28.996	.581	.493	.596	72	41	100
Sums																		
Me'ns							84°.96	58°.54				28.836	.605	.715	.649	82	64	91
Average														.659			79	

METEOROLOGICAL OBSERVATIONS FOR THE MONTH OF AUGUST, 1870.

Day of Month.	Thermometer in the open air. 7 A. M.	2 P. M.	9 P. M.	Mean.	Rain and Snow. Time of beginning of rain or snow.	Time of ending of rain or snow.	Amount of rain or melted snow in gauge, in inches.	Depth of snow, in inches.	Clouds. 7 A. M. Amount of cloudiness.	Kind of clouds.	2 P. M. Amount of cloudiness.	Kind of clouds.	9 P. M. Amount of cloudiness.	Kind of clouds.	Winds. 7 A. M. Direction.	Force.	2 P. M. Direction.	Force.	9 P. M. Direction.	Force.	Register Thermom'tr. Max.	Min.	Barometer Height Reduced to Freezing Point. 7 A. M.	2 P. M.	9 P. M.	Mean.	Force or Pressure of Vapor, in Inches. 7 A. M.	2 P. M.	9 P. M.	Relative Humid'y, or Fraction of Saturation. 7 A. M.	2 P. M.	9 P. M.
1	71	90	71	77⅓					50	Cu.	40	Cir.Cu.	90	Cir.	S W	2	S W	2		0	90	56	28.907	28.865	28.874	28.882	.682	.707	.644	90	51	86
2	67	79	68	71⅓	11 A. M.				60	Cir.Cu.	90	Cu.	100	Nim.		0	W	1		0	81	62	28.880	28.861	28.806	28.849	.604	.731	.685	94	74	100
3	70	85	78	77⅔		10 A. M.	1.820		100	Cu.	50	Cu.	00			0	S E	2		0	86	62	28.767	28.751	28.780	28.766	.733	.650	.744	100	54	78
4	70	85	64	73					10	Cu.	00		00		W	2	S W	1		9	85	67	28.851	28.865	28.833	28.849	.658	.493	.529	90	41	89
5	70	90	68	76					00		30	Cir. St.	20	Cir.	S E	1	S W	2	W	3	90	62	28.836	28.814	28.809	28.819	.658	.582	.612	90	41	90
6	70	90	68	76					40	Cir. St.	40	Cir.Cu.	00		S W	1	S W	2	W	2	90	61	28.828	28.755	28.786	28.789	.621	.542	.648	85	38	95
7	76	86	68	76⅔					100	Cu.	50	Cu. St.	100	Cu.	N E	2		0		0	86	62	28.783	28.761	28.833	28.792	.652	.687	.685	73	51	100
8	68	76	63	69	10 A. M.				90	Cu.	100	Nim.	100	Nim.	N E	2	N E	3	S W	2	77	58	28.873	28.800	28.785	28.819	.476	.897	.576	69	100	100
9	64	80	69	71		8 A. M.	.670		Mi	st.	90	Cu.	100	Cu. St.	S W	1	S W	2		0	86	58	28.824	28.832	28.885	28.847	.596	.677	.708	100	66	100
10	69	80	61	70					90	Cu. St.	50	Cir.Cu.	90	Cir. St.		0	S W	2		0	80	47	28.924	28.922	28.952	28.932	.708	.638	.537	100	62	100
11	68	79	64	70⅓					90	Cir. St.	90	Cir.Cu.	30	Cir. St.		0		0		0	80	50	28.978	28.995	28.997	28.990	.685	.651	.596	100	66	100
12	70	88	69	75⅔					50	Cir. St.	60	Cu.	50	Cir.	S W	2		0		0	89	51	28.969	29.034	28.984	28.995	.658	.609	.671	90	46	95
13	71	76	62	69⅔	Night.		.080		50	Cu. St.	70	Cir.	100	Cir.Cu.	W	1	S W	2		0	81	52	29.027	29.067	29.063	29.052	.608	.541	.556	80	60	100
14	58	62	64	61⅓					100	Nim.	100	Cu.	20	Cu.	S W	1	N E	1	E	2	66	42	29.063	29.064	28.991	29.039	.423	.429	.529	88	77	89
15	66	76	57	66⅓					30	Cir. St.	40	Cu.	00		S W	2	N	2		0	78	50	29.009	28.996	29.037	29.014	.639	.652	.466	100	73	100
16	67	81	62	70					20	Cir.	30	Cu.	00		W	2	S W	3	S W	1	81	46	29.021	28.795	28.700	28.838	.425	.474	.523	64	45	94
17	64	83	66	71					00		00		10	Cu.	S W	1	S W	4	W	2	83	50	28.645	28.610	28.683	28.646	.529	.558	.639	89	50	100
18	70	88	74	77⅓					00		60	Cu.	90	Cu.	S W	2	S W	2	S W	2	88	50	28.748	28.760	28.766	28.758	.695	.735	.680	95	56	81
19	71	82	61	71⅓					100	Cu. St.	90	Cu.	100	Cu.	S W	3	S W	4	S W	3	83	42	28.773	28.731	28.919	28.807	.682	.731	.537	90	67	100
20	60	72	53	61⅔					00		00		20	Cir. St.		0	S W	1		0	82	47	29.111	29.121	29.141	29.124	.426	.559	.403	82	72	100
21	62	74	55	63⅔					100	Cu.	50	Cu.	00		N E	1	N E	1		0	77	47	29.179	29.181	29.176	29.178	.399	.468	.433	72	56	100
22	65	77	64	68⅔					00		90	Cir. St.	10	Cu.		0	S W	2	S W	1	77	55	29.197	29.145	29.022	29.121	.549	.527	.596	89	57	100
23	64	82	68	71⅓					100	Cir. St.	100	Cu.	40	Cu.	S W	2	S W	1	S W	1	82	54	28.914	28.759	28.786	28.819	.497	.816	.577	88	75	85
24	72	79	70	73⅔					100	Cu.	20	Cir. St.	10	Cu.	S W	2	S W	3	W	2	92	65	28.708	28.707	28.698	28.702	.550	.574	.658	72	58	90
25	72	79	58	66⅓	Night.		.430		10	Cu. St.	20	Cir. St.	80	Cu. St.	S W	1	S W	3	S W	1	79	42	28.686	28.793	29.005	28.827	.745	.387	.423	95	54	88
26	53	76	58	60⅔					50	Cir. St.	50	Cu.	00			0	W	2		0	76	40	29.208	29.176	29.191	29.191	.403	.614	.348	100	68	86
27	51	79	64	64⅔					10	Cir.	50	Cu. St.	100	Cu.	S W	1	S W	1	S E	2	79	45	29.157	29.133	29.041	29.110	.374	.574	.464	100	58	77
28	65	74	65	68	6 A. M.				100	Cu. St.	100	Nim.	100	Nim.	S W	1		0	S W	3	85	61	28.918	28.800	28.745	28.82[illegible]	.583	.758	618	94	90	100
29	75	73	54	67⅓		11 P. M.	1.530		90	Cu.	50	Cu.	00		S W	3	S W	4		0	75	45	28.567	28.655	28.771	28.664	.758	.771	.418	90	95	100
30	51	85	64	66⅔					00		60	Cu.	90	Cu.		0		0	S W	2	85	54	28.850	28.800	28.811	28.820	.888	.819	.464	100	68	77
31	62	84	64	70					00		70	Cu.	00			0	W	2		2	84	53	28.794	28.772	28.791	28.752	.556	.877	.596	100	75	100
Sums							4.530																									
Me'ns				70°.11					50		56		45								82°.68	53°.06				28.890	.579	.630	.566	89	62	93
Average									50																		.594			81		

METEOROLOGICAL OBSERVATIONS FOR THE MONTH OF SEPTEMBER, 1870.

Day of Month.	Thermometer in the open air.				Rain and Snow.				Clouds.						Winds.						Register Thermom'tr		Barometer Height Reduced to Freezing Point.				Force or Pressure of Vapor, in Inches.			Relative Humid'y, or Fraction of Saturation.		
									7 A. M.		2 P. M.		9 P. M.		7 A. M.		2 P. M.		9 P. M.													
	7 A. M.	2 P. M.	9 P. M.	Mean.	Time of beginning of rain or snow.	Time of ending of rain or snow.	Amount of rain or melted snow in gauge, in inches.	Depth of snow, in inches.	Amount of cloudin's.	Kind of clouds.	Amount of cloudin's.	Kind of clouds.	Amount of cloudin's.	Kind of clouds.	Direction.	Force.	Direction.	Force.	Direction.	Force.	Max.	Min.	7 A. M.	2 P. M.	9 P. M.	Mean.	7 A. M.	2 P. M.	9 P. M.	7 A. M.	2 P. M.	9 P. M.
1	67	88	63	72⅔	3 P. M.				10	Cir.	90	Cu.	50	Cu.		0	S W	2	W	1	88	53	28.794	28.751	28.786	28.777	.556	.650	.576	84	49	100
2	58	75	60	64⅓		5 A. M.	1.300		100	Cu.	70	Cir. Cu.	90	Cir.		0	N W	2	N W	1	75	51	28.801	28.706	28.811	28.772	.483	.826	.518	100	95	100
3	56	65	60	60⅓	2 A. M.				100	Nim.	100	Cu.	00			0		0		0	65	43	28.794	28.771	28.784	28.788	.449	.549	.518	100	89	100
4	59	76	53	62⅔		6 P. M.	.200		50	Cu.	80	Cu.	00		N W	2	N W	1		0	76	42	28.799	28.804	28.865	28.822	.410	.402	.403	82	45	100
5	76	76	60	70⅔	Night.	Night.	.060		50	Cu.	50	Cu.	90	Cir. St.		0	S W	2		0	76	43	28.894	29.008	29.006	28.969	.614	.470	.518	68	52	100
6	61	84	61	68⅔	Night.	Night.	.060		90	Cu. St.	80	Cir. Cu.	00		S W	1	S	3		0	84	51	29.014	28.979	29.001	28.998	.505	.789	.505	94	68	94
7	71	89	70	76⅔					00		10	Cu.	00			0	S W	2		0	89	56	29.006	29.013	29.005	29.008	.608	.678	.733	80	50	100
8	63	90	69	74					00		30	Cir. Cu.	30	Cir. St.		0	S W	3		0	90	59	29.084	29.060	29.043	29.062	.576	.707	.671	100	51	95
9	65	74	67	68⅔					90	Cir. Cu.	90	Cu. St.	00			0	N E	3	N E	2	77	46	29.060	29.031	29.081	29.057	.618	.641	.489	100	77	75
10	57	76	59	64					50	Cir. St.	50	Cu.	10	Cir. St.	N W	2	N	2	N W	2	77	38	29.108	29.090	29.088	29.093	.407	.402	.500	87	45	100
11	56	78	52	62					20	Cir. St.	00		00		N E	1		0	N E	1	78	32	29.177	29.163	29.158	29.093	.449	.664	.361	100	69	93
12	41	73	53	55⅔					00		20	Cu. St.	10	Cir.		0	S W	1		0	73	36	29.200	29.168	29.181	29.183	.257	.442	.403	100	55	100
13	46	74	56	58⅔					10	Cir. St.	50	Cu.	00			0		0		0	74	40	29.178	29.143	29.116	29.145	.311	.680	.391	100	81	87
14	50	82	77	69⅔	Night.	Night.	.200		00		50	Cir. St.	00			0	S W	2	N W	1	88	56	29.094	29.062	28.992	29.049	.361	.355	.502	100	32	78
15	62	83	64	69⅔					100	Cu. St.	90	Cir. Cu.	50	Cir.		0	S W	2		0	83	57	28.983	28.966	28.897	28.948	.556	.677	.596	100	60	100
16	64	75	60	66⅓					100	Cu.	100	Cu.	100	Cu.	N	1	N E	2	N E	2	75	52	29.008	29.088	29.081	29.029	.596	.415	.518	100	48	100
17	58	75	49	60⅔					100	Cu.	100	Cu.	00			0	N	2		0	75	38	29.107	29.116	29.141	29.121	.483	.628	.348	100	73	100
18	49	71	54	58					70	Cir. Cu.	50	Cir. St.	10	Cu.	N E	1	N E	2	N W	1	75	35	29.190	29.163	29.161	29.171	.348	.469	.418	100	62	100
19	41	72	51	54⅔					00		10	Cir.	10	Cir.		0	N W	2		0	72	36	29.203	29.194	29.161	29.186	.257	.422	.374	100	54	100
20	44	74	51	56⅓					30	Cir. St.	50	Cir. St.	00			0	N E	1		0	75	35	29.200	29.166	29.158	29.174	.289	.463	.374	100	56	100
21	44	82	57	61					10	St.	20	Cu.	00			0	S W	1		0	82	40	29.175	29.143	29.135	29.151	.289	.289	.466	100	26	100
22	43	80	48	57					10	Cu.	50	Cu.	10	Cu.		0	S W	2		0	80	41	29.144	29.113	29.097	29.118	.278	.561	.335	100	55	100
23	49	75	46	56⅔	7 P. M.				90	Cu.	100	Cu.	100	Cu.	S W	1		0		0	80	40	29.097	29.106	29.084	29.095	.247	.591	.238	71	68	77
24	72	73	64	69⅔					100	Nim.	100	Cu.	100	Cu.		0	S W	2	N E	1	78	57	29.043	28.910	28.897	28.950	.668	.771	.596	86	95	100
25	64	56	56	58⅔		3 P. M.	.490		100	Nim.	100	Cu.	30	Cu. St.		0	S W	3		0	70	44	28.906	28.980	29.003	28.966	.563	.363	.449	94	81	100
26	46	74	58	59⅓					50	Cir. St.	60	Cir. Cu.	00			0		0		0	74	41	29.130	29.097	29.118	29.115	.289	.568	.483	100	67	100
27	47	80	56	61					50	Cu.	30	Cir. Cu.	20	Cu. St.		0	S W	3		0	80	40	29.084	29.080	29.053	29.072	.273	.599	.449	85	59	100
28	46	74	58	59⅓					60	Cu.	30	Cir. Cu.	90	Cu.		0	N E	2	S E	2	75	45	29.041	28.997	28.980	29.006	.215	.463	.483	69	56	100
29	60	74	65	66⅓	9 A. M.				100	Cu.	90	Cu.	100	Cu.		0	N W	2	N E	2	75	55	28.951	28.900	28.888	28.911	.518	.641	.549	100	77	89
30	65	75	60	66⅔		10 A. M.	.536		100	Cu.	100	Cu.	100	Cu.	N E	2	N E	2		0	75	48	28.817	28.843	28.855	28.838	.549	.666	.518	89	77	100
31																																
Sums							2.846																									
Me'ns				63°.66					54		61		88								77°.46	45°.00				29.022	.434	.561	.476	93	60	96
Average									51																		.490			83		

METEOROLOGICAL OBSERVATIONS FOR THE MONTH OF OCTOBER, 1870.

Day of Month.	Thermometer in the open air. 7 A. M.	2 P. M.	9 P. M.	Mean.	Rain and Snow. Time of beginning of rain or snow.	Time of ending of rain or snow.	Amount of rain or melted snow in gauge, in inches.	Depth of snow in inches.	Clouds. 7 A. M. Amount of cloudin's.	Kind of clouds.	2 P. M. Amount of cloudin's.	Kind of clouds.	9 P. M. Amount of cloudin's.	Kind of clouds.	Winds. 7 A. M. Direction.	Force.	2 P. M. Direction.	Force.	9 P. M. Direction.	Force.	Register Thermom'tr Max.	Min.	Barometer Height Reduced to Freezing Point. 7 A. M.	2 P. M.	9 P. M.	Mean.	Force or Pressure of Vapor, in Inches. 7 A. M.	2 P. M.	9 P. M.	Relative Humid'y, or Fraction of Saturation. 7 A. M.	2 P. M.	9 P. M.
1	60	73	65	66					100	Cu.	90	Cu.	50	Cu.		0	N E	2		0	73	41	28.894	28.928	28.931	28.917	.518	.655	.618	100	81	100
2	63	70	61	64⅔					100	Cu. St.	50	Cir.	60	Cir.	N E	1	N E	3	N E	2	70	43	28.817	28.886	28.887	28.880	.576	.551	.505	100	75	94
3	60	69	60	63					100	Cir. St.	50	Cu.	50	Cu.		0	S W	2		0	69	45	28.905	28.864	28.854	28.874	.518	.685	.518	100	90	100
4	55	62	55	57⅓					100	Cu.	90	Cu.	100	Cu.	S W	2	S W	2	N E	2	62	42	28.844	28.784	28.946	28.858	.433	.491	.433	100	88	100
5	49	58	51	52⅔	Squalls		.380		100	Cu.	100	Cu.	90	Cu.	N E	3	N E	3	N E	1	51	41	29.085	29.065	29.143	29.097	.322	.394	.374	92	82	100
6	51	62	52	55					100	Cu.	80	Cu.	0		N E	1	N E	1		0	63	37	29.204	29.201	29.201	29.202	.374	.370	.348	100	66	93
7	49	63	48	53⅓					100	Cu.	0		0			0		0		0	63	32	29.211	29.235	29.219	29.221	.348	.446	.335	100	77	100
8	41	68	58	55⅔					100	Cu.	50	Cu.	100	Cu.		0		0		0	68	30	29.254	29.282	29.185	29.240	.244	.349	.483	91	51	100
9	45	63	47	51⅔					0		100	Cu. St.	50	Cir. St.		0		0		0	63	38	29.291	29.161	29.169	29.207	.275	.446	.323	92	77	100
10	47	63	64	58	6 P. M.				90	Cir. St.	100	Cu.	100	Nim.		0		0		0	69	42	29.019	28.888	28.775	28.894	.323	.446	.596	100	77	100
11	54	59	54	55⅔		2 A. M.	.200		50	Cu.	100	Cu.	100	Cu.	S W	2	S W	3	S W	1	59	41	28.615	28.574	28.576	28.588	.418	.439	.362	100	88	87
12	54	54	45	51					100	Cu.	80	Cu.	90	Cu.		0	S W	3	S W	3	54	40	28.605	28.574	28.620	28.599	.418	.308	.251	100	74	84
13	45	60	36	47					80	Cu.	50	Cu.	0		S W	2	S W	3		0	60	31	28.760	28.841	29.059	28.886	.275	.288	.212	92	54	100
14	44	54	51	49⅔	Night.		.160		90	Cir. St.	100	Cu.	30	Cu. St.	S W	2	S W	2	S W	2	55	42	28.778	28.775	28.783	28.778	.241	.110	.348	84	26	93
15	53	72	50	58⅓					100	Cu.	90	Cir. St.	0		S W	2	S W	3	W	2	72	47	28.794	28.997	29.030	28.940	.362	.422	.258	87	54	71
16	60	70	58	62⅔					10	St.	100	Cu. St.	100	Cu.	S W	3	S W	3		0	70	45	29.025	28.943	29.014	28.994	.426	.449	.453	82	61	100
17	50	55	50	51⅔	6 A.M.				100	Nim.	100	Nim.	100	Nim.		0		0	W	1	56	31	29.025	28.934	28.999	28.986	.361	.405	.258	100	94	71
18	38	53	32	41		3 A. M.	.492		30	Cu.	20	Cu.	0		N W	1	N W	1		0	53	25	29.019	29.011	29.100	29.043	.221	.269	.181	100	67	100
19	38	44	40	40⅔					80	Cir. Cu.	100	Nim.	100	Nim.	S W	2	S W	4		0	50	32	28.912	28.737	28.676	28.775	.208	.241	.248	91	84	100
20	38	50	41	43					40	Cu.	50	Cu. St.	40	Cu.		0	W	2	S W	1	50	31	28.709	28.641	28.615	28.655	.229	.283	.257	100	78	100
21	36	52	42	43⅓					10	St.	60	Cu.	80	Cu.		0	W	1	S W	1	52	30	28.652	28.688	28.745	28.695	.229	.257	.267	100	66	100
22	36	58	36	43⅓					90	Cu.	50	Cu.	30	Cu.		0	S W	3		0	50	33	29.039	29.149	29.195	29.127	.212	.309	.212	100	64	100
23	37	72	60	56⅓					0		0		0			0	S W	3	S W	3	72	45	29.203	29.153	29.109	29.155	.221	.358	.518	100	46	100
24	53	72	62	62⅓					30	Cir. St.	30	Cir. St.	10	St.	S W	2	S W	5	S W	2	72	47	29.111	29.072	29.041	29.074	.348	.422	.370	86	54	66
25	59	58	44	53⅔	Shower		.090		100	Cu.	100	Nim.	100	Cu.	S W	3		0	N E	2	59	35	29.021	29.021	29.086	29.042	.500	.483	.289	100	100	100
26	41	63	48	50⅔					100	Cu.	100	Cu.	100	Cu.	N E	1	W	3	S W	4	63	36	29.178	29.094	29.089	29.120	.244	.243	.285	91	42	85
27	62	72	48	60⅔					100	Cu.	30	Cir. Cu.	10	St.	S W	3	S W	5	S W	3	72	26	28.810	28.761	28.891	28.820	.556	.422	.285	100	54	85
28	32	67	49	49⅓					50	Cir. St.	60	Cu.	70	Cir. St.		0	W	2		0	67	22	29.046	29.053	29.094	29.064	.181	.190	.322	100	29	92
29	28	55	49	44					30	Cir. St.	90	Cir.	100	Cu.		0	S W	2		0	57	27	29.167	29.175	28.791	29.044	.153	.218	.322	100	50	92
30	42	52	42	45⅓	3 A. M.	10 P. M.	.970		100	Nim.	100	Nim.	100	Nim.	S E	1		0	W	3	52	30	28.713	28.494	28.483	28.563	.267	.308	.244	100	79	91
31	36	53	28	39					100	Cu.	0		0		S W	4	W	1	S W	1	53	30	28.801	28.791	28.898	28.830	.149	.295	.153	71	73	100
Sums							2.292																									
Me'ns				52°.45					73		68		57								61°.25	32°.80				28.939	.327	.370	.343	95	68	93
Average											66																	.346			85	

METEOROLOGICAL OBSERVATIONS FOR THE MONTH OF NOVEMBER, 1870.

Day of Month.	Thermometer in the open air. 7 A. M.	2 P. M.	9 P. M.	Mean.	Rain and Snow. Time of beginning of rain or snow.	Time of ending of rain or snow.	Amount of rain or melted snow in gauge, in inches.	Depth of snow, in inches.	Clouds. 7 A. M. Amount of cloudin's.	Kind of clouds.	2 P. M. Amount of cloudin's.	Kind of clouds.	9 P. M. Amount of cloudin's.	Kind of clouds.
1	33	67	31	43⅔					10	St.	50	Cu.	30	Cu.
2	34	69	30	44⅓					00		00		00	
3	29	53	33	38⅓					10	Cu.	70	Cu. St.	90	Cu.
4	23	50	34	35⅔					100	Cu.	50	Cu.	100	Nim.
5	27	45	32	34⅔					00		00		00	
6	26	53	33	37⅓					50	Cir.	100	Cu. St.	00	
7	24	54	42	40	In	night.	.160		10	St.	90	Cir. St.	90	Cir. St.
8	53	64	48	54⅓	In	night.	.330		30	Cir. St.	100	Cu. St.	00	
9	38	46	34	39⅓					100	Cu. St.	90	Cu.	100	Cu. St.
10	19	45	34	32⅔					10	St.	100	Nim.	50	Cir. St.
11	32	53	34	39⅔					50	Cir. St.	10	Cu.	30	Cir. St.
12	29	58	42	43					90	Cir. St.	100	Cu.	00	
13	36	50	36	40⅔	Rain in	night.	.400		100	Cu. St.	50	Cu.	100	Nim.
14	37	50	32	39⅔	Snow	squalls.			100	Nim.	100	Cu.	100	Cu. St.
15	23	50	31	34⅔					50	Cir.	10	Cu.	10	Cu.
16	30	52	30	37⅓					10	Cir.	100	Cir.Cu.	50	Cu.
17	32	45	32	36⅓					100	Cu.	50	Cu.	100	Cu.
18	26	32	18	25⅓	Snow	squalls.			50	Cu.	70	Cu.	00	
19	13	35	32	26⅔					20	Cu.	100	Cu. St.	100	Cu. St.
20	34	48	38	40					100	Cu. St.	90	Cu.	100	Cu. St.
21	28	40	27	31⅔					90	Cu. St.	70	Cu.	10	Cu.
22	20	36	31	29					10	Cu.	90	Cu.	100	Cu. St
23	30	43	35	36					100	Cu. St.	80	Cu. St.	100	Cu. St.
24	35	41	31	35⅔					100	Cu. St.	10	Cu.	00	
25	29	34	30	34⅓					100	Cu. St.	90	Cu.	00	
26	28	46	46	40					50	Cir. St.	100	Cu. St.	20	Cu.
27	42	61	47	50					50	Cir.Cu.	00		10	St.
28	43	60	57	53⅓	In	night.	.015		00		100	Cu. St.	100	Cu. St.
29	42	47	31	40					100	Cu. St.	100	Cu. St.	10	St.
30	31	49	35	38⅓					10	St.	00		00	
31														
Sums							.905							
Me'ns				38°.40					58		65		46	
Average											54			

Day of Month.	Winds. 7 A. M. Direction.	Force.	2 P. M. Direction.	Force.	9 P. M. Direction.	Force.	Register Thermom'tr. Max.	Min.	Barometer Height reduced to freezing point. 7 A. M.	2 P. M.	9 P. M.	Mean.	Force or Pressure of Vapor, in Inches. 7 A. M.	2 P. M.	9 P. M.	Relative Humid'y, or Fraction of Saturation. 7 A. M.	2 P. M.	9 P. M.
1		0		0	W	1	67	20	28.892	28.824	28.885	28.867	.188	.425	.174	100	64	100
2	S W	1	W	2	W	1	69	24	28.918	29.017	29.110	29.013	.155	.398	.167	79	56	100
3		0	W	2	S W	1	53	22	29.095	29.121	29.009	29.108	.142	.219	.188	88	54	100
4		0		0	S W	2	50	21	29.038	29.084	29.173	29.098	.123	.186	.158	100	51	79
5		0	E	3		0	50	17	29.200	29.185	29.190	29.191	.147	.138	.143	100	46	79
6		0		0		0	53	20	29.203	29.196	29.209	29.202	.141	.295	.168	100	73	89
7		0	S E	1	S E	3	54	20	29.244	29.075	28.935	29.084	.129	.206	.267	100	49	100
8	S	3	S W	4	W	1	64	33	28.605	28.577	28.584	28.588	.375	.529	.209	93	89	75
9	S W	3	S W	4	W	2	46	14	28.602	28.844	29.087	28.844	.186	.262	.155	81	84	79
10		0	E	2	S E	2	45	15	29.197	29.164	29.194	29.185	.103	.160	.196	100	53	100
11	S W	1	W	2	S W	1	53	23	29.068	29.019	28.886	28.991	.143	.186	.155	79	67	79
12		0	W	1	W	1	58	20	28.890	28.849	28.839	28.859	.142	.104	.222	88	22	88
13		0		0		0	50	20	28.937	28.916	28.863	28.905	.170	.258	.212	80	71	100
14	N W	2	W	1	W	2	50	20	28.827	28.713	28.713	28.751	.195	.139	.162	82	33	89
15		0		0		0	50	17	28.668	28.610	28.669	28.649	.123	.186	.174	100	51	100
16	S W	1	W	2		0	52	18	28.679	28.708	28.739	28.708	.167	.232	.167	100	60	100
17	S W	2	W	1	N W	1	46	17	28.838	28.813	28.919	28.856	.181	.182	.162	100	61	89
18	W	1	N W	2		0	32	10	29.054	29.007	29.061	29.040	.141	.162	.098	100	89	100
19		0	S W	2	S	2	34	13	29.008	28.934	28.838	28.926	.078	.142	.181	100	70	100
20	S W	3	S W	3	W	2	48	25	28.704	28.707	28.827	28.746	.196	.143	.229	100	43	100
21		0	W	1		0	40	12	28.941	28.979	29.050	28.990	.153	.182	.147	100	73	100
22		0	N E	2	N E	3	36	16	28.941	28.733	28.597	28.757	.108	.149	.136	100	71	73
23	N W	2	N W	2	W	2	43	25	28.610	28.776	28.882	28.756	.167	.079	.162	100	28	80
24	W	3	W	3	W	2	41	25	28.918	28.967	28.937	28.939	.162	.147	.174	80	57	100
25	S W	3	S W	3	S W	4	35	22	28.877	28.708	28.561	28.715	.160	.155	.148	100	79	89
26	S W	2	S W	3	S W	3	46	28	28.505	28.463	28.496	28.488	.158	.109	.169	100	54	54
27	W	1		0	W	1	62	36	28.694	28.750	28.805	28.753	.222	.269	.202	83	59	62
28	W	2	W	1	W	1	62	40	28.811	28.813	28.877	28.833	.231	.426	.378	83	82	81
29	N E	2	N E	3	N E	1	47	20	28.989	29.064	29.160	29.071	.244	.156	.174	91	48	100
30		0	S W	2	S W	2	50	25	29.246	29.219	29.115	29.193	.123	.223	.168	100	64	89
31																		
Sums																		
Me'ns							49°.58	21°.26				28.903	.164	.216	.181	93	60	89
Average														.187			80	

METEOROLOGICAL OBSERVATIONS FOR THE MONTH OF DECEMBER, 1870.

Day of Month.	Thermometer in the open air.				Rain and Snow.				Clouds.						Winds.						Register Thermom'tr		Barometer Height Reduced to Freezing Point.				Force or Pressure of Vapor, in Inches.			Relative Humid'y, or Fraction of Saturation.		
	7 A. M.	2 P. M.	9 P. M.	Mean.	Time of beginning of rain or snow.	Time of ending of rain or snow.	Amount of rain or melted snow in gauge, in inches.	Depth of snow, in inches.	7 A. M. Amount of cloudin's.	7 A. M. Kind of clouds.	2 P. M. Amount of cloudin's.	2 P. M. Kind of clouds.	9 P. M. Amount of cloudin's.	9 P. M. Kind of clouds.	7 A. M. Direction.	7 A. M. Force.	2 P. M. Direction.	2 P. M. Force.	9 P. M. Direction.	9 P. M. Force.	Max.	Min.	7 A. M.	2 P. M.	9 P. M.	Mean.	7 A. M.	2 P. M.	9 P. M.	7 A. M.	2 P. M.	9 P. M.
1	31	45	40	38⅔					90	Cu. St.	90	Cu.	100	Cu. St.	S W	2	S W	2	S W	3	45	19	29.003	28.811	28.716	28.843	.155	.160	.208	89	53	82
2	32	47	33	37⅓					00		10	Cu.	00		W	3	W	4	W	1	47	28	28.689	28.742	28.759	28.730	.181	.156	.168	100	48	89
3	29	44	42	38⅓					50	Cu. St.	100	Cu. St.	30	Cu.	W	1	W	1	W	3	44	29	28.806	28.755	28.508	28.689	.160	.196	.181	100	68	100
4	40	52	42	44⅔					80	St.	60	Cir. St.	100	Cu. St.	S W	2	S W	2		00	52	35	28.498	28.594	28.625	28.570	.248	.282	.244	100	78	91
5	42	46	39	42⅓	8 A. M.	12 M.	.50		100	Cu. St.	100	Cu. St.	100	Nim.	E	2		00	S W	4	46	30	28.359	28.171	28.208	28.244	.267	.311	.238	100	100	100
6	36	39	32	35⅔		10 P. M.	.95		100	Nim.	50	Cu.	100	Cir. St.	S W	3	S W	2	S W	1	39	28	28.711	28.794	28.779	28.761	.212	.178	.162	100	73	89
7	34	35	34	34⅓	7 A. M.	5 P. M.	.09		100	Nim.	100	Nim.	100	Cu. St.		00	S W	1	W	4	35	25	28.465	28.391	28.613	28.489	.196	.204	.204	100	100	100
8	28	28	27	27⅔					100	Cu. St.	100	Cu. St.	100	Cu. St.	N W	2	N W	1		00	28	18	28.828	28.890	28.869	28.862	.158	.153	.147	100	100	100
9	25	39	29	31					90	Cu. St.	70	Cu.	00		W	2	S E	1		00	39	20	28.995	28.983	29.009	28.662	.135	.152	.160	100	63	100
10	23	42	32	32⅓					10	St.	30	Cir.Cu.	30	St.		00	E	2	S E	2	42	30	29.048	29.084	29.008	29.046	.123	.184	.181	100	50	100
11	30	34	32	32	5 P. M.		.50	4	100	Cu. St.	100	Cu. St.	100	Nim.	E	4	E	4	N E	2	34	28	28.883	28.708	28.619	28.736	.167	.155	.181	160	79	100
12	33	34	34	33⅔					100	Nim.	100	Nim.	100	Nim.	E	1	S W	1	W	3	34	30	28.555	28.603	28.727	28.628	.188	.196	.196	100	100	100
13	32	32	32	32					100	Nim.	100	Nim.	100	Nim.	W	4	W	2	W	2	32	27	28.781	28.746	28.732	28.736	.181	.181	.181	100	100	100
14	27	29	29	28⅓		9 P. M.	.20		100	Nim.	100	Nim.	100	Nim.	W	3	W	3	W	4	29	22	28.726	28.855	29.021	28.867	.147	.160	.160	100	100	100
15	25	27	26	26					90	Cu. St.	100	Cu. St.	100	Cu. St.	W	1	W	2	W	1	27	24	29.131	29.154	29.205	29.163	.135	.147	.141	100	100	100
16	25	30	26	27					100	Cu. St.	100	Cu. St.	100	Cu. St.		00		00		00	30	22	29.191	29.106	28.919	29.072	.135	.120	.141	100	78	100
17	27	30	30	29					100	Cu. St.	100	Cu. St.	100	Cu. St.	W	2	W	2	W	1	30	22	28.845	28.766	28.734	28.781	.147	.167	.167	100	100	100
18	29	30	27	28⅔					100	Cu. St.	100	Cu. St.	100	Cu. St.	W	1		00	W	3	30	22	28.714	28.732	28.774	28.740	.160	.148	.147	100	89	100
19	30	30	27	29	1 P. M.				100	Cu. St.	100	Nim.	100	Nim.		00	N E	2	N E	2	30	24	28.710	28.438	28.212	28.453	.167	.174	.147	100	100	100
20	30	30	15	25					100	Nim.	100	Cu. St.	100	Cu. St.	W	1	W	2	W	3	30	-3	28.405	28.451	28.572	28.476	.167	.148	.086	100	89	100
21	5	9	2	5⅓		7 P. M.	.30	8	100	Cu. St.	90	Cu. St.	50	Cu. St.	S W	2	S W	4	S W	3	9	-9	28.752	28.927	28.983	28.887	.055	.065	.048	100	100	100
22	3	12	-3	4					100	Cu. St.	30	Cu.	30	Cu.	N W	1	S W	2	W	1	12	-15	29.039	29.036	28.946	29.007	.050	.075	.038	100	100	100
23	-11	5	-3	-3					30	St.	30	Cu.	10	Cu.		00	W	1	S W	2	6	-12	28.891	28.919	28.978	28.929	.027	.055	.028	100	100	100
24	-3	2	-3	-1⅓					100	Cu. St.	100	Cu. St.	100	Cu. St.	S W	2	S W	2	S W	2	2	-13	29.108	29.288	29.219	29.205	.028	.048	.038	100	100	100
25	-4	7	15	6	8 A. M.		.07	1	90	Cu. St.	100	Cu. St.	100	Nim.	N E	2	N E	2	S W	3	15	-5	29.173	28.944	28.819	28.978	.036	.045	.086	100	76	100
26	9	11	0	6⅔		3 P. M.	.11	1½	100	Nim.	100	Nim.	20	St.	S W	3	S W	3	S W	2	11	-5	28.919	29.062	29.078	29.019	.065	.057	.044	100	79	100
27	10	25	23	19⅓					70	Cir. St.	10	St.	100	Cu. St.	S W	2	S W	2	S W	2	25	7	28.876	28.791	28.708	28.791	.068	.135	.106	100	100	86
28	18	20	11	16⅓					100	Nim.	100	Cu. St.	100	Cu. St.	N E	1	N E	3	N E	2	20	-14	28.707	28.751	28.887	28.781	.098	.075	.071	100	70	100
29	-11	12	11	4					10	Cu.	30	Cu.	100	Cu. St.		00	S W	1	S W	4	12	-11	28.924	28.909	28.702	28.845	.027	.060	.071	100	80	100
30	25	34	32	30⅓	4 A. M.				100	Nim.	100	Nim.	100	Nim.	S W	4	S W	3		00	34	25	28.238	28.068	28.238	28.178	.135	.196	.181	100	100	100
31	30	32	24	28⅔		10 A.M.	.35	4	100	Nim.	100	Cu. St.	00		W	3	W	2	W	2	32	18	28.588	28.794	28.952	28.778	.167	.162	.129	100	89	100
Sums							2.57	18½																								
Me'ns				24°.80					82		80		76								29°.06	14°.38				28.772	.135	.145	.138	99	85	98
Average									79																		.189			94		

WINDS.

This is for the record of the direction *from* which the wind is blowing, as indicated by a vane, and its force by estimation. The direction is entered in eight points of the compass: N., N. E., E., S. E., S., S. W., W., N. W. The force is estimated and registered by the following table, in figures from 1 to 10:

1. Very light breeze	2	miles per hour.
2. Gentle breeze	4	" " "
3. Fresh breeze	12	" " "
4. Strong wind	25	" " "
5. High wind	35	" " "
6. Gale	45	" " "
7. Strong gale	60	" " "
8. Violent gale	75	" " "
9. Hurricane	90	" " "
10. Most violent hurricane	100	" " "

ABSTRACT OF METEOROLOGICAL OBSERVATIONS FOR 1870.

	Average for 1870.	Average for 7 years.
Thermometer in open air	49°.10	47°.03
Barometer at 32°	28.876	28.887
Rain-fall	37.76	31.45

CONTENTS.

ILLUSTRATIONS.

www.ingramcontent.com/pod-product-compliance
Lightning Source LLC
LaVergne TN
LVHW011300110826
845149LV00001B/206

* 9 7 8 1 4 1 8 1 8 8 6 1 0 *